D0204823

BIOLOGICAL STRUCTURE AND FUNCTION 7

BIOLOGY OF CARTILAGE CELLS

BIOLOGICAL STRUCTURE AND FUNCTION

BIOLOGY OF CARTILAGE CELLS

R. A. STOCKWELL
Reader in Anatomy, University of Edinburgh

CAMBRIDGE UNIVERSITY PRESS
CAMBRIDGE
LONDON · NEW YORK · MELBOURNE

Published by the Syndics of the Cambridge University Press
The Pitt Building, Trumpington Street, Cambridge CB2 1RP
Bentley House, 200 Euston Road, London NW1 2DB
32 East 57th Street, New York, NY 10022, USA
296 Beaconsfield Parade, Middle Park, Melbourne 3206, Australia

First published 1979

Printed in Great Britain at the
University Press, Cambridge

Library of Congress Cataloguing in Publication Data
Stockwell, Robert Amos, 1933–
Biology of cartilage cells.
(Biological structure and function; 7)
Bibliography: p.
Includes index.
1. Cartilage cells. 2. Cartilage. I. Title.
II. Series.
QP88.23.S84 591.1′852 78-11704
ISBN 0 521 22410 1
(ISSN 0308-5384)

To Dr Jill Fyfield, Robert, Stephen and Elizabeth

CONTENTS

PREFACE

When asked to write a book about cartilage, it was perhaps inevitable that I should choose the chondrocyte as the theme. First, cartilage exists because of the cells. Secondly, it serves to emphasise that cartilage is a living tissue, even though in the adult it may appear inactive and can be regarded in some ways as an inert engineering material. Thirdly, in the last decade or so, there has been a tremendous expansion of knowledge in the various aspects of cartilage research – for example, cell differentiation, macromolecule synthesis, the structure and function of the epiphysial growth plate and of articular cartilage. The results of this research can be integrated only if the chondrocyte is taken as the common denominator.

Advances in knowledge have been possible because cartilage lends itself to investigation by individuals in many different disciplines – chemists, physicists, mechanical engineers as well as biological scientists and those in clinical practice. It is obviously difficult to satisfy the requirements of such a diversity of creatures in one short book; thus the adoption of the cellular approach to the subject has rendered inappropriate a full quantitative account of the mechanical properties of cartilage nor has it been possible to discuss fully the pathology of the chondrocyte. Nevertheless I hope that this monograph will assist all workers, whatever their scientific background or status, to be aware of progress in fields of cartilage research other than their own, especially regarding the chondrocyte and the interaction with its environment.

The book falls naturally into two parts. The first deals with the structure, metabolism and environment of the cartilage cell, predominantly in normal adult tissue. The interaction of the cell with matrix macromolecules, hormonal influences, and other factors controlling chondrocyte metabolism, structure and numbers are described as fully as possible. An entire chapter has been allotted to chondrocyte nutrition in different forms of cartilage, a subject which has been controversial but has seen considerable advances in the last decade. The second part of the book outlines chondrogenesis, cartilage growth and degeneration. The factors relating to chondrocyte differentiation are discussed and there is a full account of the role of chondrocyte proliferation in cartilage growth and repair, with special reference to long bone development. The concluding chapter surveys briefly the various forms of cartilage degeneration and the involvement of the chondrocyte in cartilage calcification and in articular cartilage fibrillation.

I have been fortunate in my friends and colleagues who have contributed in various ways to this book, although any errors of omission or commission are entirely mine. In particular, I have benefited from numerous discussions about the chemistry and pathology of cartilage with Dr Alice Maroudas, Dr George Meachim, Professor John Scott, and Dr Rudolph Sprinz, who has in addition allowed me to use some of his valuable material. I am indebted to Professors George Romanes and David Van Sickle for specimens and helpful comments relevant to cartilage development; to Dr Helen Barrett for guidance in embryological matters and to Mr J. Connolly for permission to use some of his unpublished results. Dr Ruth Bellairs kindly provided the electron micrograph of the notochord used in Fig. 91, Dr Barry Longmore supplied the valuable photographs of the articular surface obtained by reflected light interference microscopy for Fig. 76; Dr Peter Gould produced enlargements of his electron micrographs for use in Figs. 90 and 94. It is a pleasure to express my gratitude to Mr R. MacDougall and Mr J. Cable for their indispensable technical assistance with histological preparations and photographs in this book and to Mrs G. Liddle who persevered with an often baffling manuscript. Finally I thank the editors for their encouragement and the publishers for their patience.

Acknowledgements for permission to reproduce illustrations from my earlier publications are due to the editor of the *Journal of Anatomy* for Figs. 3, 5, 14, 15, 16, 45, 60, 83 and 106; to the Royal Society for Fig. 12; to the editor and publishers of the *Annals of the Rheumatic Diseases* for Fig. 13; to the editor and publishers of the *Journal of Clinical Pathology* for Figs. 8 and 37; and to the editor and publishers of the *Transactions of the Ophthalmological Society of the United Kingdom* for Fig. 30.

I am grateful to Dr R. P. Gould and the editor and publishers of *Experimental Cell Research* for their kind permission to publish Figs. 90 and 94; to Dr N. F. Kember and the editor and publishers of the *Journal of Bone and Joint Surgery* for Fig. 96; to Dr G. Meachim and the editor and publishers of the *Journal of Bone and Joint Surgery* Fig. 103; to Professor D. C. Van Sickle and the editor and publishers of the *Anatomical Record* for Fig. 62; and to Professor L. Wolpert and the Ciba Foundation for Fig. 93. I thank Dr A. J. Bailey and the editor and publishers of *Protides of the Body Fluids* for permission to publish a redrawn and modified copy of an original diagram as Fig. 20*e*; Dr Dick Heinegard and the editor and publishers of the *Journal of Biological Chemistry* for part of Fig. 33 and for Fig. 34; Dr Helen Muir, FRS and the Institute of Orthopaedics for part of Fig. 33; Professor S. M. Partridge and the editor and publishers of *Federation Proceedings* for Fig. 27; Professor J. Paul and the editor and publishers of *Nature* for Fig. 95; and Professor J. W. Smith and the editor and publishers of *Nature* for Fig. 26*a*.

August 1978 R.A.S.

1

INTRODUCTION

Cartilage or gristle (Old Teutonic *grinstu*, to grind) is a connective tissue, relatively resistant to compressive, tensile and shearing forces. It is more deformable than bone, which explains why bone forms the major part of the skeleton. In most situations, cartilage forms large masses of tissue devoid of blood vessels, nerves or lymphatics except at its periphery. The absence of these vulnerable structures and the physical characteristics of cartilage make it an ideal cushioning material to distribute the load on bones articulating in cartilaginous and synovial joints. It is sufficiently rigid to provide skeletal support for structures such as the ear, respiratory system and part of the rib cage where elastic deformability is also required. These mechanical properties of cartilage are a consequence of the physico-chemical organisation of the matrix which makes up the bulk of the tissue. The matrix itself exists as the result of the activity of the cartilage cell or chondrocyte (χονδρος, a grain).

Despite the importance of the mechanical role of cartilage, it is the cells and their response to environmental factors which determine the nature of the tissue. Chondrocytes are remarkable because they can produce unique matrix components and maintain the orderly form of the tissue relatively remote from nutritional sources or obvious control mechanisms. Although they are few in number, the chondrocytes are the only living agents available to adapt the tissue to local change. The regulation of the activity of these cells, their mode of interaction with the environment and the reasons why their response to pathological lesions may be of little avail are problems of great interest.

Cartilage has a deceptive appearance of simplicity and homogeneity, but in reality is quite variable. This is true of the cells as well as the matrix. All chondrocytes have a deeply crenated contour which increases the surface area:volume ratio, but they differ in general shape, size and distribution in different parts of the tissue. Functionally, the cells of adult cartilage have a low respiratory activity and exhibit some regional and perhaps local diversity also with respect to rates of synthesis and types of secretory product. The cells in their lacunae are separated from one another by variable amounts of matrix, lying either singly or in small groups. The cellularity of different cartilages and of the regions within a single specimen shows a wide range of variation, the cells occupying 1–10% of the tissue volume (Hamerman & Schubert, 1962).

Apart from the cell content, about three-quarters of cartilage (Table 1) is water, collagen fibres forming about half of the dry matter, with smaller variable amounts of proteoglycans (protein–glycosaminoglycan complexes). Other components are non-collagenous protein, lipid and inorganic material. The three varieties of cartilage (Fig. 1) found in the body are distinguished by their fibre content. In hyaline cartilage, the proportions of collagen and proteoglycan are such that the tissue has a glassy, sometimes translucent, appearance and no fibres can be seen either macroscopically or with routine histological staining techniques. Fibrocartilage has a higher proportion of collagen and a very low proteoglycan content. Hence the fibres are visible to the naked eye and accessible to routine stains. Elastic cartilage is characterised by elastic fibres in addition to collagen and proteoglycan (Table 1).

A fibrous membrane, the perichondrium, surrounds cartilaginous tissue proper. Though it lacks a perichondrium on its superficial and deep surfaces, articular cartilage in many respects resembles one half of the other hyaline cartilages (Fig. 1). Nevertheless, the different cartilages vary considerably in terms of chemical content and cellularity (Table 1). In articular cartilages, where comparison is not complicated by the presence of a perichondrium, the chemical content and cell density vary even in different parts of the same joint and at different depths in the tissue. In addition, since the chondrocyte forms a discontinuity in the matrix and is a centre of synthetic and degradative activity, it is not surprising that it imposes biochemical heterogeneity on the matrix in its vicinity, as readily shown by local variation in histological and ultrastructural features.

During development much of the skeleton is cartilaginous. This is associated with the small scale of the rudiments and related stresses. An important factor in the requirement for cartilage during development is its capacity for interstitial growth. While bone can grow only on its surface (appositional growth), chondrocytes can divide and produce more matrix in the interior of cartilage. Thus chondrocyte proliferation and hypertrophy in cartilaginous epiphysial growth plates (Fig. 1) permit growth in length of a bone without disturbing joint organisation. However, apart from its growth characteristics the pre-eminence of cartilage in the embryonic skeleton prior to ossification must be due to other qualities. Possibly it is a consequence of the simple, rapid and biochemically economical method by which chondrocytes produce a rigid tissue. The cells have a relatively small nutrient requirement; strength and resilience depend only on the retention of water within the collagen–proteoglycan gel.

The palaeontological origin of cartilage as a tissue is not known, partly because uncalcified cartilage leaves no fossils. Cartilage is found in invertebrates as well as vertebrates. Schaffer (1930) and others believed 'true' cartilage to be present only in vertebrates. He excluded invertebrate cartilage partly on histological grounds but principally because such tissues would not yield gelatin. However, Person & Philpott (1969) consider that tissue which satisfies criteria for cartilage in terms of consistency, histology, ultrastructure and chemical composition (i.e. the presence of collagen and chondroitin

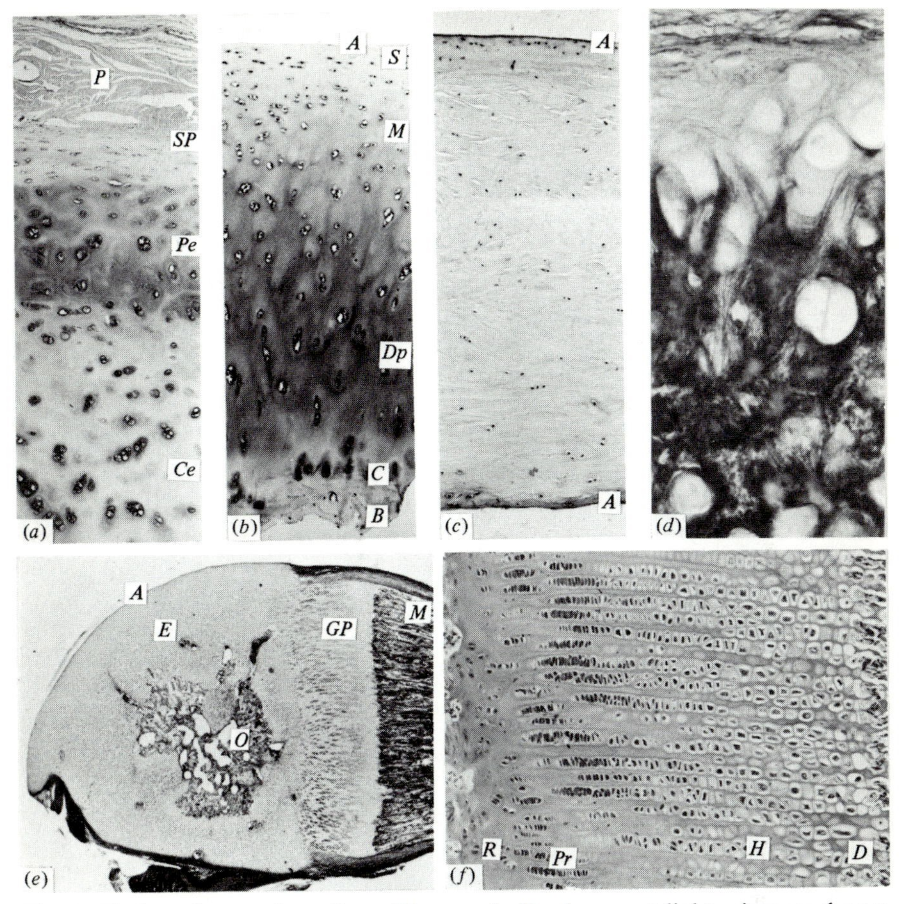

Fig. 1. Various forms of cartilage. These and all subsequent light micrographs are stained with haematoxylin and eosin unless otherwise stated. (*a*) Hyaline (human costal), magnification × 32. The perichondrium (*P*) surrounds the cartilage which may be subdivided into subperichondrial (*SP*), peripheral (*Pe*) and central (*Ce*) zones. (*b*) Articular hyaline (human, capitulum of humerus), magnification × 40. The articular surface (*A*) lacks a perichondrium. Superficial (*S*), middle (*M*) and deep (*Dp*) zones are described in the uncalcified cartilage. This joins the calcified cartilage (*C*) adjacent to the subchondral bone (*B*) at a faint basophilic lamina, the 'tidemark'. (*c*) Fibrocartilage (human sterno-clavicular intra-articular disc), magnification × 50. Both articular surfaces (*A*) and fibrous laminae are visible. (*d*) Elastic (human epiglottis). Stained with orcein, magnification × 250. Note the densely stained elastin mesh in the matrix. (*e*) Immature hyaline (rabbit lower femur), magnification × 10. Epiphysial cartilage (*E*) surrounds the ossification centre (*O*) and merges imperceptibly with the articular cartilage (*A*). The growth plate (*GP*) separates the epiphysial bony centre from the growing end or metaphysis (*Me*) of the bone shaft. (*f*) Epiphysial cartilaginous growth plate (dog upper tibia), magnification × 64. The metaphysis lies to the right. The columns of cartilage cells can be subdivided into reserve (*R*), proliferative (*Pr*), hypertrophic (*H*) and degenerative (*D*) zones.

TABLE 1

Chemical composition of various forms of cartilage in two species. Non-collagenous proteins, lipids and other materials form the percentage of dry weight not accounted for by collagen and glycosaminoglycans

Type of cartilage	Water content (%)	% dry weight: Collagen	Glycosamino-glycan	Elastin	Cell density (cells per mm³ tissue)
Human					
*Epiphysial (1, 5)	81	37	15	–	–
*Bronchial (4)		37	11	–	–
*Costal (2, 5)	61	38	6	–	11000
*Articular, femoral head (6)	72	66	18	–	10000
Fibrocartilage, meniscus of knee joint (3)	74	78	2.4	0.6	12000
Bovine					
Epiphysial (3)	76	51	22	0	–
Nasal (3)	74	35	43	0	–
Articular (3)	78	72	14	0	20000
Fibrocartilage, meniscus (3)	66	82	2	0	–
Elastic, auricular (3)	71	53	12	19	–

* Calculated from authors' data.

References: (1) Lindahl (1948); (2) Linn & Sokoloff (1965); (3) Peters & Smillie (1971); (4) Saltzman, Sicker & Green (1963); (5) Sewell & Pennock (1976); (6) Venn & Maroudas (1977). Cell densities: author's data.

sulphates, although the collagen type has not been characterised) occurs in various invertebrate species. These include the 'feather-duster'worm (*Eudistyla polymorpha*) in the annelida, the horseshoe crab (*Limulus polyphemus*) in the arthropoda and the squid (*Loligo pealii*) in the mollusca. Squid head cartilage in particular resembles vertebrate hyaline cartilage (Philpott & Person, 1970). Similarly, tissue closely resembling hyaline cartilage has been identified with the electron microscope in insects (*Locusta migratoria*) in the anterior wall of the ejaculatory duct (Martoja & Bassot, 1965). Invertebrate cartilage is not mineralised but calcification (by hydroxylapatite) can be induced by raising the temperature of the water to 37 °C (Eilberg, Zuckerberg & Person, 1975). The possibility that cartilage originated in the invertebrates is considered by Person & Philpott but its occurrence there and in the vertebrates is attributed to convergent evolution, a view shared by Mathews (1975).

In the vertebrates, the relationship of cartilage and bone has been controversial. Romer (1942) believed cartilage to be ancient but bone to be the more primitive tissue. He considered the presence of cartilage in the adult an example of neoteny, the fossil record indicating that the persistence of cartilaginous skeletons in the adult cyclostomata and chondrichthyes resulted from secondary reduction of bone during evolution. However, Mathews (1975) affirms that the cartilaginous state of shark vertebrae may be primitive since although most of the cartilaginous skeleton is heavily calcified, the centra in a primitive species (*Notorhynchus maculatum*) are uncalcified; the more remote placoderms also had little or no ossification of the vertebral skeleton. Whatever the status of shark cartilage, however, it appears that Romer's view requires modification in the light of evidence of calcified cartilage in an early vertebrate (*Eriptychius Americanus*) of the Middle Ordivician, about 500 million years old (Denison, 1967). This finding substantiates the idea that cartilage preceded endoskeletal, though not dermal, bone (Halstead, 1969; Mathews, 1975).

There seems litle doubt that the major chemical constituents of the tissue are very old indeed although evidence of polysaccharides in interstellar space (Hoyle & Wickramasinghe, 1977), while provocative, is unlikely to be related to the evolution of cartilage. Mathews (1975) considers that collagen is one of several fibrous proteins which may have a common ancestry; fibres with the characteristic collagen periodic cross-banding have been found in protozoa and the protein is common in the sponges and coelenterates. Since oxygen is required for certain stages of synthesis and fibre stabilisation, collagen production (and its consequences in terms of the development of rigid tissues) may have been affected by the marked increase in atmospheric oxygen in the Palaeozoic from Pre-Cambrian conditions (Towe, 1970). The low level of oxygen in the Pre-Cambrian atmosphere has again been invoked as the stimulus for polyanion synthesis and secretion (Scott, 1975). A pericellular sheath of sulphated and carboxylated material would have provided the cells with an electrostatic shield against the ravages of negatively charged hydrated electrons, produced in abundance by radiation in an oxygen-free environment. Whether or not the Pre-Cambrian fossil bacteria of 3000 million years ago (Banghoorn & Schopf, 1966) had polysaccharide capsules is not known but undoubtedly glycocalyces must have been acquired in early metazoan forms. Moores & Partridge (1974) speculate that protein secretion in general is a modification of the primitive function of producing a glycoprotein cell-coat.

2

CHONDROCYTE STRUCTURE

Structure gives information about function and modifications of cell activity resulting from intrinsic or extrinsic causes. Ideally any investigation of the biochemical or biomechanical properties of cartilage should be accompanied by adequate histological control; in fact such multidisciplinary approaches are becoming standard practice. Unavoidably, much of our knowledge of chondrocyte morphology and ultrastructure in particular is derived from histologically 'fixed' (i.e. dead) tissue. It is reassuring, therefore, to find that observations of living chondrocytes substantiate results obtained on fixed tissue, as far as this is possible (Arlet, Mole & Barriuso, 1958; Kawiak, Moskalewski & Darzynkiewicz, 1965; Chesterman & Smith, 1968).

Typical chondrocytes are ovoid cells and have an irregular surface with projecting cell processes (Fig. 2) but their overall shape can vary from spheroidal to a flattened form. Cell size also varies considerably, the largest diameter ranging from 10 to 30 μm or more. Many factors appear to influence the shape, size and other characteristics of the chondrocyte, including the type of cartilage, the position of the cell within the tissue, the cell density, the age of the organism and so on. However, chondrocytes have many features in common with each other and with other cells. Indeed it is surprisingly difficult to identify morphological characteristics which can be used to define a chondrocyte, although this may be possible using biochemical and functional criteria.

CELL MEMBRANE

The chondrocyte surface is limited by the cell membrane or plasmalemma. No exhaustive studies have been carried out on chondrocyte membranes *per se* but the plasmalemma evidently has the usual structure of a bimolecular leaflet of phospholipids and cholesterol coated with protein. Enzymes typically associated with the plasmalemma, such as 5′-adenosine monophosphatase, are found in chondrocyte membranes (Ali, Sajdera & Anderson, 1970).

The effects of substances such as vitamin A and oleic acid, which in cartilage eventually result in degradation of matrix macromolecules, are said to be mediated through their action on the cell membrane, as in erythrocyte preparations (Lucy & Dingle, 1964).

Cells of most tissues possess a coating of glycoprotein on their outer surface. It is not known whether the chondrocyte has such a 'glycocalyx' and it is

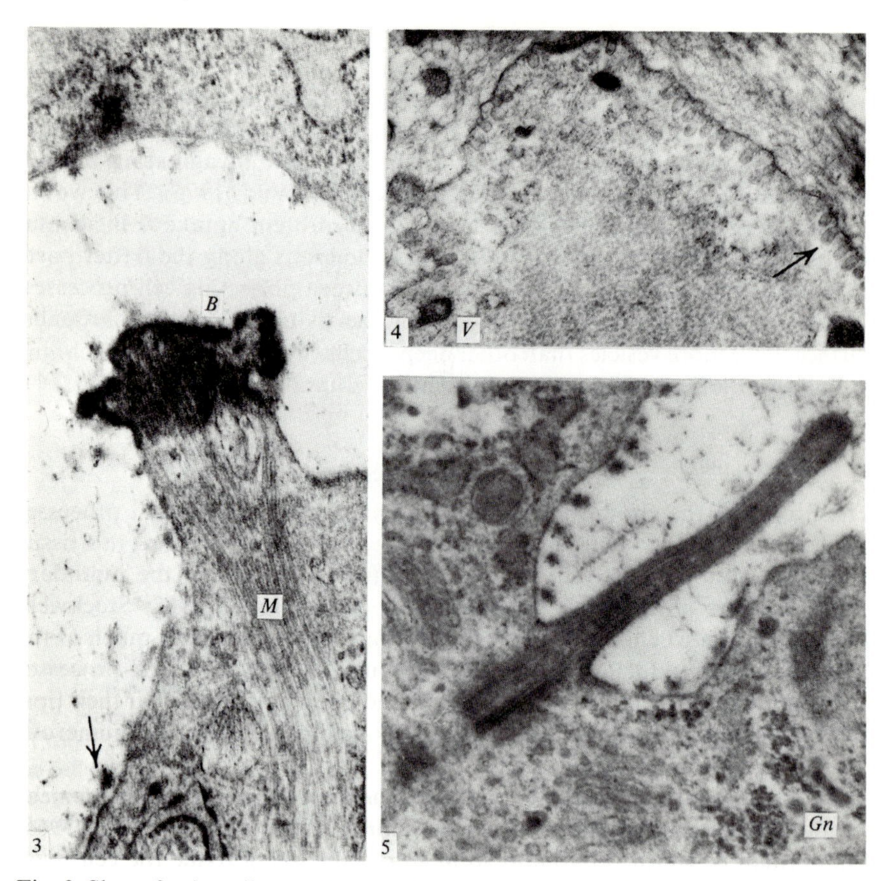

Fig. 3. Sheep fetal cartilage. Glutaraldehyde fixation, magnification ×28 500. Section through mid-body (*B*) of cell in terminal stages of mitosis. Numerous microtubules (spindle fibres) can be seen (*M*). The surfaces of the separating daughter cells exhibit amorphous material (arrow). (From Stockwell, 1971*a*.)
Fig. 4. Rabbit articular cartilage (three months). Osmium tetroxide fixation, magnification ×23 500. Most of the cytoplasmic area visible is occupied by glycogen. A coated vesicle (*V*) and numerous micropinocytotic vesicles are present (arrow).
Fig. 5. Sheep fetal cartilage. Glutaraldehyde fixation, magnification ×30 000. A single cilium projecting from a 'bay' in the cell margin lined by amorphous material. Glycogen (*Gn*) granules and Golgi membranes are present. (From Stockwell, 1971*a*.)

in general appears to decline with growth, a chondrocyte with a single cilium may be found in aged cartilage reacting to pathological damage. Such cilia have a basal body which is one of the centrioles of the cell and, although exhibiting eight or nine peripheral pairs, have no central fibres in the shaft. They are generally considered to be non-motile but their function is conjectural. By analogy with the cilium-like structure of the middle part of the retinal rod, a role as a chemoreceptor has been suggested but there appears to be no constant orientation of the solitary cilium in fetal epiphysial cartilage either with respect to vascular canals or to neighbouring cells (Stockwell, 1971*a*).

Even in the earliest stages of development of the cartilage blastema, the pre-cartilage cells only occasionally make extensive contacts with adjacent cells, though they possess long processes and are relatively closely packed (Gould, Day & Wolpert, 1972). Such contacts show 20-nm intervals between the plasma membranes, containing electron-dense material; gap junctions also occur. In older, more mature cartilage, the contiguous membranes of adjacent chondrocytes may be apposed at sites which exhibit a demosome-like structure (Palfrey & Davies, 1966); such an appearance may indicate recent cell division.

CELLULAR ORGANELLES

The cell substance consists of a protein-rich fluid – the cytosol – in which a number of structures are embedded. These include the nucleus of the cell and the various tubular and membranous organelles which, with the cytosol, form the cytoplasm. Various inclusions of the cytoplasm – glycogen and lipid droplets – are prominent features in chondrocytes.

CELL NUCLEUS

The chondrocyte nucleus (Fig. 2) is about 4 μm in diameter in mature tissue although larger in fetal cartilage.

In general, nuclear shape conforms to that of the cell; it is usually ovoid but in a 'flattened' superficial articular chondrocyte the nucleus also is discoidal. With increasing age, the nucleus becomes more irregular (Fig. 6*a*) and indented and 'dumb-bell' forms may be found in aged cartilage (Barnett, Cochrane & Palfrey, 1963). The nuclear envelope consists of two membranes separated by the perinuclear space, occasionally in continuity with the cisterna of the granular endoplasmic reticulum. The envelope membranes become continuous at the margins of the nuclear pores. These openings (about 0.1 μm in diameter) in the envelope allow the passage of large molecules such as messenger RNA. A 'nuclear fibrous lamina' on the internal aspect of the inner nuclear membrane (Fig. 6*b*) is thicker (20–30 nm) in older chondrocytes (Oryschak, Ghadially & Bhatnagar, 1974).

The nucleus contains most of the DNA complement of the cell. Together with basic histone and acidic non-histone proteins, DNA forms the chromosomes visible at mitosis. In the interphase nucleus, this material (the chromatin) is dispersed in the clear nucleoplasm but chondrocyte nuclei display a large amount of condensed chromatin (hetero-chromatin) especially in adult tissue. This tends to abut on the nuclear fibrous lamina. One or more nucleoli per nucleus are found in immature chondrocytes but are less frequent in adult cells. These bodies contain RNA and the DNA coding for the ribosomes of the cytoplasm.

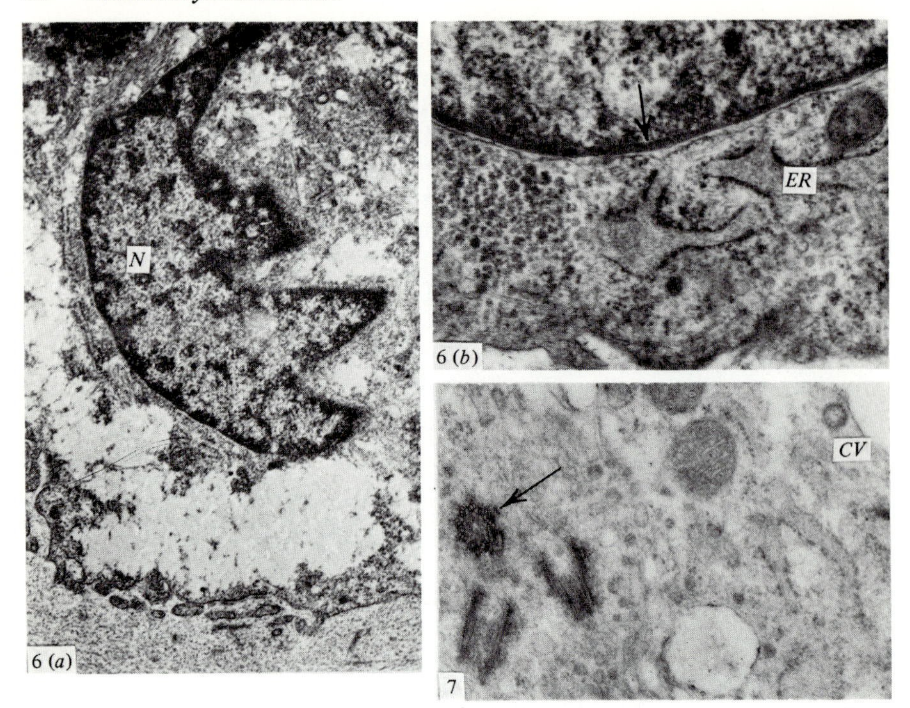

Fig. 6. (*a*) Human articular cartilage (56 years). Osmium tetroxide fixation, magnification × 12000. This ageing chondrocyte has a deeply crenated nucleus (*N*). (*b*) Sheep fetal cartilage. Glutaraldehyde fixation, magnification × 25000. A 'nuclear fibrous lamina' (arrow) lines the inner aspect of the nuclear envelope. Glycogen and granular endoplasmic reticulum (*ER*) are present.

Fig. 7. Rat upper tibial growth plate (four weeks). Glutaraldehyde fixation, magnification × 28000. Three centrioles are present in this section, possibly of a cell about to divide: note the appearance in transverse section (arrow). A coated vesicle (*CV*) lies near the cell margin.

The DNA content of the tissue may be used as an index of the number of cells present in a specimen, as, for example, in relating metabolic studies to cell content. Studies of DNA content and cell density of tissue in the rabbit (Mankin & Baron, 1965) and man (Stockwell, 1967*a*) have shown that the amount of DNA per chondrocyte (about 7×10^{-12} g) conforms to that of other tissue cell types in the species.

CYTOPLASMIC TUBULES

Centrioles

A single centriole (Fig. 7) or a pair are commonly observed in the chondrocyte in immature cartilage where the mitotic index is relatively high (Mankin, 1964) and solitary cilia not infrequent. In normal adult tissue where mitosis has ceased, centrioles are uncommon and are found only when reactive or regenerative cell multiplication occurs in the locality. They are short cylindrical bodies approximately 0.35 μm by 0.15 um, with a wall composed of nine sets of triplet fibres or tubules. Centrioles are associated with the production of two morphologically distinct and functionally separate structures although both contain fibres classed as microtubules; these are the mitotic spindle and the cilium.

Microtubules and cytoplasmic filaments

Microtubules are frequently observed in immature cells (Fig. 3) and occasionally short lengths may be found in adult chondrocytes (Palfrey & Davies, 1966; Weiss *et al.*, 1968). The microtubule is about 25 nm in external diameter with walls 5 nm thick, and has a less dense core 15 nm in diameter. The wall is composed of longitudinally running protofilaments; in transverse section it shows 13 subunits each 5 nm in diameter made of the protein tubulin (Burnside, 1975). An outer component of the microtubule wall (Behnke, 1975) which increases the outer diameter to 36 nm can be visualised with various polycations and is also thought to contain tubulin. Tubulin binds alkaloids such as colchicine, vinblastine and vincristine which thereby act as microtubule inhibitors. The protein in some ways resembles actin but has a different molecular weight and is also immunologically distinct. The subunit (molecular weight 55000–60000) exists as a dimer (6S) or as a large aggregate (36S) when not assembled into the microtubule. Recent work shows that a very high molecular weight (360000) protein is also required for microtubule formation (Keates & Hall, 1975). This forms 15% by weight of the intact tubule and each molecule interconnects 40 subunits. Formation of microtubules is believed to be a process of tubulin polymerisation on a template, extension of the tubule possibly stemming from the inner ring of a short duplex tube (Jacobs, Bennett & Dickens, 1975). In the special cases of the spindle and the cilium, the templates must be associated with centrioles or similar structures.

Several functions have been ascribed to microtubules apart from their structural and contractile role in the mitotic spindle and cilium. There is evidence that they have a role in the maintenance and regulation of cell shape and in the intracellular organisation and movement of materials. It has been postulated that they are involved in collagen secretion since colchicine and vinblastine depress collagen secretion (Bornstein, 1974), the procollagen accumulating in the Golgi region of the cell (Harwood, Grant & Jackson, 1976). If so this might account for the scarcity of microtubules in ageing chondrocytes where collagen turnover is very low or absent.

Cytoplasmic filaments were first noted in elastic cartilage (Sheldon & Robinson, 1958) where they are a prominent feature of the mature cells (Cox & Peacock, 1977). They are a common constituent of the chondrocyte, lying in the cytosol. In hyaline cartilage cells they are normally only a minor component but large masses and whorls of filaments (Fig. 8) are found in ageing or degenerate cartilage and under certain experimental conditions

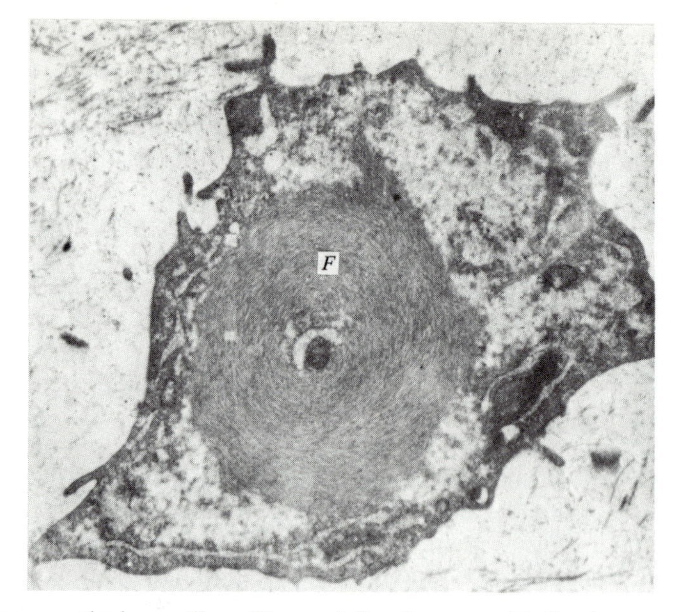

Fig. 8. Human articular cartilage (21 years). Osmium tetroxide fixation, magnification × 13000. A large whorl of fine filaments (*F*) is present in this deep zone chondrocyte. (From Stockwell, 1978*b*.)

(Barnett *et al.*, 1963; Meachim & Roy, 1967; Sprinz & Stockwell, 1976). Most filaments are about 10 nm in diameter but several classes of filaments may be present. Cultured chondrocytes contain 5–8 nm diameter filaments of actin or similar protein capable of binding heavy meromyosin, in addition to the 10-nm filaments (Ishikawa, Bischoff & Holtzer, 1969) which do not react with anti-actin sera (Kurki, Linder, Virtanen & Stenman, 1977). The chemical nature of the 10-nm filaments is not known. There is no evidence that they are collagenous since although they occasionally exhibit periodic beading (Sprinz & Stockwell, 1976) there is no typical collagen periodicity, nor do they contain hydroxyproline (Threadgold, 1975, cited by Dickson, 1977).

The relationship of filaments to microtubules is uncertain. It has been held that the two structures are interconvertible. Thus following incubation with vincristine, the cells lose their complement of microtubules and instead exhibit microtubule 'crystals' which bind colchicine-^{3}H; after post-incubation in normal media the cells gain large aggregates of filaments (Krishnan & Hsu, 1969, 1971). However, de Brabander, Aerts, Van de Veire & Borgers (1975), using fibroblasts, find that the 10-nm filaments are unaffected by anti-tubulins even though co-existing microtubules disappear; although large masses of filaments appear subsequently, the microtubules begin to re-form earlier. These results suggest an enhanced synthesis of filaments rather than microtubule interconversion. Similarly, tubulin crystals derived from microtubules disrupted by vinblastine do not react with antisera specific to 10-nm filaments

(Kurki *et al.*, 1977). The filaments appear to be a much more stable element of the cell than microtubules, which are susceptible to increased pressure and low temperature as well as colchicine and other alkaloids.

Although the cytoplasmic filaments may confer structural rigidity on the cytoplasm, the evidence suggests that they are redundant features of the cell and a sign of cell deterioration when present in abundance (Meachim & Roy, 1967). Possibly they are products of synthetic and/or secretion pathways which normally produce and secrete useful proteins (perhaps related *inter alia* to tubulin) but which become blocked under abnormal conditions. This might be a slow process in ageing or much more rapid following treatment of the cell with anti-tubulins. Whatever the role of the filaments, their intimate spatial relationship to lipid droplets is difficult to explain (Kostovic-Knezevic & Svalger, 1975; Sprinz & Stockwell, 1976).

CYTOPLASMIC MEMBRANOUS ORGANELLES

These are compartments of the cytoplasm enclosed by a single or double membrane, in which physico-chemical conditions and metabolic processes can differ from those in the cytosol or in other compartments. The membranes of certain organelles may show characteristic differences but the various organelles are not completely isolated from one another. Although it is unlikely that there is continuity between membranes of all organelles at any one time, nevertheless the smooth functioning of the cell requires that there should be intermittent but frequent connections between organelles.

Granular endoplasmic reticulum

This structure and the Golgi complex are normally the most prominent cytoplasmic organelles in the chondrocyte (Figs. 2, 9) since they produce the macromolecules. The cavity of the granular or 'rough' endoplasmic reticulum (ER) is extensive and highly irregular in form and, with its limiting membrane, forms a three-dimensional network. As seen in thin sections, it consists of short lengths of pairs of membranes, or oval membrane profiles enclosing spaces called cisternae. The membranes are studded on their outer (cytoplasmic) surface with electron-dense granules 15 nm in diameter, the ribosomes, which are also found unattached to membrane in the cytosol. Ribosomes contain RNA and protein and are complex structures where amino acids selected by transfer RNA are assembled into new polypeptide chains. Chondrocytes stain for RNA with the methyl green–pyronin technique (Shaw & Martin, 1962) and incorporate ^{3}H-cytidine, indicating active RNA synthesis (Mankin, 1963*b*). The prominence of the granular ER is an indication of protein synthetic activity in the cell.

Transcription of parts of the inherited DNA sequence in the chromosomes produces messenger RNA. These long, thread-like molecules act as inter-

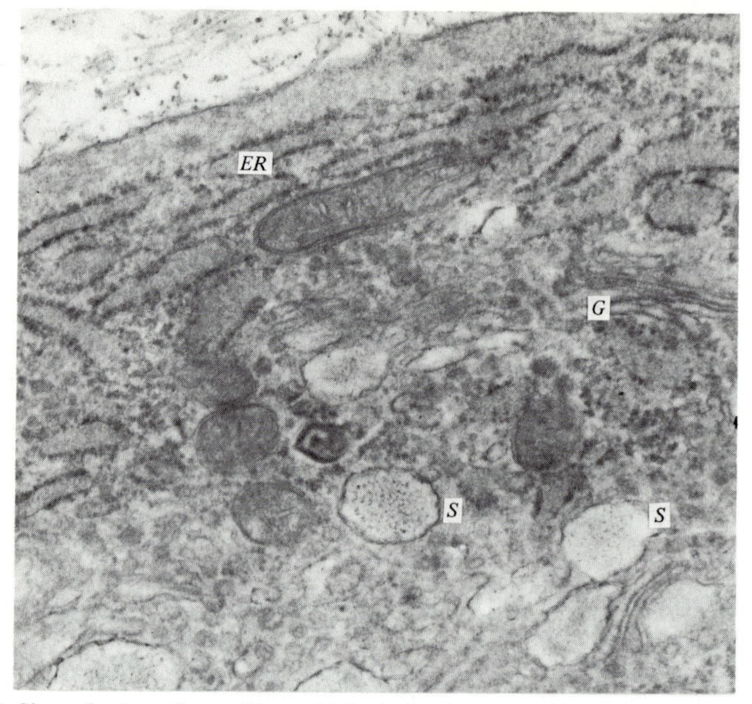

Fig. 9. Sheep fetal cartilage. Glutaraldehyde fixation, magnification ×33000. Part of a cell containing Golgi membranes (*G*) and granular endoplasmic reticulum (*ER*). Several secretion vacuoles (*S*) containing punctate and finely filamentous material are present.

mediaries between the nuclear DNA and the cytoplasmic ribosomes, where translation of the messenger RNA determines the appropriate sequence of amino acids for the protein synthesised. Post-translational modification and additions to the newly synthesised polypeptides and protein may occur, particularly in chondrocytes. The material enters the cisternae of the rough ER and is secreted into the matrix via the Golgi complex, in contrast with proteins synthesised on free ribosomes, which are retained within the cell. However, not all protein synthesised by the attached ribosomes is destined for export, for example the outer mitochondrial membrane, soluble proteins (e.g. cytochrome c) of the mitochondrion and lysosomal enzymes are synthesised at this site (Beattie, 1968; Campbell & Von der Decken, 1974).

Golgi complex

This organelle consists of smooth (agranular) membranes characteristically forming flattened sacs often dilated at their margins, the swellings detaching as vacuoles about 0.5 μm in diameter. The parallel membranes and vacuoles are associated with microvesicles about 50 nm in diameter. Golgi membranes may occur in more than one site in the chondrocyte cytoplasm and are particularly well developed in immature chondrocytes.

The Golgi complex has several functions but its primary role appears to be the provision of membrane for 'packaging', usually for secretion. As in other secretory cells, proteins synthesised in the rough ER are transported in microvesicles to the flattened sacs, where they are condensed and packaged into Golgi vacuoles (Revel & Hay, 1963). In the epithelial cell, these become the familiar and plentiful secretion granules. However, while the Golgi itself is a prominent organelle in the chondrocyte, secretion vacuoles may not be observed and if present are often difficult to identify with certainty. They are relatively few in number compared with an epithelial gland cell and contain only a small amount of rather tenuous electron-dense material (Fig. 9). Tangential sections through the scalloped cell margin of the chondrocyte can present a similar appearance. Golgi vacuoles are sparse in elastic cartilage (Serafini-Fracassini & Smith, 1974; Cox & Peacock, 1977), although vacuoles discharging numerous small vesicles into the matrix are observed.

A second major function of the Golgi complex concerns polysaccharide synthesis. It appears that secreted proteins normally contain a small carbohydrate moiety which is initially attached in the rough ER, completion of synthesis occurring in the Golgi complex. In the chondrocyte (and other cells secreting large amounts of polysaccharide) this function of the Golgi complex is accentuated (Godman & Lane, 1964; Neutra & Leblond, 1966; Horwitz & Dorfman, 1968). It makes a significant contribution to the synthesis and sulphation of the polysaccharide side chains of proteoglycan molecules secreted into the matrix. Hence in the growth cartilage where synthesis is very active, secretion vacuoles are plentiful and their polyanionic content stainable with bismuth nitrate (Smith, 1970) and ruthenium red (Thyberg, Lohmander & Friberg, 1973).

A third role of the Golgi complex is in the production of lysosomes, vesicles containing degradative enzymes, which are concerned with intracytoplasmic digestion. It is probable that the essential function of the Golgi in this case is to provide the wrapping membrane of the lysosome.

Lysosomes

The lysosome was first detected during biochemical cell fractionation procedures, its properties indicating a body similar in size to small mitochondria, bounded by a relatively impermeable membrane and containing hydrolytic enzymes, particularly acid phosphatase. Lysosomal enzymes are synthesised in the rough ER and pass to the Golgi region of the cell for packaging. A specialised set of smooth membranes associated with the concave surface of the Golgi complex may be involved (Novikoff, Novikoff, Quintana & Hann, 1971). The structures produced are known as primary lysosomes and are usually only of moderate electron density, possibly because many lysosomal enzymes are glycoproteins (Lloyd & Beck, 1974). They may be as small as 50 nm in diameter but also occur as larger, darker 'dense bodies' (0.4–0.8 μm). Morphologically, primary lysosomes (Fig. 10*a*) differ from mitochondria in possessing a single rather than a double membrane but have few characteristics which distinguish them from other vacuoles of similar size. A narrow electron-lucent interval may separate the membrane from the contents of the lysosome (Daems & Van Rijssel, 1961). The presence of acid phosphatase activity is an essential criterion but for various reasons such enzyme histochemical studies are not usually performed in routine electron microscopic studies of chondrocytes and consequently only a tentative identification of lysosomes may be permissible.

Primary lysosomes are destined to fuse with other cytoplasmic vacuoles or with the plasmalemma, thereby discharging their enzymes into the vacuole or extracellularly. After fusion with a vacuole, they form secondary lysosomes (Figs. 2, 10*b*). 'Autolysomes' result from fusion of a primary lysosome with a vacuole in which part of the cytoplasm (usually containing an effete organelle or excess secretion product) has been sequestered: the process of digestion is termed autophagy. 'Heterolysomes' are formed by fusion with a vacuole arising by pino- or phagocytosis: this process of 'endocytosis' facilitates the digestion of extracellular substances. Digestion in the secondary lysosome may be incomplete and the heterogeneous indigestible residues may then form complex bodies. These often contain a lipid globule as well as granular and fibrillar material and are occasionally found in chondrocytes (Meachim, 1967). When all hydrolytic enzyme activity has been lost, the aggregate may be classified as a post-lysosome or residual body (de Duve & Wattiaux, 1966; Lloyd & Beck, 1974).

Lysosomes obviously have an important role in cell metabolism and the turnover of cytoplasmic organelles. In cartilage, however, lysosomes could be involved also in the major function of the chondrocyte, i.e. the local control of the matrix. Ideas concerning lysosomal involvement stem from the work of Fell and her colleagues, who showed that hypervitaminosis A leads to dissolution of matrix in cartilage explants in culture (Fell & Mellanby, 1952), by potentiation of lysosomal proteases (Fell & Dingle, 1963). Thus exteriorisation of lysosomal enzymes by fusion of lysosomes with the plasmalemma followed by endocytosis could be of considerable significance in normal degradation and turnover of the matrix (Fell, 1969; Dingle, 1975). Such mechanisms are relevant to the degradative changes in growth plate cartilage matrix just before calcification (Campo & Dziewiatkowski, 1963; Jibril, 1967) and to matrix depletion prior to cartilage degeneration as, for example, in articular cartilage fibrillation, i.e. roughening and splitting of the articular surface (Matthews, 1953; Mankin & Lippiello, 1970). Lysosomes which could facilitate these processes have been demonstrated ultrastructurally in chondrocytes both in the growth plate (Thyberg & Friberg, 1970;

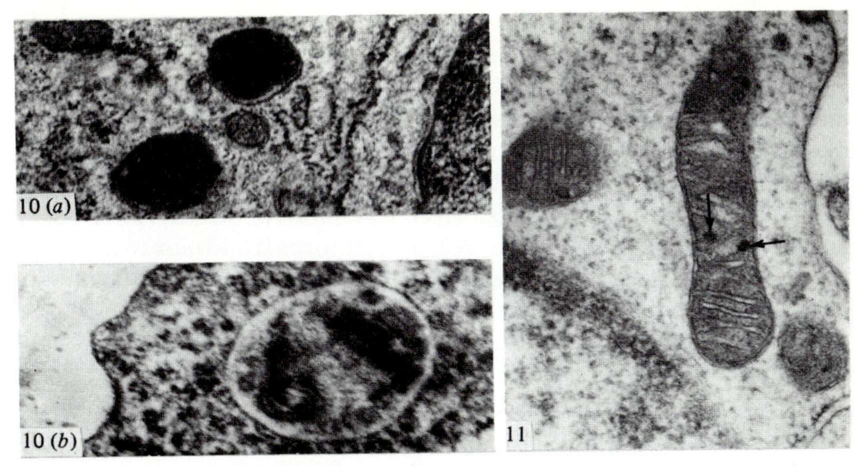

Fig. 10. Rabbit articular cartilage (three months). Osmium tetroxide fixation. (*a*) Primary lysosomes, magnification ×30000. (*b*) Secondary lysosome, magnification ×49000. (*b*) by courtesy of Dr R. Sprinz.
Fig. 11. Sheep fetal cartilage. Glutaraldehyde fixation, magnification ×36000. Mitochondria containing several dense granules (arrows).

Serafini-Fracassini & Smith, 1974) and in articular cartilage (Chrisman, 1967).

Mitochondria

It is well known that mitochondrial morphology is affected physiologically by the ambient metabolic state and artefactually by histological fixation conditions. When adequately fixed these organelles vary in shape but in chondrocytes are usually ovoid with the lesser diameter ranging from about 0.25–0.5 μm (Fig. 11). Characteristically the mitochondrial wall is formed by two membranes. The outer membrane appears to be little more than an enclosing vacuolar membrane although it has certain metabolic functions. The inner membrane is more complex and is invaginated into the interior in the form of shelf-like cristae although tubular and vesicular forms can occur in other cells. In older chondrocytes the cristae tend to be few and difficult to distinguish in the electron-dense mitochondrial matrix enclosed by the inner membrane.

Mitochondria have a number of functions, including fatty acid synthesis, steroid hydroxylation, amino acid metabolism and so on. However, their principal role is the production from foodstuffs of reducing equivalents such as reduced nicotinamide-adenine dinucleotide (NADH); these can subsequently be aerobically oxidised. The oxidation steps are coupled to the production of ATP in the process of oxidative phosphorylation; ATP is used to drive chemical reactions required for cellular activities. Enzymes (e.g. dehydrogenases) involved in producing NADH are located in the mitochondrial matrix or in the cytosol, while the respiratory 'chains' of cytochromes required for aerobic oxidation form integral units of the inner mitochondrial membrane. ATP generation is associated with the inner membrane or elementary particles, spheroidal structures 8–9 nm in diameter which project into the mitochondrial matrix and are attached to the membrane next to the cytochrome chain units (Baum, 1974). Thus large numbers of mitochondria with profuse cristae indicate a high level of respiratory activity. Although in young chondrocytes mitochondria display well-formed and numerous cristae, in older chondrocytes the poor development of these organelles is compatible with the low respiratory activity found in ageing cartilage.

The mitochondrial matrix possesses dense granules (Fig. 11) containing calcium phosphate. This salt accumulates as the result of calcium ion uptake by transport systems located in the inner membrane which can be activated by ATP hydrolysis (Lehninger, 1970). Prior to calcification, chondrocyte mitochondria contain increased numbers of dense granules and take up large amounts of calcium (Martin & Matthews, 1969) but in non-calcifying cartilage the granules are not prominent. The mitochondrial matrix also contains a small amount of DNA and RNA. This is thought to be sufficient to code for some proteins of the inner membrane. However, mitochondrial biogenesis is probably largely dependent on the usual cytoplasmic ribosomal systems (Beattie, 1968).

CYTOPLASMIC INCLUSIONS

Glycogen

Glycogen was first observed in chondrocytes in epiphysial cartilage by Rouget (1859). Large amounts accumulate in the growth plate but disappear in the lower hypertrophic zone with the onset of calcification. Ultrastructurally, glycogen takes the form of particles 20–30 nm in diameter in the cytosol (Figs. 4, 5). The particles are composed of 3–10-nm subunits and become interconnected to form aggregates 40–70 nm in diameter in the hypertrophic chondrocytes of the growth plate (Daimon, 1977). In the degenerative zone the number but not the size of the aggregates is reduced. Glycogenolysis was once thought to be an essential step in the calcification mechanism by providing a source of sugar phosphates to act as substrates for alkaline phosphatase (Robison, 1923). This interpretation has been modified over the years, a more indefinite role being assigned to glycogen as a raw material for matrix synthesis (Pritchard, 1952), but current views suggest that sugar phosphates together with other organic phosphates are directly involved in the calcification process (Ali, 1976).

Glycogen is also abundant in the cells of permanent cartilages including those that do not normally calcify. It is more profuse in the deeper cells of the tissue and increases during the later stages of growth until full maturity is reached (Montagna, 1949; Stockwell, 1967*b*). A simple storage function is

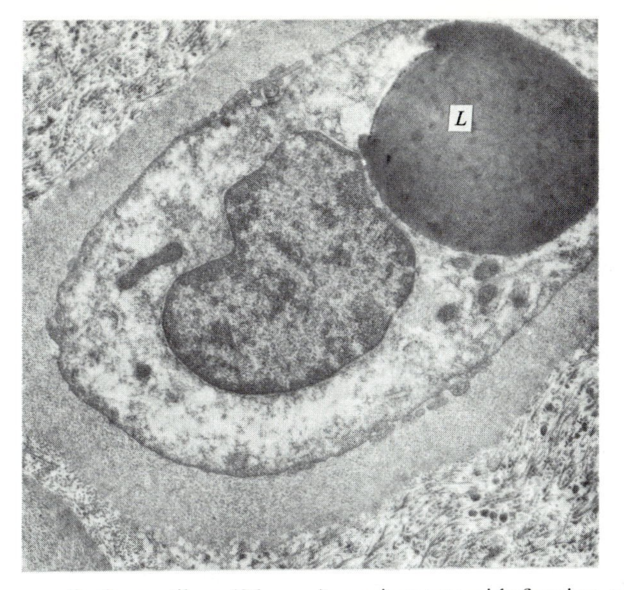

Fig. 12. Human articular cartilage (56 years), osmium tetroxide fixation, magnification × 8000. Middle zone chondrocyte containing large lipid globule (*L*). (From Stockwell, 1975*b*.)

indicated by the diminution in glycogen content as ear cartilage cells are stimulated to synthesise glycosaminoglycan rapidly, following dissolution of the matrix by intravenous papain (Sheldon & Robinson, 1960).

Lipid

In cartilage, lipid occurs both in the cells and in the matrix (Putschar, 1931). Chondrocytes usually contain one or more fat globules (Figs. 12, 16*a*) in addition to the lipid forming an integral part of the cell membranes. Fat globules in chondrocytes were first described by Leidy (1849) and have long been considered to be a normal constant inclusion (Sacerdotti, 1900; Putschar, 1931). The amount varies with the type of cartilage, its age and the position of the chondrocyte in the tissue (Sacerdotti, 1900; Collins, Ghadially & Meachim, 1965). These factors appear to be more important than species differences although fat globules (10–15 μm in diameter) are larger in rabbit than in human chondrocytes.

In the central region of non-articular cartilage the intracellular fat increases in amount from fetal to middle life and then retrogresses (Fig. 13); in the more peripheral (subperichondrial) cells, fat is detectable earlier but the globules remain discrete and small. In articular chondrocytes the fat content is less than in other hyaline cartilage but is also most prominent in the middle zone. The fat globules in the central cells of adult cartilage react histochemically for neutral (triglyceride) lipid (Montagna, 1949; Stockwell, 1967*b*) although the

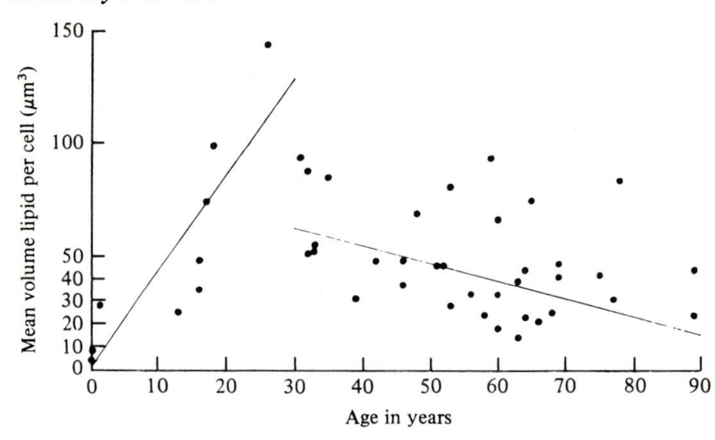

Fig. 13. Age-related changes in intracellular lipid of human costal cartilage. 0–30 years: $y = 0.4 + 4.27x$; 31–90 years: $y = 86.2 - 0.68x$. (From Stockwell, 1967*c*.)

small droplets in the subperichondrial cells stain for phospholipid only. In newborn cartilage also, only phospholipid can be detected histochemically in the chondrocytes (Stockwell, 1967*b*) but birefringence (Sheehan, 1948) and electron microscopy studies (Stockwell, 1971*a*) of epiphysial cartilage demonstrate small globules of triglyceride. Electron microscopy of chondrocytes often reveals cytomembranes and mitochondria in close proximity to fat globules (Davies *et al.*, 1962; Collins *et al.*, 1965) although the globule itself is not membrane-bound. The phospholipid present in these membranes might effectively mask the histochemical reaction of triglyceride in the very small fat globules found in immature cartilage.

Cytoplasmic filaments also abut on lipid droplets especially in ear cartilage where the chondrocytes contain large globules (Kostovic-Knezevic & Svalger, 1975; Cox & Peacock, 1977); this is also found in fat cells and subsynovial tissue. The small droplets accumulating in chondrocytes after experimental lipoarthrosis characteristically lie in the middle of a filamentous zone (Fig. 14) or exhibit a thin rim of cytoplasmic filaments. However, in the early stages of lipid formation there are few filaments at the periphery of the globule (Fig. 15), suggesting that the spatial relationship may be the consequence of passive adsorption or of packing within the cytoplasm (Sprinz & Stockwell, 1976).

Many investigators have been intrigued by the large amount of intracellular fat and its role in the chondrocyte. In other tissues, fat globules represent either a reserve energy store for the cell itself or for the whole organism. There may be exceptions, for example the fat pads of synovial joints (Davies & White, 1961) where a mechanical function may be served. However, as Collins *et al.* (1965) remark, while the amount stored in some chondrocytes appears to be excessive to local requirements, when regarded as a form of depot fat for the whole organism it would seem to be negligible. Quantitative data (Stockwell, 1966) suggest that per cm^3 cartilage, chondrocyte lipid droplets

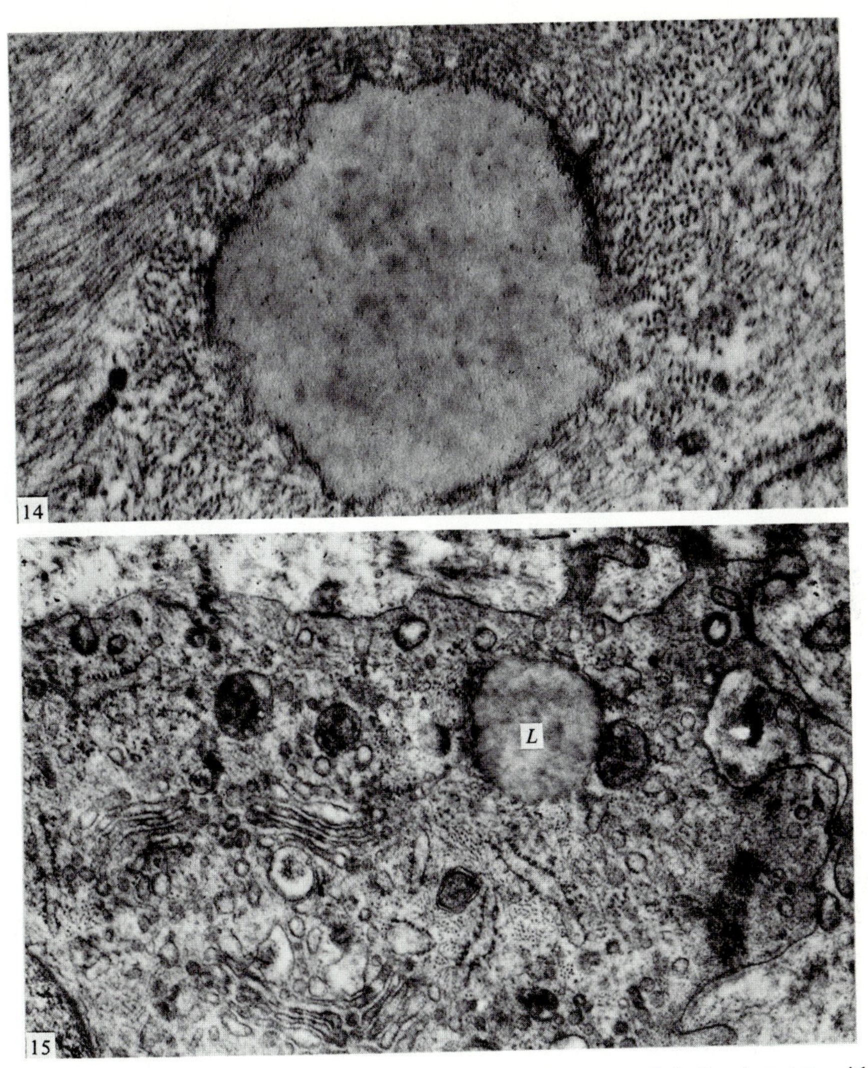

Fig. 14. Rabbit femoral condylar articular cartilage (three months). Osmium tetroxide fixation, magnification ×45000. Lipid globule surrounded by intracytoplasmic fine filaments. Specimen taken four days after intra-articular injection of glyceryl trioleate. (From Sprinz & Stockwell, 1976.)

Fig. 15. Rabbit femoral condylar articular cartilage (three months). Osmium tetroxide fixation, magnification ×26000. Specimen taken one day after intra-articular lipid injection. Intracytoplasmic fine filaments few or absent. Lipid globule (*L*) next to mitochondrion. Note profuse Golgi membranes. (From Sprinz & Stockwell, 1976.)

contain only about 0.1% of the amount of fat stored in the same volume of adipose tissue.

A possible role as a source of depot fat has been investigated. Sacerdotti (1900) found that the extractable lipid of rabbit ear cartilage does not change during starvation and Rabe (1910) reported a slight increase. However, in human hyaline and elastic cartilage, Wlassics (1930) and Bacsich and Wlassics (1931) concluded that the size of the intracellular lipid globule is related to the nutritional status of the individual and is unaffected by ageing. A somewhat contradictory finding in this investigation was that tuberculous cases, who of course were in the group of wasted individuals, tend to have as much intracellular lipid as the well-nourished cases; moreover, in the Tb group the lipid content shows the normal age-related decline. More recently, quantitative studies of human costal cartilage (Stockwell, 1967*c*) show that although there is a tendency for the total lipid content of the tissue to be reduced in thin individuals the mean content of intracellular lipid is not diminished; this suggests that only the extracellular lipid is affected. Similar investigations of articular cartilage also suggest that extracellular lipid (spread diffusely in the superficial zone) may be related to the nutritional status of the individual (Stockwell, 1967*c*). Since post-mortem material was used in all these studies, many of the thin individuals could have suffered from a relatively short terminal wasting illness; naturally thin individuals might contain less intracellular lipid but this requires further investigation.

Alternatively, the intracellular lipid might act as an energy store for the chondrocyte itself. However, there appear to be no studies of the chondrocyte lipid under conditions where synthetic activity is stimulated. Thus observing the effects of papain-induced depletion of ear cartilage matrix, Sheldon & Robinson (1960) did not comment on the state of the intracellular lipid during the recovery (matrix replenishment) period although, as mentioned above, glycogen rapidly disappears from the cells.

Although there is little evidence that the amount of intracellular lipid is related to the nutritional status of the whole organism, it is not known whether it is used by the chondrocyte itself as an energy reserve. Why then does it occur in chondrocytes in relatively large amounts? Possibly it is the result of macrophage-like activity. If fat is injected into the rabbit joint cavity, the articular chondrocytes accumulate lipid droplets (Ghadially, Mehta & Kirkaldy-Willis, 1970) although it appears to be the fatty acid rather than the glycerol moiety of the triglyceride which is incorporated (Sprinz & Stockwell, 1976), re-esterification occurring in the chondrocytes (Fig. 16*a–d*). However, zonal localisation of the intracellular lipid found in experimental lipoarthrosis is abnormal and does not reflect the natural process of accumulation of fat in chondrocytes. Thus it is the more superficial cells of the cartilage, normally relatively devoid of fat, which become engorged with lipid rather than those of the middle zone in which the cells normally have one or more large lipid globules. Furthermore the lipid-rich stage is transitory

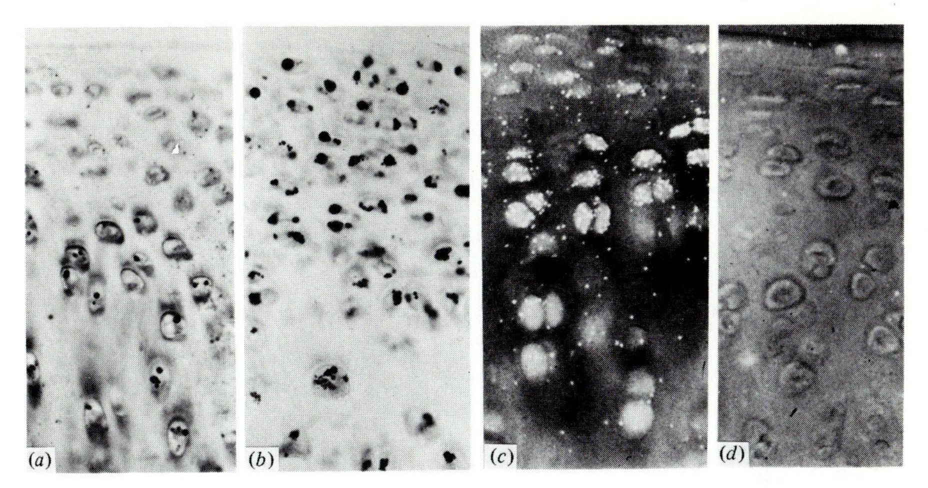

Fig. 16. Rabbit femoral condylar articular cartilage (three months), magnification (*a*) and (*b*) ×205; (*c*) and (*d*) ×410. Specimens taken four days after commencement of experimental lipoarthrosis. (*a*) Control (intact) joint. Frozen section, sudan black B. (*b*) Injected joint showing chondrocytes with large lipid inclusions. Frozen section, sudan black B. (*c*) Joint injected with glyceryl tri-(oleate-9,10-H^3). Frozen section autoradiograph, incident light. (*d*) Joint injected with glyceryl-2-^3H trioleate. The grain density over the cells is high in (*c*) but very low or absent in (*d*). Frozen section autoradiograph, incident light. (From Sprinz & Stockwell, 1976.)

and many of the cells degenerate and die (Sprinz & Stockwell, 1977). It is probable that fat droplets normally form during the course of intermediary metabolism rather than through macrophage-like activity.

CHONDROCYTE SIZE

Chondrocytes at the centre of a cartilage specimen are larger than those near the periphery (Beneke, 1973; Stockwell & Meachim, 1973). However, the difference in size is artificially accentuated by the gradation in shape from flattened discoidal and spindle-shaped cells near the perichondrial and articular surfaces to the deeper spheroidal cells. Hence in the usual transverse histological section, the peripheral cell exhibits only a narrow profile. It should be noted that cell shape is unaltered near osseochondral boundaries in the adult although profound changes occur in the growth plate. Modifications of chondrocyte shape also occur *in vitro*, dependent on the conditions and mode of culture (Sokoloff, 1976).

Interferometric methods give values for the dry mass of isolated calf nasal chondrocytes of $2.5 \pm 0.6 \times 10^{-10}$ g per cell (Kawiak *et al.*, 1965). There are considerable difficulties in estimating the linear dimensions of chondrocytes *in situ* using either light or electron microscopy. Cell boundaries are difficult to determine in light microscopy and it is unwise to utilise the classical lacunar

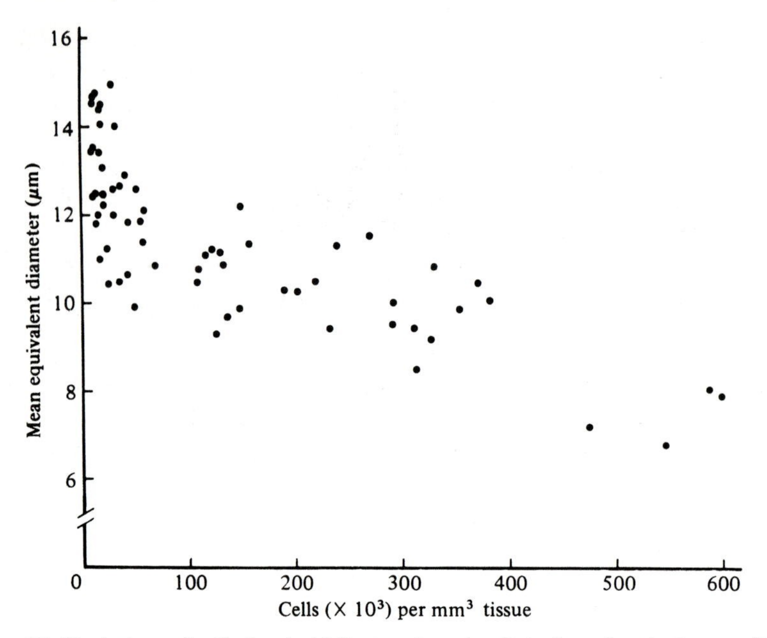

Fig. 17. Variation of cell size (middle zone) and cellularity of articular cartilage.

'rim' for this purpose. In electron microscopy there is the problem of selectivity of cell profiles in ultra-thin sections, coupled with the considerable shrinkage (about 50% by volume in cartilage) associated with tissue preparation (Anderson & Sajdera, 1971). Consequently, the few estimates of chondrocyte size vary widely, even within a single type of cartilage. Thus, based on light microscopy measurements of human articular cartilage, Leidy (1849) gives a value equivalent to 28 × 11 μm, while more recent reviews suggest larger (30–40 μm) and smaller (19 × 12 μm) sizes (Barnett, Davies & MacConaill, 1961; Stockwell & Meachim, 1973). In general, electron micrographs from a large number of sources indicate a smaller size for articular chondrocytes, with a maximum diameter of 10–12 μm. However, non-articular hyaline and elastic cartilages undoubtedly have much larger cells at their centre: even in electron micrographs, cells with diameters of 15–20 μm are commonly found.

Chondrocyte size increases during maturation of permanent cartilage (Beneke, 1973). This may be associated with the inverse relationship between tissue cellularity and cell size (Fig. 17), which seems to apply in adult cartilages of different species and dimensions (Stockwell, 1971*c*), since cell density falls during maturation. However, many of the enlarged cells in the central regions of cartilage contain fat globules up to 10 μm in diameter. These inclusions may partly account for the increase in cell size. Cell hypertrophy is a cardinal feature of the temporary cartilage of the growth plate (Figs. 18, 19) where the height of the juxta-metaphysial cell lacuna is a key factor in the rate of bone elongation (Sissons, 1956). Measurements of the intact tissue, using either light or electron microscopy, agree in confirming a massive increase ($\times$ 5–10) in cell volume from the proliferative to the hypertrophic zone. This is substantiated by observations of isolated proliferative and hypertrophic chondrocytes (Bourret & Rodan, 1976*a*). Maximum cell dimensions vary according to age up to 40 μm in the lower hypertrophic zone (Sissons, 1956; Walker & Kember, 1972*b*).

While in the permanent cartilages cell enlargement is insidious, with a time scale of decades in some species, the onset of hypertrophy in the growth plate seems to be rather abrupt, adjacent cells presenting very different features. Hypertrophy is not due to a 'ballooning' of the cell accompanied by a smoothing of its contour since projecting cell processes remain at least as numerous. Consistent with the high metabolic activity of these cells (Kuhlman, 1960) the mitochondria are plentiful and contain many dense granules. The cisternae of the granular ER become dilated although the Golgi complex and lysosomes are reduced in late hypertrophy. Extreme dilatation of ER cisternae is also a feature of randomly distributed chondrocytes in fetal epiphysial cartilage but such cells do not become hypertrophied (Stockwell, 1971*a*). Although chondrocyte hypertrophy in the growth plate is not associated with lipid accumulation, cytoplasmic organelles are more widely spaced (Holtrop, 1972) or even reduced in amount (Serafini-Fracassini & Smith, 1974); the cytosol is voluminous and pale. This suggests that the cells become increasingly hydrated.

HORMONAL EFFECTS ON CHONDROCYTE STRUCTURE

Many hormones cause ultrastructural changes in chondrocytes. Investigations have been carried out principally on epiphysial or articular cartilage. As might be expected the effects though similar are more pronounced in immature than in adult tissue. All hormones investigated appear to cause ultrastructural signs of cell regression (mitochondrial swelling and the presence of autophagic vacuoles) but this may be partly attributed to dosage. Although their physiological and biochemical actions vary considerably, hormones may be classified broadly into two groups as to whether they cause overdevelopment or underdevelopment of chondrocytes and their cytoplasmic organelles (Silberberg, 1968).

HORMONES CAUSING OVERDEVELOPMENT

Growth hormone (somatotrophin) acting via its intermediary somatomedin (see Chapter 4) and the hormones working synergistically with it in the body though not perhaps at the level of the cell (thyroxin, insulin, oestrogen and testosterone) belong to this group. As exemplified by growth hormone (Silberberg, Silberberg & Hasler, 1964; Silberberg & Hasler, 1971) there is

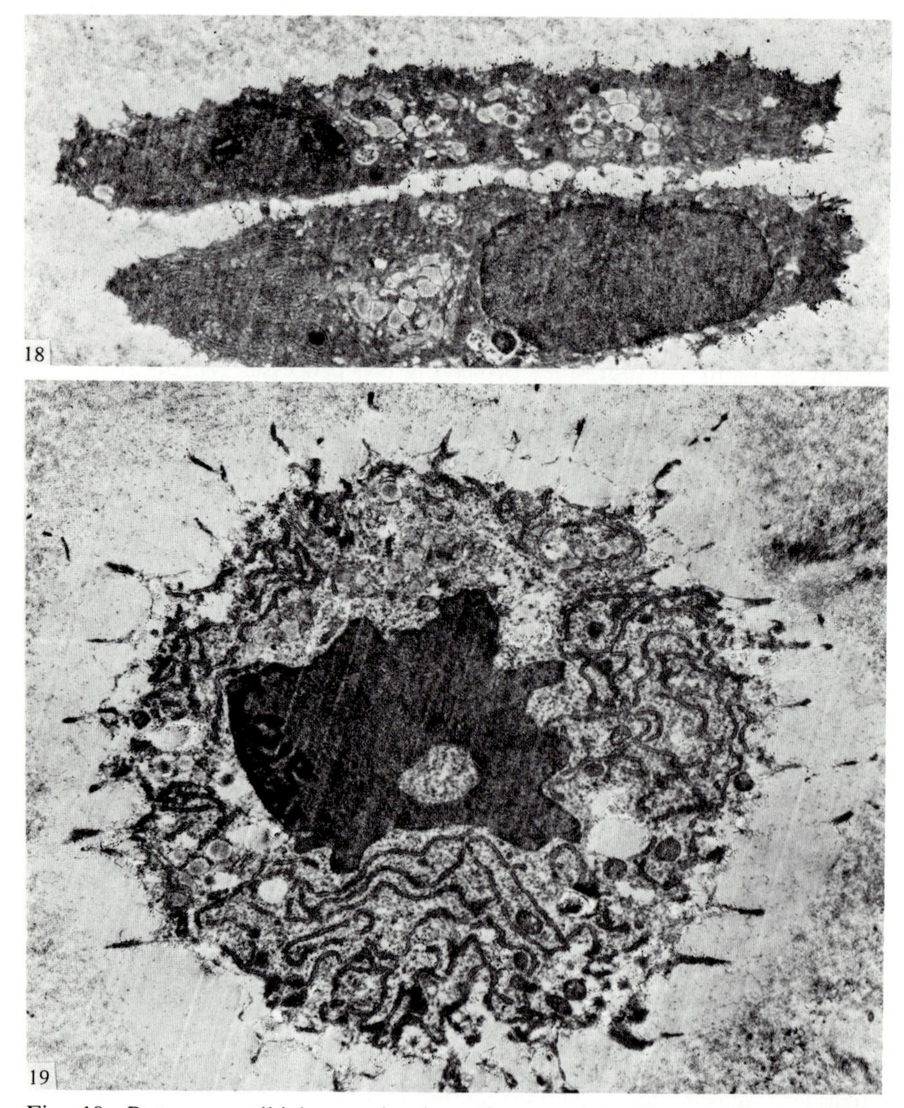

Fig. 18. Rat upper tibial growth plate (four weeks). Glutaraldehyde fixation, magnification × 6000. Two chondrocytes in proliferative zone. Note nuclei at opposite ends of the cells and the profuse secretion vacuoles.

Fig. 19. Rat upper tibial growth plate (four weeks). Glutaraldehyde fixation, magnification × 6000. Chondrocyte in hypertrophic zone. The nucleus is distorted, cytoplasmic organelles are unusually widely separated although there are large numbers of mitochondria and cisternae of the endoplasmic reticulum. Note the long radiating cell processes.

an increase in cell size, the amount of granular ER and Golgi complex; mitochondria and micropinocytotic vesicles become more numerous. Glycogen accumulates in cells of immature cartilage. Accompanying these cytoplasmic changes, the number of cells increases although there is also some evidence of cell degeneration and death. The pericellular matrix becomes more fibrous. These effects are entirely consistent with the action of the hormone in enhancing growth. The other hormones are broadly similar in their actions although exhibiting characteristic differences. The actions of testosterone (Silberberg & Hasler, 1972) and of insulin (Silberberg, Hasler & Silberberg, 1966) most closely resemble those of growth hormone. With testosterone, lipid as well as glycogen accumulates in the cell. However, although testosterone causes cell multiplication both in articular and in epiphysial cartilage, it is considered that its effects on maturation are more significant than on cell proliferation, resulting in a net acceleration of the normal maturation sequence (Fahmy, Lee & Johnson, 1971; Silberberg & Silberberg, 1971). The actions of oestrogen (Silberberg & Hasler, 1971) are somewhat at variance with the rest since although the ER and mitochondria become well developed and glycogen accumulates, the Golgi complex does not enlarge nor do the cells hypertrophy or proliferate; cytoplasmic filaments also accumulate. This conforms with the known action of oestrogens in inhibiting growth and glycosaminoglycan synthesis and in accelerating skeletal maturation (Silberberg & Silberberg, 1971). Thyroxin, while causing cell hypertrophy, etc. has little effect on cell proliferation and it is said that cell death is more common with this hormone (Silberberg, 1968; Silberberg & Hasler, 1971). However, propylthiouracil also causes considerable cell degeneration in the epiphysial plate (Dearden, 1974), where matrix vesicle formation and mineralisation is enhanced.

HORMONES CAUSING UNDERDEVELOPMENT

Cortisol and related substances belong to this group; there appears to be no data for aldosterone. Large doses of corticosteroids (*c.* 20 mg kg^{-1} body weight) cause profound ultrastructural changes in articular (Silberberg, Silberberg & Hasler, 1966) and in epiphysial plate chondrocytes (Dearden & Mosier, 1972). Cell size is reduced and the granular ER diminishes, becoming vesiculated, and later disintegrates; the Golgi complex shows some initial enlargement in articular chondrocytes but later retrogresses. Mitochondrial swelling is followed by vacuolation; cytoplasmic dense bodies, vacuoles and fine filaments appear. After three to five days of treatment, many dead cells are observed but cell proliferation in the growth plate is not totally impaired. Although histochemically glycogen diminishes in growth chondrocytes (Balogh & Kunin, 1971) ultrastructural observations indicate an increase in glycogen granules. In the growth plate, matrix vesicles are smaller and less numerous. These structural changes suggest a decrease in cell and matrix protein synthesis consistent with the inhibition of synthesis found biochemically.

ANTIGENICITY OF CHONDROCYTES AND CARTILAGE GRAFTS

Compared with most other tissues in the body, cartilage enjoys a considerable degree of immunological privilege. In man and other animals, homografts can persist for several years with the survival of many chondrocytes, as judged by their normal lipid and glycogen content (Craigmyle, 1955), respiration and glycolysis (Laskin, Sarnat & Bain, 1952) and ^{35}S-sulphate incorporation (Gibson, Curran & Davis, 1957). There have been two schools of thought concerning this phenomenon. The earlier view held that chondrocytes and cartilage are only weakly antigenic (Loeb, 1926; Gibson *et al.*, 1957), while alternatively it has been proposed that the matrix forms an immunological barrier between the cell and the lymphoid system (Bacsich & Wyburn, 1947). The elucidation of the problem has been complicated by many conflicting results and, as Heyner (1973) and Elves (1976) comment, has awaited the development of appropriate techniques and an adequate knowledge of immunological mechanisms.

Although only weakly immunogenic, immune responses to intact cartilage grafts do occur (Craigmyle, 1958; Stjernsward, 1964, quoted by Elves, 1976). The response is much more pronounced when preparations of chondrocyte suspensions isolated from whole cartilage by enzymatic digestion of the matrix (Kawiak *et al.*, 1965; Smith, 1965) are administered as homograft inocula. The cartilage nodules produced undergo lymphocytic infiltration at about 28 days, with loss of matrix metachromasia, peripheral chondrocyte necrosis and resorption of the tissue (Moskalewski & Kawiak, 1965). Heyner (1973) notes that while intact cartilage allografts exhibit lymphocytic infiltration only if transplanted into a sensitised host, nodules formed from suspensions of allogeneic chondrocytes undergo resorption whether placed in sensitised or non-sensitised hosts. These observations suggest that the intercellular matrix prevents rejection and that the immunological privilege enjoyed by cartilage cannot be the result of lack of antigenicity of the chondrocytes.

That chondrocytes do possess major histocompatibility antigens has been demonstrated unequivocally by Elves (1974) who has shown that isolated sheep articular chondrocytes stripped of their matrix have an antigen profile similar to that of autologous lymphocytes. Similarly, rat chondrocytes bear species-specific, strain-specific and tissue-specific antigens (Malseed & Heyner, 1976). Thus in an immunological context, the chondrocyte can no longer be regarded as a unique cell.

The protective role of the matrix lies in the prevention both of egress of chondrocyte antigens to the immunity system of the body and of access of cytotoxic antibodies to the chondrocytes (Elves, 1976). Thus the antigenicity of isolated chondrocytes increases as the amount of matrix remaining around each cell is reduced by successively longer periods of papain digestion (Elves, 1974). Moskalewski & Kawiak (1965) also found that that the degree of initial nodule formation by allogeneic isolated chondrocytes depended on the amount of 'jelly-like' substance present around the cells at transplantation. The presence of matrix also impedes antibody access. Immunological studies of articular cartilage fragments exposed to antiserum indicate that antibodies do not enter the tissue (Poole, Barratt & Fell, 1973) unless the matrix is first depleted either by growth of the cartilage explant in contact with soft connective tissue (Fell & Barratt, 1973) or by prior treatment with trypsin (Millroy & Poole, 1974). In the latter circumstances, the chondrocytes near the surface of the explant are killed but death of the more deeply placed cells does not occur owing to the protective barrier produced by new synthesis of ground substance (Millroy & Poole, 1974). Both aspects of the protective function of the matrix are consistent with permeability studies, showing that high molecular weight proteins (60000–80000) either do not enter cartilage or else diffuse through it only with difficulty (Peacock, Weeks & Petty, 1960; Maroudas, 1970), largely owing to steric hindrance.

3

THE PERICELLULAR ENVIRONMENT

The matrix in which the cells are embedded forms the bulk of the tissue and is responsible for many of its properties, especially the mechanical characteristics. At the same time it forms the environment of the chondrocytes. It is the medium through which any chemical agent or mechanical stimulus reaches the cell. Hence it regulates the life of the cell though the cell in turn can modify its environment. The structure of the matrix is of considerable interest. It is heterogeneous, functional characteristics depending on the nature of its components, mainly water, collagen, proteoglycan and other proteins, and their interaction. Its architecture has to be considered both at the molecular and the histological level. The chondrocyte is an isolated cell, perhaps more so even than the cells of most other connective tissues. However, it is affected by the presence of other cells in the tissue, the distribution and density of the cell population forming part of the pericellular environment.

COLLAGEN STRUCTURE

Collagen fibres are remarkable for their inextensibility and high tensile strength, about 15–30 kg mm^{-2}, weight for weight equivalent to steel (Harkness, 1968). They can form ropes as in ligaments or tendons, sheets as in aponeuroses or fascial membranes and three-dimensional nets as in cartilage. In all connective tissues the fibres are intimately associated with the protein-polysaccharides of the ground substance.

The fibres in cartilage range from 5–200 nm in diameter and are made up of rod-like molecules called tropocollagen each about 300 nm long and 1.5 nm wide. The tropocollagen molecule consists of three polypeptide α-chains extending colinearly throughout the length of the molecule and wound round each other to form a triple helix. Thus collagen is a complex material and exhibits several levels of organisation, from the primary structure of the α-chain to the mode of interaction of fibres with the molecules of the ground substance. The structure and biosynthesis of collagen are of considerable interest and many reviews related to its biological and pathological significance may be consulted (Ramachandran, 1967; Traub & Piez, 1971; Perez-Tamayo & Rojkind, 1973; Serafini-Fracassini & Smith, 1974; Mathews, 1975).

TABLE 2

Molecular species of collagen. Fibrocartilage contains Type I collagen; Type II may be only a minor component (20% in pig anulus fibrosus) or absent (pig menisci). Type I is also found in human fibrillated articular cartilage, and 'dedifferentiated' cultured chick chondrocytes synthesise a Type I trimer $(\alpha_1 I)_3$. Recent evidence suggests that the α_1II chain exists in differing 'major' and 'minor' forms in bovine nasal cartilage. References: Deshmukh & Nimni (1973); Miller & Matukas (1974); Eyre & Muir (1975); Gay *et al.* (1976); Mayne, Vail, Mayne & Miller (1976*a*); Butler, Finch & Miller (1977)

Form of α-chain	Molecular type	Tissue source
α_1I	Type I: $(\alpha_1 I)_2\alpha_2$	Bone, skin, tendon
α_2		
α_1II	Type II: $(\alpha_1 II)_3$	Hyaline and elastic cartilage
α_1III	Type III: $(\alpha_1 III)_3$	Cardiovascular system
α_1IV	Type IV: $(\alpha_1 IV)_3$	Basement membranes

THE α-CHAIN

Although there are marked differences in some lower species, in mammalian collagen the various α-chains are rather similar in their amino acid composition, form and molecular weight (95000), function and mode of synthesis. Until recently, only two forms of the polypeptide chains, α_1 and α_2, were distinguished. Most mammalian collagens were believed to consist of molecular units containing two α_1 chains and one α_2 chain, i.e. $(\alpha_1)_2\alpha_2$, as in rat skin and tendon, although in codfish skin, three types of chain had been identified (Mathews, 1975). However, Miller & Matukas (1969) were able to differentiate between two types of α_1 chain in chick sternal cartilage: they referred to these as Type I or α_1I, which was similar to that found in skin, and Type II or α_1II. Further investigation revealed at least five forms of α-chain in different tissues characterised on the basis of their amino acid content and other factors; four types of mammalian collagen containing them have been distinguished (Table 2). It appears that the collagens are more tissue- than species-specific in their α-chain content although immunological differences exist (Mathews, 1975). Collagen in elastic and hyaline cartilage either temporary or permanent and in intervertebral discs (Muir, 1977) is of the Type II or $(\alpha_1 II)_3$ variety. However, cultured chondrocytes may produce different types of collagen according to the environmental conditions and, *in vivo*, Type I collagen may be produced in cartilage under pathological conditions (Nimni & Deshmukh, 1973). Before considering the peculiarities of the cartilage collagen it is helpful to consider first the common basic structure of the α-chains.

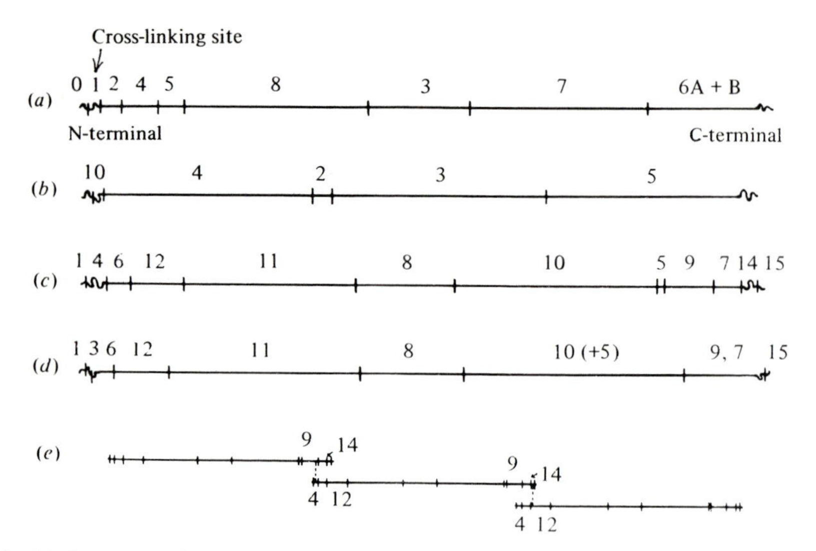

Fig. 20. Sequence of cyanogen bromide peptides and intramolecular cross-linking sites in: (*a*) α_1I, (*b*) α_2, (*c*) chick α_1II, (*d*) mammalian α_1II. Intermolecular cross-linking sites in chick type II collagen are shown in (*e*): for clarity, only one α-chain per molecule is included. References: Serafini-Fracassini & Smith (1974); Miller, Woodall & Vail (1973); Bailey *et al.* (1975).

The α-chain is made up of a unique sequence of about 1000 amino acids in which glycine (Gly) forms about one-third and the imino acids proline and hydroxyproline about one-quarter of the whole. Analysis of the primary structure is facilitated by cyanogen bromide (CB) which cleaves the polypeptide chain at its methionyl residues into a number of relatively short peptides which are remarkably constant. The CB peptides were originally numbered according to their place in the fractionation procedure, which does not necessarily correspond to their order in the molecule (Fig. 20). Other methods employing hydroxylamine and collagenase degradation have also been used in such investigations. Although the amino acid sequences of some of the CB peptides have been known for some time it is only quite recently that this goal has been achieved for the whole chain (Hulmes *et al.*, 1973). There appear to be no large blocks of repeating sequences of identical amino acid composition: the chain is synthesised as a single polypeptide. However, a very important feature of the structure is that the greater part of the length of the chain is made up of the general repeating tripeptide: Gly-*X*-*Y*, where *X* and *Y* can be any of a number of different amino acids, although in most vertebrate collagens cysteine is absent (Piez, 1967). When it occurs, hydroxyproline (Hyp) is restricted to position *Y*; proline (Pro) may occur in either the *X* or the *Y* position, a common tripeptide (35% of the chain) consisting of the sequence Gly-Pro-*Y*; about 10% of the chain consists of the sequence Gly-Pro-Hyp (Ramachandran, 1967; Traub & Piez, 1971).

The unique sequence with its high content of imino acids cannot fit into the common α-helix conformation and instead the α-chain has the polyproline type of left-handed helix. Most of the polypeptide is helical; however, certain regions within the left-handed helix may be replaced by a less common (2_7) type of helix (Hopfinger & Walton, 1970) where there are groups of polar amino acids. In addition, at both ends of the polypeptide there is a short length in which glycine forms less than one-third of the amino acid content and does not occupy its regular repeat position. These 'telopeptides' are in a random coil state and are referred to as N-(amino) and C-(carboxyl) terminal non-helical regions (Fig. 20). These regions are the only parts of the sequence which are significantly antigenic, although complement-fixing activity is dependent on an intact molecular triple helix conformation (Traub & Piez, 1971; Mathews, 1975). The non-helical regions are important in intra- and inter-molecular cross-linking and are related to the additional non-helical extensions of the newly synthesised molecule (procollagen).

CARTILAGE COLLAGEN α-CHAINS

These are designated α_1II and their differences from other types of α-chains are reviewed by Miller & Matukas (1974). The amino acid composition differs from that of other collagens, but the number of glycine, proline and hydroxyproline residues is about the same as in the α_1I and α_2 chains. However, the number of peptide fragments produced by cyanogen bromide is greater (Fig. 20) and hence there are more methionyl residues (Miller, 1971*a*, *b*). The most striking difference lies in the much higher frequency of hydroxylysine, about 25 residues per chain in α_1II compared with 12 in α_1I and α_2 (Miller & Matukas, 1974). Disaccharide units are attached to one or two hydroxylysines per chain in skin collagen (Butler & Cunningham, 1966; Spiro, 1970). This is accentuated in cartilage collagen, which has about 10 units per chain associated with the greater content of hydroxylysine. Histologically, Types I and II collagen may be differentiated by immunofluorescent techniques (von der Mark, von der Mark & Gay, 1976).

MOLECULAR STRUCTURE

Before the chemistry of the α-chain was elucidated, X-ray diffraction (Herzog & Gonell, 1925; Astbury, 1938; Cowan, North & Randall, 1953) provided a basis for acceptable models of the molecular structure (Ramachandran & Kartha, 1955). The diffraction pattern shows that there are 10 equivalent scattering units in three complete turns of the helix (Traub & Piez, 1971). In the model proposed by Ramachandran (1967), which accords most closely with the X-ray data, the molecule consists of three α-chains winding about a common central axis (Fig. 21). Each chain is a coiled coil, in which the primary coil or minor helix is left-handed with a unit twist of $-110°$ per amino

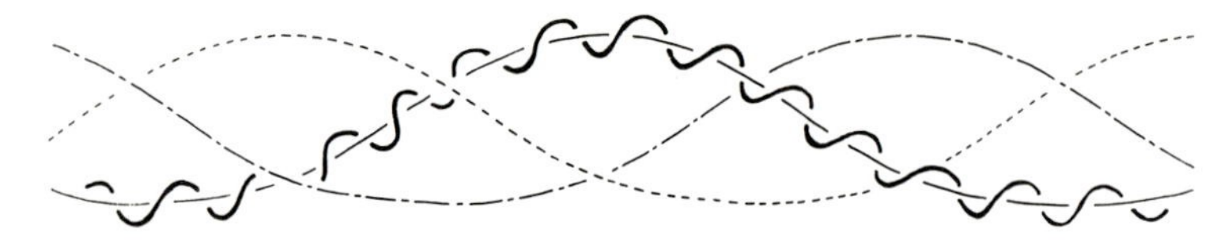

Fig. 21. Schematic diagram of the helical portion of the tropocollagen molecule. The left-handed helix of only one of the three α-chains is included.

acid residue. If the axis of this minor helix was straight, the α-chain would have 3.3 residues per turn and every third residue (i.e. the α-carbon atoms of residues 1, 4, 7, 10...) would be out of register by 30°. Instead, the axis of the minor helix itself follows a helical course. This secondary coil or major helix has a right-handed twist of +30° for each three residues, with a common axis for all three α-chains. This permits just three residues per turn so that the α-carbon atoms of the repeating glycine residues are located at the same radius from the central axis of the molecule. Hence in the triple helix the chains are interrelated by a helical symmetry of −110° and a rise of 2.9 Å in going from one α-carbon atom in one chain to the corresponding α-carbon in the next chain.

Although there are covalent cross-linkages between the α-chains, the triple helix of the tropocollagen molecule owes its stability to the presence of a large number of imino acids and extensive interchain hydrogen bonding. Proline and hydroxyproline confer stability by virtue of the rotational constraint imposed on the C—C bonds of the polypeptide backbone by the pyrrolidine rings. Thus the temperature at which the triple helix of collagen is disorganised to a random coil (denaturation temperature) is related to the total imino acid content (Piez & Gross, 1960). Hydrogen bonds occur between the amide hydrogens and the carbonyl oxygens of the peptide linkages in adjacent chains. In the collagen α-chain the peptide linkages lie across the helical axis rather than parallel to it as in the more usual α-helix. The glycine residues of the repeating tripeptide sequences lie towards the central axis of the triple helix and since they lack side groups, permit the three chains to come close together, facilitating hydrogen bonding.

The number of hydrogen bonds per tripeptide is uncertain. In the Collagen II model of Rich & Crick (1961) there is one hydrogen bond, while in the Ramachandran model there are two bonds. While various physico-chemical data favour the two-bonded structure, a serious criticism has been that the presence of an imino acid in position II of the triplet precludes the formation of a second bond. Study of a large number of synthetic polytri- and polyhexapeptides related to collagen suggested that a modified Collagen II model (N_IH_I—O_{II}, Fig. 22) was substantially correct (Traub, Yonath & Segal, 1969). However, it appears that the energy of denaturation is too high (12.6 kJ mol^{-1} triplet) for one bond only (McClain & Wiley, 1972); investigations using hydrogen–tritium exchange methods also suggest two

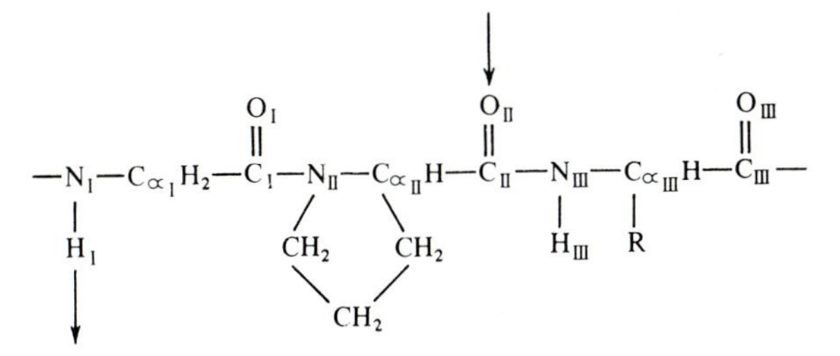

Fig. 22. Pattern of hydrogen bonding for the tripeptide sequence -Gly-Pro-*Y*- according to the collagen II model. The presence of an imino acid, e.g. proline, in position 2 precludes a second H bond similar to that found in position 1.

bonds (Yee, Englander & Hippel, 1974). The hypothesis (Gustavson, 1955) that hydroxyproline contributes to helix stability by hydrogen bonding via the hydroxy group is well known. However, in all molecular models of collagen the side chain projects out radially from the triple helix, presenting difficulties with interchain bonding, so instead it was believed that both imino acids (hydroxyproline and proline) contributed equally due to their pyrrolidine rings. Recent work supports a unique role for hydroxyproline in helix stability, however. It appears that protocollagen (an abnormal non-hydroxylated form) has a lower melting point (i.e. the helix is less stable) by some 15 °C than procollagen, the newly synthesised form of normal collagen (Berg & Prockop, 1973). Hence there is probably a second hydrogen bond in the triplet, via hydroxyproline when present. A direct interchain bond Hyp_{III}—O_I) has been suggested by Berg *et al.* (1973) although this requires a *cis*-peptide linkage between Gly_I and Pro_{II}. However, Ramachandran, Bansal & Bhatnagar (1973) reject an interchain linkage direct to hydroxy-proline and suggest instead that its γ-hydroxy group acts via a water bridge. It appears that several bond structures are possible and that different parts of the α-chains may form hydrogen bonds in different ways.

CROSS-LINKAGES

Covalent bonding occurs between the three chains of each tropocollagen molecule and also between individual molecules during fibrogenesis. In mature collagen, very stable irreducible cross-linkages are involved and their chemistry is not yet understood. In young collagen, covalent linkages of the Schiff-base type confer a high tensile strength on the fibre, the degree of solubility of the fibre varying with the type of tissue and with the predominant type of cross-linkage. In immature collagen the linkages are reducible with borohydride to yield a stable form susceptible to chemical investigation. The non-reduced form in the native fibre may be heat- and acid-labile and hence not stable enough for direct chemical investigation of the tissue. The precise nature or even the existence of the linkage in the unstable non-reduced form is therefore conjectural in some cases.

Linkages in immature collagen

The intramolecular linkage is an aldol which is acid stable. It is formed from two allysine residues produced by oxidative deamination of two lysine residues on adjacent peptide chains (Fig. 23). The lysine residues are situated at position 9 on the α_1 and at position 5 on the α_2 chain in the non-helical portion (Fig. 20) at the N-terminal ends (Bornstein & Piez, 1966).

A number of components have been isolated from borohydride-reduced collagen fibres which have been thought to be involved in intermolecular cross-linking. It is believed that some of the non-reduced forms proposed earlier do not exist as cross-links *in vivo* (Bailey, Robins & Balian, 1974). Intermolecular linkages are formed from lysine or hydroxylysine and their derivatives. Lysine or hydroxylysine residues in the non-helical regions at both ends of the tropocollagen molecule undergo oxidative deamination to allysine or hydroxyallysine. Condensation of these aldehyde forms with the ε-amino group of lysine or hydroxylysine situated in the helical region of an adjacent molecule produces a bond of the aldimine type.

(i) Hydroxylysinonorleucine. The non-reduced (dehydro-) form is labile to dilute acids and to heat and is produced from the condensation of allysine with hydroxylysine (Fig. 23). A more stable form occurring *in vivo* is derived from hydroxyallysine and lysine. Hydroxylysinonorleucine is present in reduced fibres but cannot be isolated from solutions of monomeric collagen.

(ii) Hydroxylysinohydroxynorleucine (di-hydroxylysinonorleucine). This is a major component of the reduced collagen of tendon, bone and cartilage. Its non-reduced form, hydroxylysino-5-keto-norleucine, is stable to heat, dilute acids and D-penicillamine and is derived from hydroxyallysine and hydroxylysine (Fig. 23). The stable 5-keto form is produced *in vivo* by a spontaneous Amadori rearrangement at the aldimine double bond, and can be isolated from non-reduced cartilage collagen where it is the major reducible intermolecular cross-link. Bonds occur between CB peptide 4 in the N-terminal telopeptide and peptide 9 in the helical region of an adjacent molecule, and between peptide 14 in the C-terminal telopeptide and the helical peptide 12 (Fig. 20); these arrangements are consistent with the 1D stagger of the collagen fibre (Bailey *et al.*, 1975).

(iii) Histidino-hydroxymerodesmosine. This structure isolated from reduced collagen is theoretically capable of linking three or four molecules, since its constituents, derived from an aldol, histidine and hydroxylysine, could each be situated in separate peptide chains. However, investigation of native fibres suggests that the presumed non-reduced form does not exist in life, the compound probably being formed during the reduction and isolation procedure.

(iv) Other linkage compounds. Other components have been found in small amounts in reduced collagen. In the case of aldol-histidine and hydroxymerodesmosine the evidence suggests similar conclusions as with histidino-

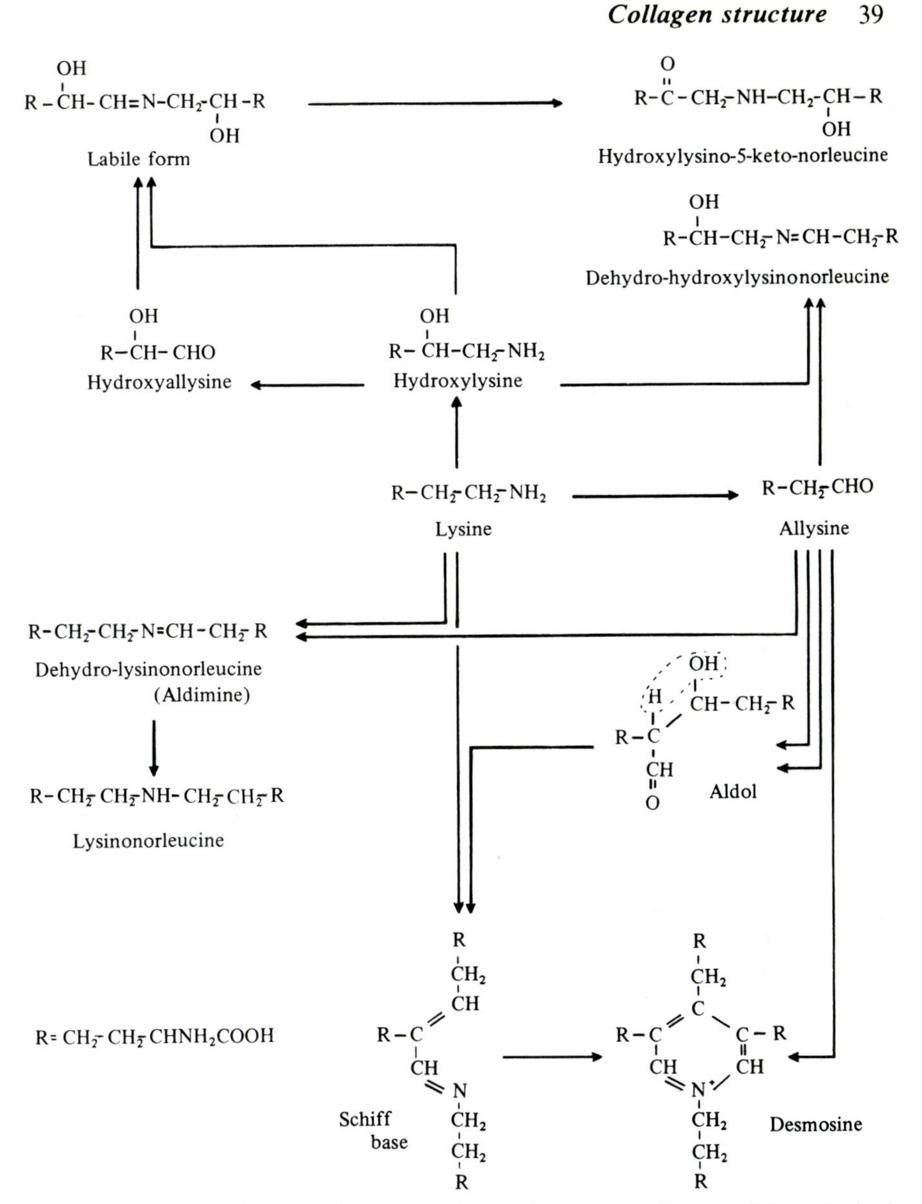

Fig. 23. Schematic diagram of the formation and structure of some of the principal cross-links found in collagen and elastin. Lysine in all cases forms one of the amino acid residues of the polypeptides of the collagen α-chains or of elastin. References: Partridge (1973); Bailey *et al.* (1974).

hydroxymerodesmosine. The presence of a third minor component, lysinonorleucine (Fig. 23) may reflect incomplete hydroxylation of lysine residues. In any case the small proportions of these substances which are detectable suggests that they are unlikely to be physiologically important in cross-linking.

Thus Bailey *et al.* (1974) consider that only mono- and di-hydroxylysinonorleucine are derived from cross-links present *in vivo*. Since in cartilage collagen hydroxylation of lysine residues in the non-helical regions is almost complete the di-hydroxy form is the major linkage.

Linkages in mature collagen

In mature collagen the nature of the intermolecular cross-linkages is uncertain. The reducible cross-linkages of the Schiff-base type found in young tissue increase to a maximum in infancy but are almost absent by the end of the growth period, with no further change during adult life. The diminution in the reducible linkages does not appear to be time-dependent since the change is associated with physiological rather than temporal age. The fibres become less soluble, less prone to thermal denaturation and less susceptible to enzymic degradation. Hence non-reducible cross-links must replace those of the Schiff-base type and even those of the stable 5-keto form predominant in young cartilage collagen. Probably the juvenile types undergo spontaneous conversion so that they become more stable and irreducible. Although it is suggested that stable bonds may be formed by chemical reduction *in vivo* (Mechanic, Gallop & Tanzer, 1971) as in elastin, others claim that there is no supportive evidence (Bailey & Peach, 1971; Robins, Shimokomaki & Bailey, 1973). Minor reducible components increasing in amount with age consist of condensation products derived from lysine and hydroxylysine with glucose or mannose (Robins & Bailey, 1972; Tanzer, Fairweather & Gallop, 1972). However, these could not effect linkages either between collagen peptide chains or between the peptide chains and glycoproteins, since the latter have no reducing endgroups available.

ORGANISATION OF THE COLLAGEN FIBRE

Collagen fibres vary enormously in diameter and can aggregate into bundles visible with the light microscope or the unaided eye. Native fibres positively stained with heavy-metal salts, such as uranyl acetate, or with phosphotungstic acid at acid pH, exhibit major electron-dense cross bands with a periodicity ranging from 64 nm to 70 nm, depending on the hydration state of the collagen and the method of preparation. Each period (D) is made up of at least 10 fine dark cross-striations (Fig. 24) identified conventionally as a_1–e_2 (Gross & Schmitt, 1948). When a solution of tropocollagen in physiological saline at neutral pH is warmed from 4 to 37 °C, the monomeric collagen

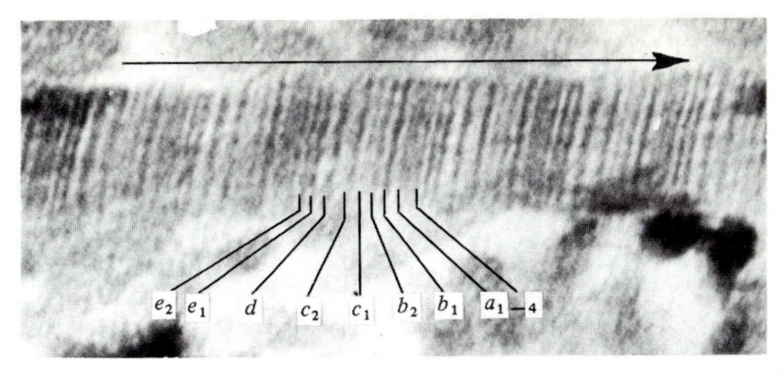

Fig. 24. Human articular cartilage (46 years). Glutaraldehyde fixation, magnification × 225000. Collagen fibre in a section stained with uranyl acetate and lead citrate. The tropocollagen molecule (arrow) extends over 4.4 periods with the N-terminal end at the c_2 band (at left) and the C-terminal end at the a_2 band.

aggregates into fibres with the characteristic periodicity (67 nm). If, however, ATP is added to an acid solution of tropocollagen, short fibrous segments are formed with the same length as the tropocollagen molecule (about 300 nm), known as 'segment long spacing' (SLS) collagen. In this type of preparation the molecules are orientated in the same direction and lie side by side with similar portions of each molecule arrayed in precise lateral alignment. Thus the cross-striations of SLS represent specific portions of the tropocollagen molecule. The cross-striations of SLS and the banding of the native fibre can be matched in position but not in intensity; furthermore, the length of SLS collagen (and therefore of a tropocollagen molecule) extends over 4.4D (Fig. 24) of a native fibre (Hodge & Schmitt, 1960). This suggests that laterally adjacent molecules in the fibre are longitudinally displaced by 67 nm or 1D (Fig. 25), i.e. approximately 'quarter-staggered' (Gross, Highberger & Schmitt, 1954). Thus the periodicity of the native fibre is the summation in many 1D-staggered tropocollagen units of portions of the molecule represented by cross-striations at 1D intervals in SLS (Fig. 25).

The cross-banding pattern of collagen was early postulated to be associated with the location of polar amino-acids in the molecule (Hodge & Schmitt, 1960; Kuhn, Grassmann & Hofmann, 1960). Using skin collagen, good correlations have now been obtained between the SLS striation pattern and the distribution of acidic and basic side groups in the amino acid sequence of partial (von der Mark, Wendt, Rexrodt & Kuhn, 1970) and of whole α_1-chains (Chapman, 1974; Doyle *et al.*, 1974*b*). Computer-generated patterns from the amino acid sequence compared with the natural cross-striations indicate that there are 232–234 amino acids per period (Hulmes *et al.*, 1973; Chapman & Hardcastle, 1974). Chapman (1974) considers that the tropocollagen molecule extends over rather more than 4.4D, the N-terminal end of the molecule commencing at the C-terminal side of the c_2 band and the C-terminal end finishing at a_2–a_3 (Fig. 24). Most (90%) of the charged amino acid residues are located in the cross-striations of the native fibre and

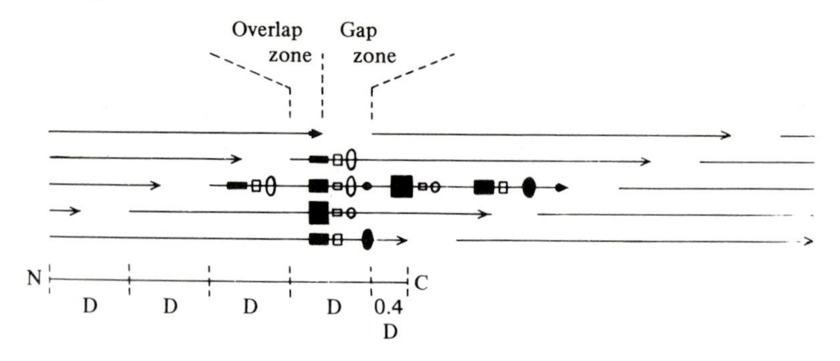

Fig. 25. Diagram showing orientation and overlap of tropocollagen units in the fibre. Note the resulting transverse alignment of sequential molecular regions which have similar periodic spacing but dissimilar staining intensity. Key: ■ = $a+b$; ● = c; ○ = d; □ = e.

it is noteworthy that the a_2–a_4 and c_2 bands in particular carry a net positive charge (Doyle *et al.*, 1975).

Since the tropocollagen molecule is about 4.4D long, it follows that there must be a 0.6D space between successive molecules so that the orderly 1D stagger can be preserved. Therefore each 1D period (Fig. 25) consists of an 'overlap' zone 0.4D long and a 'gap' zone 0.6D long (Hodge & Petruska, 1963; Olsen, 1963). The gap zone is made up of only 80% of the molecules while in the overlap zone the N- and C-terminal ends of certain tropocollagen molecules are contiguous, facilitating intermolecular cross-linking. In negatively stained preparations (phosphotungstic acid at neutral pH) where the metal 'stains' spaces in the tissue, the gap zone forms a darker wider *B* zone than the pale narrow *A* zone.

While a regular 1D stagger of tropocollagen units can be maintained indefinitely in two dimensions, it cannot so readily be extended in three dimensions, as required in the native collagen fibre. A number of ideas have been put forward but the best model is that of Smith (1968). This is based on the observation that negatively stained fibres exhibit longitudinal striations, indicating that the fibre is made up of filaments about 3 nm in diameter. In two dimensions, a 1D stagger results in a laterally repeating pattern at every fifth tropocollagen unit. Smith points out that a monolayer rolled into a hollow cylindrical filament containing five units in cross-section (Fig. 26*a*) satisfies the requirements of a 1D stagger in three dimensions. The diameter of the hypothetical primary filament would be about 3 nm in the negatively stained dry state and 4–5 nm in embedded material. X-ray diffraction studies provide confirmation of the model yielding filament spacings of 3.8 nm (Miller & Wray, 1971; Miller & Parry, 1973). Such primary filaments have been visualised in reconstituted cartilage collagen (Doyle *et al.*, 1974*a*) as well as in skin collagen. Smith notes that assembly of primary filaments to form thick cross-banded fibres requires only that the filament periods should be in register.

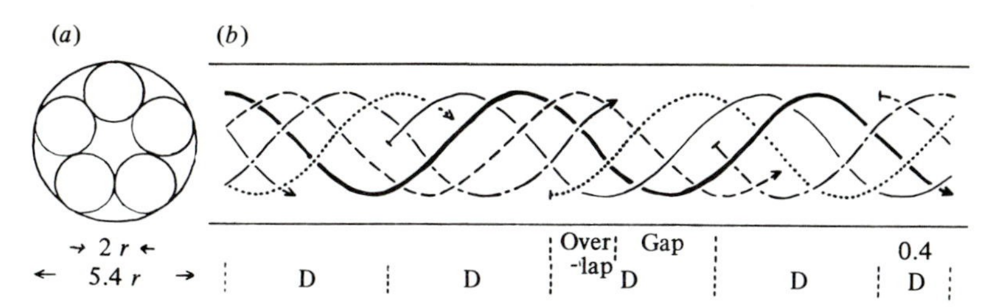

Fig. 26. The pentameric primary filament of collagen according to the concepts of Smith (1968) and Piez & Trus (1977), shown in (*a*) transverse section and (*b*) longitudinal aspect. The axes of the tropocollagen units form left-handed helixes with a pitch of 122 nm. The diameter of the filament is about 3 nm. ((*a*) redrawn from Smith, 1968.)

Although structures have been suggested containing four (Veis, Anesey & Mussell, 1967), seven (North, Cowan & Randall, 1954) or eight units (Hoseman, Dreissig & Nemetschek, 1974), they lack the elegance of the Smith pentamer which has gained wide acceptance as the basic supra-molecular unit of assembly of the collagen fibre. X-ray diffraction suggests that the primary filament itself is a super-coiled structure (Fig. 26*b*) with a four-fold screw axis, the tropocollagen units following a super-helical path around the filament with a pitch of 60 to 120 nm (Miller & Parry, 1973; Frazer, Miller & Parry, 1974; Piez & Trus, 1977). A left-handed helix is suggested by Piez & Trus (1977), who point out that this would comply with the principles of rope-building. Thus at successive levels of organisation, the α-chains form a left-handed helix, the chains in the molecules follow a right-handed path and the molecules in the filament form a left-handed helix again. Alternation of the 'handedness' prevents the 'rope' uncoiling and converts an axial into a lateral (compressive) stress. In the fibre the primary filaments are packed together on a square lattice (Miller & Parry, 1973; Frazer *et al.*, 1974). Both electrostatic and hydrophobic interactions are thought to facilitate binding of neighbouring filaments (Piez & Trus, 1977).

ELASTIN STRUCTURE

A few cartilages, for example those of the ear, arytenoid and epiglottis, contain elastic fibres in addition to collagen. Elastic fibres stain with orcein and show no periodic cross-striations as found in collagen. The fibre consists of moderately electron-dense amorphous material, elastin, associated with peripheral microfibrils 10 nm in diameter. In cartilage the fibres tend to coalesce forming irregular networks between the cells.

Elastin is a very stable, chemically inert protein, resistant to collagenase but degraded by pancreatic and leucocyte elastase. It is a cross-linked, three-dimensional isotropic polymer made up of globular polypeptide units. It shows about 70% extensibility and recoil when fully swollen with water, which forms two-thirds of its volume (Partridge, 1973). Somewhat unexpectedly, perhaps, a large proportion (95%) of its amino acids are non-polar, many of them with hydrophobic side chains. The hydroxyproline content is much lower (10–20 residues per 1000) than in collagen; although the tripeptide sequence Gly-Pro-*Y* is uncommon or absent, hydroxyproline residues tend to occur in the sequence -Gly-*X*-Hyp-Gly-. The polypeptide unit may consist of long sequences of non-polar amino acids containing relatively large amounts of alanine separated by polar regions where the cross-linkages occur (Mathews, 1975). The cross-linkages in mature elastin are due to two isomers, desmosine and isodesmosine (Partridge, Elsden & Thomas, 1963; Thomas, Elsden & Partridge, 1963); after reduction with borohydride, lysinonorleucine is also found.

Desmosine and isodesmosine are formed from four lysine residues by successive oxidation and condensation:

$$\text{2 Lysine} \rightarrow \text{2 Allysine} \rightarrow \text{Aldol} + \text{Lysine} \rightarrow$$
$$\text{Schiff base} + \text{Allysine} \rightarrow \text{Desmosine or Isodesmosine}$$

These are C_{24} compounds, containing positive charged pyridinium rings with the structure shown in Fig. 23. Since they have four side chains ($CHNH_2COOH$), they are capable of cross-linking four polypeptides. For a short period before the formation of these irreversible cross-links, there exists a dynamic equilibrium of aldols and Schiff bases. Orientation of the lysyl residues prior to bond formation probably depends on the tertiary structure of the globular polypeptide unit (Partridge, 1973).

A corpuscular structure for elastin was suggested by Robert & Poullain (1963) to account for the physical properties of the material. The corpuscle has a hydrophobic centre (Fig. 27) and the intercorpuscular space is filled with water; hence in water the polypeptide tends to a compact globular shape to reduce the interfacial area. In good swelling solvents, however, such as aqueous formamide, the structure assumes a random coil form. Weis-Fogh & Anderson (1970) showed by calorimetry that the deformation of elastin in water causes a reversible change in internal energy several times greater than the mechanical work done. The change in internal energy is little affected by pH (hence charged groups are not involved) but is reduced in organic solvents. When elastin is stretched the spherical corpuscular units are distorted into ellipsoids with an accompanying increase in their surface area. Weis-Fogh & Anderson find that the measured change in energy can be accounted for by that required to shift sufficient hydrophobic amino acid side chains from the internal to the external aspect of the corpuscular shell to provide for the increase in surface area. It has also been shown that during stretching the elastic energy is stored as a change in water structure (Partridge, 1973).

According to Gotte & Volpin (1975), there are three possible models for the organisation of elastin at the ultrastructural level: amorphous, filamentous

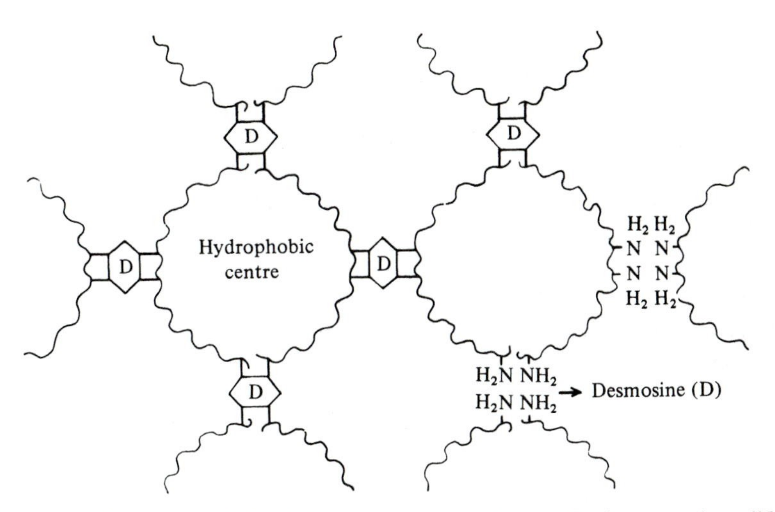

Fig. 27. Schematic diagram of the globular structure of elastin. (Redrawn and modified from Partridge (1966), *Fedn Proc. Fedn Am. Socs exp. Biol.*, **25**, 1023–9.)

and corpuscular. An isotropic pattern of elements 11–14 Å in diameter and at 30–40 Å centres noted by Cox & O'Dell (1966) is consistent with a corpuscular arrangement. On the other hand, following extraction with hot alkali, long slender filaments 3–3.5 nm in diameter and with a 3-nm periodicity can be observed lying parallel to the long axis of the main fibre (Gotte, Giro, Volpin & Horne, 1974). This could result from a linear array of globules 2.8 nm in diameter, joined by desmosine or isodesmosine cross-links.

ELASTOGENESIS

A soluble non-cross-linked protein precursor of elastin has been named tropoelastin. It has a minimum molecular weight of 74000 and does not contain desmosine or other lysine-derived amino acids (Smith, Brown & Carnes, 1972). However, it has about 38 lysine residues, more than the 20 or so required to form intermolecular cross-links. Within the cell, coacervate formation may be prevented by the very high positive charge (Partridge, 1973).

In fetal tendons and ligaments, microfibrils 11–12 nm in diameter are formed in the matrix close to the cell surface, and elastin subsequently appears among them. They are chemically distinct from elastin and can be selectively degraded by trypsin and chymotrypsin. The microfibrils consist of glycoproteins and can be extracted with 6-M guanidine following reduction of disulphide bonds with dithiothreitol. The amino acids are predominantly polar and desmosine and isodesmosine are absent (Greenlee, Ross & Hartman, 1966; Greenlee & Ross, 1967; Ross & Bornstein, 1969). Extra-

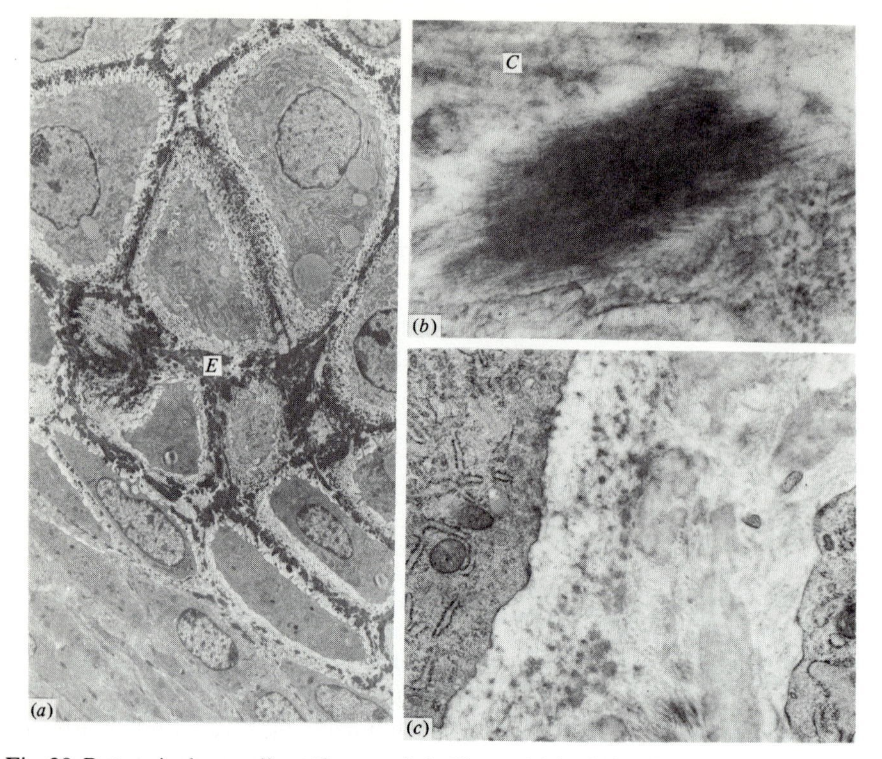

Fig. 28. Rat auricular cartilage (four weeks). Glutaraldehyde fixation. (*a*) Chondrocytes with elastic tissue (*E*) forming in the matrix. Magnification ×2000. (*b*) Aggregate of microfibrils in peripheral zone of cartilage prior to deposition of elastin. *C* = collagen fibre. Magnification ×30000. (*c*) Presence of amorphous elastin between cells. Specimen taken nearer base of ear. Magnification ×14500.

cellular precipitation of tropoelastin may occur either on nucleating sites provided by the microfibrils (Kadar & Robert, 1975) or on an elastin surface with the required conformation properties. It is possible in either case that a regulatory proteolytic enzyme at the site of fibre formation may be required to adjust the iso-electric point of the precursor protein (Partridge, 1973).

In rabbit ear cartilage, elastic tissue is stainable in the 26-day fetus and has been observed from birth onwards with the electron microscope (Cox & Peacock, 1977). Although in mature rat cartilage no microfibrils can be detected at the periphery of the elastic fibres, in young rats (Fig. 28) the amorphous components show a microfibrillar structure where the fibre runs out of the plane of section (Serafini-Fracassini & Smith, 1974). In rabbit ear cartilage, microfibrils are always (birth to three years) associated with the amorphous component (Cox & Peacock, 1977).

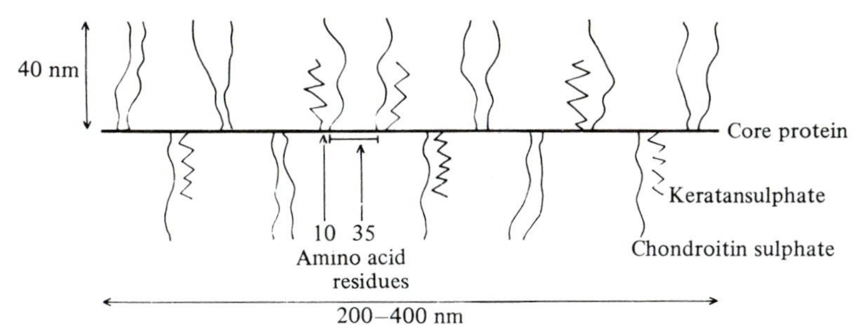

Fig. 29. Diagram of proteoglycan molecule, molecular weight 1–2 × 10^6. The doublet structure is indicated.

PROTEOGLYCAN STRUCTURE

These are high molecular weight molecules consisting of a small amount of protein covalently bound (Schatton & Schubert, 1954) to carbohydrate polymers. The polysaccharides are each made up of negatively charged disaccharide repeating sequences and are named glycosaminoglycans (synonym mucopolysaccharides), such as chondroitin sulphate and keratan-sulphate. In most proteoglycans (Fig. 29), the glycosaminoglycans form long side chains attached by their potentially reducing ends to a protein core (Mathews & Lozaityte, 1958; Partridge, Davis & Adair, 1961). A major class of cartilage proteoglycans appear to be unique to the tissue in that they can form very large aggregates in which about 30 or more proteoglycans are linked to a 'backbone' of hyaluronic acid, itself a glycosaminoglycan (Hardingham & Muir, 1972*a*; Hascall & Heinegard, 1974*a*).

GLYCOSAMINOGLYCANS

A number of different glycosaminoglycans are known although not all of them occur in mammals or in cartilage. In cartilage only chondroitin sulphate, keratansulphate and hyaluronic acid have so far been identified, all forming part of the same macromolecule in combination with protein.

Chondroitin sulphate

Chondroitin sulphate is the most plentiful glycosaminoglycan in cartilage. It has been subject to investigation since 1837 when Müller obtained 'chondrin' solutions by autoclaving cartilage in water. As distinct from the solutions obtained from bone and tendon which yielded gelatin, 'chondrin' gave a precipitate when treated with acetic acid and on hydrolysis yielded inorganic sulphate (Mulder, 1838) and reducing sugar (Fischer & Boedecker, 1861). Chondroitin sulphate was first isolated by Krukenberg in 1884 using extraction with alkali which splits the protein from the polysaccharide. Subsequently a number of workers including Schmiedeberg (1891) and Levene &

La Forge (1914) identified the sugars present in chondroitin sulphate. However, it was not until 1955 that Davidson & Meyer established that chondroitin sulphate (Fig. 30) has the structure of a linear polymer with repeating disaccharide units linked β-(1,4), each containing glucuronic acid and *N*-acetylgalactosamine linked β-(1,3). The hexosamine component is sulphated and hence there are normally two negative charges per unit. The length of the polymer chain varies but can contain up to about 30 disaccharide units, giving a molecular weight of about 20000 (Muir, 1973).

In mammalian cartilage there are two different forms of chondroitin sulphate, differing in their infra-red absorption spectra and other physical properties (Mathews, 1958) according to the position of the sulphate group on the hexosamine. Chondroitin sulphate A is sulphated at C4 (chondroitin-4-sulphate) and chondroitin sulphate C at C6 (chondroitin-6-sulphate) of the galactosamine moiety. Both forms are present in cartilage but chondroitin-6-sulphate predominates with age (Mathews & Glagov, 1966). Undersulphation (molar ratio sulphate: hexosamine < 1) of the polymer may occur; near the protein linkage region, sulphate groups are less numerous (Wasteson & Lindahl, 1971). Chondroitin sulphates of lower vertebrate (particularly shark cartilage) and invertebrate cartilage tend to be oversulphated. Hagfish and squid cartilage exhibit a type containing galactosamine sulphated both at C4 and C6. The comparative biochemistry of cartilage proteoglycans is reviewed by Mathews (1975).

Keratansulphate

Keratansulphate also owes its name to the anatomical structure from which it was first isolated – the cornea, in which it is the principal glycosaminoglycan (Meyer, Linker, Davidson & Weissmann, 1953). Skeletal keratansulphate (KS-II) is essentially similar to corneal keratansulphate (KS-I), a major part of both forms consisting of the same disaccharide repeating unit. The structure of keratansulphate (Fig. 30) differs from chondroitin sulphate principally in the nature of the repeating sequence but the molecule as a whole is more complex and less well understood.

The larger part of the polymer consists of repeating units, linked β-(1,3), containing galactose and *N*-acetylglucosamine linked β-(1,4). It is the only acid glycosaminoglycan to lack uronic acid. The glucosamine residue is normally sulphated at C6 but this may be absent; alternatively, extra sulphate groups may be present at C6 of the galactose residue (Bhavanandan & Meyer, 1968). Chain lengths vary (Laurent & Scott, 1964), published values for molecular weight ranging from 4000 to 18000 (Bray, Lieberman & Meyer, 1967; Hopwood & Robinson, 1974*b*).

The major differences between KS-I and KS-II lie in the protein linkage region and in the presence of carbohydrate residues other than those of the main polymer sequence (Mathews, 1975). KS-II contains appreciable

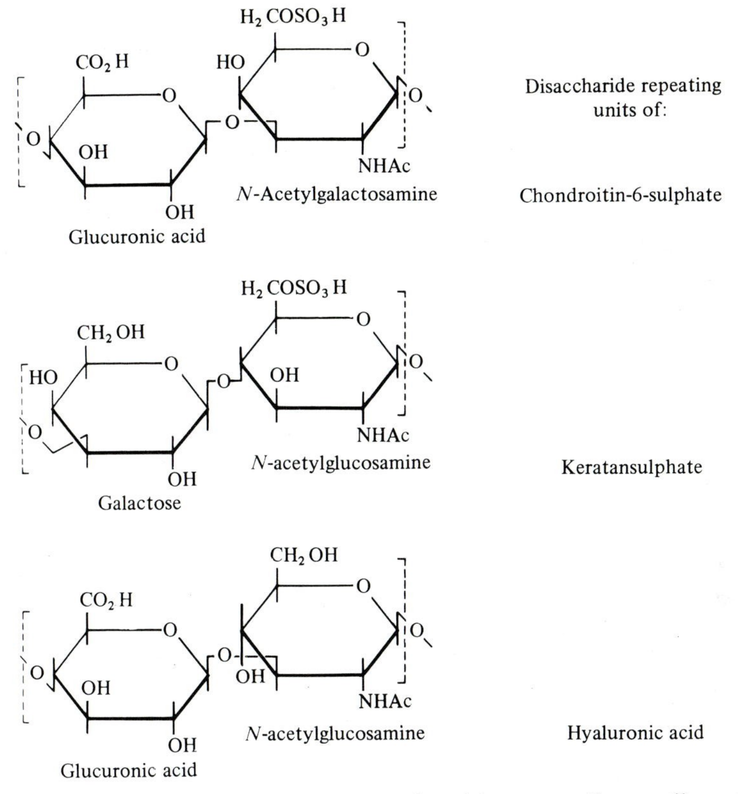

Fig. 30. Structure of the glycosaminoglycans found in mammalian cartilage. (From Stockwell, 1974.)

quantities of galactosamine, sialic acid, fucose, mannose and additional galactose; KS-I contains no galactosamine and little of the other sugars (Mathews & Cifonelli, 1965; Bray *et al.*, 1967; Bhavanandan & Meyer, 1968; Meyer, 1970). According to Hopwood & Robinson (1974*b*) there are three residues of sialic acid and one each of mannose, fucose and galactosamine per KS-II chain. KS-II is a branched structure with fucose and sialic acid at the non-reducing ends of its side branches; these side chains probably contain the extra galactose (Luscombe & Phelps, 1967; Bhavanandan & Meyer, 1968; Hirano & Meyer, 1971); the position of mannose is uncertain and may be close to the alkali-labile region or separate from it (Baker, Cifonelli, Mathews & Roden, 1969; Hopwood & Robinson, 1974*b*).

In cartilage, keratansulphate was first detected by chemical analysis of human costal cartilage. This revealed an excess of hexosamine over uronic acid with advancing age (Shetlar & Masters, 1955) accounted for by increasing amounts of *N*-acetylglucosamine (Stidworthy, Masters & Shetlar, 1958). Hence the presence of keratansulphate was inferred and confirmed by a number of subsequent studies (Kuhn & Leppelmann, 1958; Meyer, Hoffman

& Linker, 1958; Kaplan & Meyer, 1959). The amount of keratansulphate in cartilage is usually less than that of chondroitin sulphate, forming only a small proportion of the total glycosaminoglycan in immature tissue.

Hyaluronic acid

This glycosaminoglycan is widespread in normal mammalian connective tissues, and other characteristic and plentiful sources include rooster comb, certain tumours, notably mesothelioma, and streptococcal cultures. In the body, hyaluronic acid is almost the only glycosaminoglycan of synovial fluid and the vitreous body of the eyeball and it derives its name from the latter structure in which it was first detected (Meyer & Palmer, 1934). A comprehensive summary of pre-1965 analytical and other data is given by Preston, Davies & Ogston (1965) and more recent work is reviewed by Laurent (1970).

The repeating disaccharide units (Fig. 30) linked β-(1,4) consist of glucuronic acid and *N*-acetylglucosamine linked β-(1,3) (Meyer, Smyth & Dawson, 1939; Meyer, 1958) and contain no sulphate. Hyaluronic acid is usually considered to be a flexible, long-chain polymer which does not branch; the electron microscope reveals this structure as having a mean length of 2.4 μm (Fessler & Fessler, 1966). Its physico-chemical properties and a molar ratio of uronic acid: glucosamine in excess of unity suggest that there is some branching and cross-linking (Preston *et al.*, 1965). In solution, it takes a single stranded left-handed helical form and further three-dimensional organisation may be found (Sheehan, Atkins & Nieduszynski, 1975). The whole molecule adopts a spheroidal random coil structure kept expanded by its negatively charged carboxyl groups. In this state, the molecule occupies a volume (domain) of 1.4×10^{-13} cm^3 (Preston *et al.*, 1965), several orders greater than that of the molecular chain itself (Laurent, 1970). The amount of protein associated with hyaluronic acid varies but the molecule contains at least 0.35% of peptide (Swann, 1968; Laurent, 1970) chemically bound to the polysaccharide. In the tissues several molecules may be non-covalently linked together with much larger amounts of protein (Silpananta, Dunstone & Ogston, 1968).

Very small amounts of hyaluronic acid were reported in whale (Seno & Anno, 1961) and bovine nasal cartilage (Szirmai, Van Boven-de Tyssonsk & Gardell, 1967) and were not considered to be of much consequence. It is now known that the small quantity of hyaluronic acid, representing about 0.7% of the total uronic acid in laryngeal cartilage, for example (Hardingham & Muir, 1973*b*, 1974*a*), is a key factor in the aggregation of proteoglycans into complexes which are unique and essential to the structure and stability of the matrix.

LINKAGE TO PROTEIN

The structure of the linkage region is established for chondroitin sulphate. A significant advance was made by Muir (1958) who found that the glycosaminoglycan is bound to protein via the amino acid serine. The linkage amino acid is spared by proteolytic treatment but destroyed by alkali as the result of β-elimination, so degrading serine and threonine residues involved in linkages to carbohydrate (Anderson, Hoffman & Meyer, 1965). Gregory, Laurent & Roden (1964) demonstrated that chondroitin-4-sulphate is linked to protein by its potentially reducing end via xylosyl-serine and that galactose is also involved. The nature of the sugar 'bridge' (Fig. 31) between the repeating polymer and serine was characterised by Roden and his colleagues (reviewed by Roden, 1968) and is similar for both chondroitin-4- and -6-sulphate (Helting & Roden, 1968). The linkage serine is thought to form part of a specific sequence of amino acids at the attachment region of chondroitin sulphate (Mathews, 1971; Johnson & Baker, 1973).

(Disaccharide)$_x$ –Glucuronic acid–Galactose–Galactose–Xylose–Serine

Fig. 31. The protein–polysaccharide linkage region of chondroitin sulphate.

The linkages of KS-I and KS-II keratansulphates to protein are quite different. KS-I, which contains no galactosamine, has an alkali-stable bond between *N*-acetylglucosamine and asparagine (Seno, Meyer, Anderson & Hoffman 1965; Baker, Cifonelli & Roden, 1975). KS-II has both alkali-labile and alkali-stable linkages. The alkali-labile bond (Fig. 32) is between *N*-acetylgalactosamine and serine or threonine, the main keratansulphate chain substituting at the C6 position of galactosamine (Mathews & Cifonelli, 1965; Seno *et al.*, 1965; Bray *et al.*, 1967). Galactosamine is further substituted at the C3 position by the disaccharide *N*-acetylneuraminyl-galactose (Hopwood & Robinson, 1974*a*). The alkali-stable bond is less well characterised but involves glutamic acid (Seno *et al.*, 1965; Heinegard, 1972). According to Hopwood & Robinson (1974*b*), there is no evidence that the linkage glutamate is part of the proteoglycan core protein; they consider that a structure whereby the two bonds involve the same keratansulphate chain is the most likely model.

Little is known of the hyaluronate–protein linkage.

PROTEOGLYCAN AGGREGATES

Improved methods of proteoglycan extraction have permitted the characterisation of the very large proteoglycan aggregates which are unique to cartilage. The older disruptive techniques, e.g. high-speed homogenisation, produced shearing stresses which led to denaturation and splitting of the macromolecules. However, the method of 'dissociative' extraction introduced by Sajdera & Hascall (1969) eliminates these defects by employing gentle

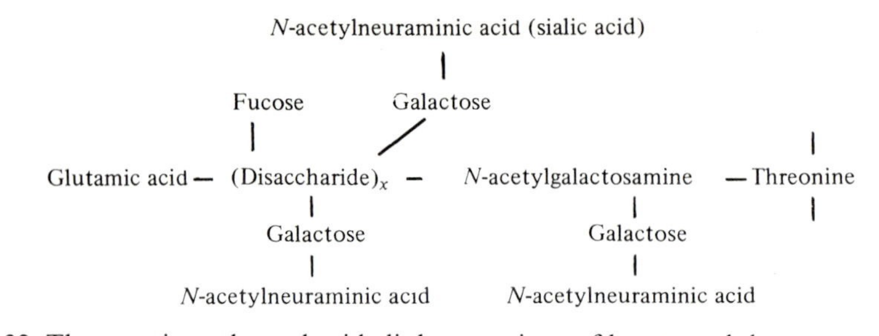

Fig. 32. The protein–polysaccharide linkage regions of keratansulphate according to Hopwood & Robinson (1974*a*, *b*).

agitation of the tissue in solvents of high ionic strength, e.g. 4-M guanidinium chloride. High-speed homogenisation produced a number of protein-polysaccharides (PP) which were grouped into two general classes. A light fraction (PPL) is a mixture and can be subfractionated in the ultracentrifuge. A heavy, rapidly sedimenting class (PPH), also heterogeneous, contains hydroxyproline and is of higher molecular weight. Homologies between PPL and PPH fractions and the proteoglycans produced by dissociative extraction are discussed by Hamerman, Rosenberg & Schubert (1970).

Dissociative extraction depends on disaggregation of proteoglycan complexes in the tissue. Reaggregation of the proteoglycan 'subunits' takes place *in vitro* at low ionic strength provided that certain other 'link' components (also dissociated during extraction) are restored (Hascall & Sajdera, 1969). These components were thought to be entirely glycoprotein in nature but, investigating apparent differences between bovine nasal and pig laryngeal cartilage, Tsiganos, Hardingham & Muir (1972) observed that an essential 'link' fraction in both forms of cartilage was largely carbohydrate in composition with significant quantities of uronic acid. Further work has led to the identification of the link material in the cartilage extracts as hyaluronic acid (Gregory, 1973; Hardingham & Muir, 1973*b*, 1974*a*) and the demonstration of the specific interaction of hyaluronic acid with cartilage proteoglycans (Hardingham & Muir, 1972*a*, 1973*a*; Hascall & Heinegard, 1974*a*, *b*). However, aggregation between hyaluronic acid and proteoglycan also requires the glycoprotein 'link' (Hascall & Sajdera, 1969; Gregory, 1973) which stabilises the linkage region to enzymatic digestion (Hascall & Heinegard, 1974*b*) and increases the sedimentation rate of the aggregate in the ultracentrifuge (Hardingham & Muir, 1974*a*).

About 30 proteoglycans are associated with one hyaluronic acid molecule (Hardingham & Muir, 1972*a*) although more can be accommodated on long hyaluronic acid chains (Fig. 33). Each proteoglycan interacts by a single binding site on the core protein with a segment of the hyaluronic acid chain corresponding to a decasaccharide unit (4–5 nm long). However, the proteoglycans are spaced at longer intervals (25 nm), corresponding to units

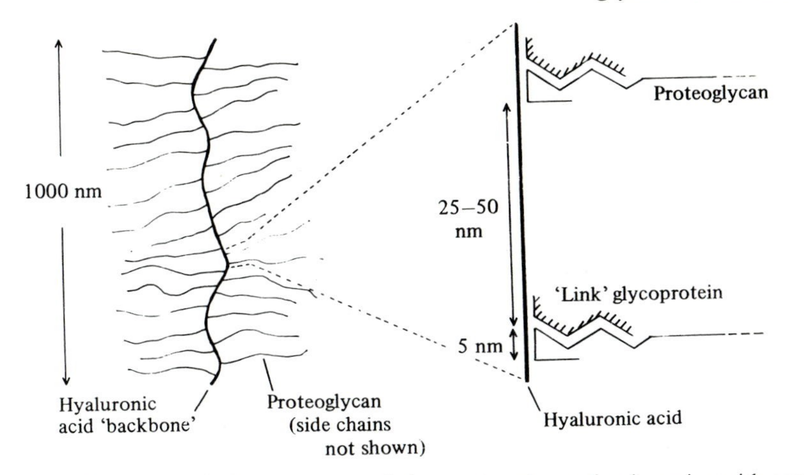

Fig. 33. Diagram of the structure of the proteoglycan–hyaluronic acid complex according to Hardingham & Muir (1974*b*) and Heinegard & Hascall (1974*a*).

with a molecular weight of 10000, as a consequence of steric effects between proteoglycan subunits (Hardingham & Muir, 1973*a*; Hascall & Heinegard, 1974*a*). These biochemical data agree closely with electron microscopic observations of isolated proteoglycan complexes (Rosenberg, Hellman & Kleinschmidt, 1970, 1975), suggesting the model of the proteoglycan–hyaluronate complex shown in Fig. 33 (Hardingham & Muir, 1974*b*; Heinegard & Hascall, 1974*a*).

Not all proteoglycans in cartilage are capable of aggregating (Tsiganos & Muir, 1969; Brandt & Muir, 1971; Mayes, Mason & Griffin, 1973). Indeed, sequential extraction techniques suggest that there are three populations of proteoglycans in addition to unextractable material released only after proteolytic digestion of the tissue (Rosenberg, Pal & Beale, 1973; McDevitt & Muir, 1976).

(i) Non-aggregating proteoglycans are extracted by non-dissociative solvents such as 0.15-M sodium acetate. They are of small hydrodynamic size, of low protein and keratansulphate content and do not interact with hyaluronic acid. They form only a few per cent of the total.

(ii) Aggregating proteoglycans are extracted by either 2-M calcium chloride or 4-M guanidinium chloride. These form the bulk of the proteoglycan content of nasal cartilage but only about half of the total in articular cartilage.

(iii) Articular cartilage which has been thoroughly extracted with 2-M calcium chloride yields a further fraction when treated with 4-M guanidinium chloride. This brings into solution a proteoglycan complex of much larger hydrodynamic size (600S) than the aggregate, reminiscent of PPH. It contains roughly 50% proteoglycan, a small amount of collagen (2%) and a 2.7S protein (40%) not detectable in hyaluronate–proteoglycan aggregate. It accounts for about a third of the total proteoglycan in articular cartilage but appears to be only a minor component in nasal cartilage.

PROTEOGLYCAN SUBUNITS

Identification of the proteoglycan–hyaluronate aggregate has necessitated a re-appraisal of the proteoglycan structure (Fig. 29) proposed by Mathews & Lozaityte (1958), although this has been a valuable model correct in its main features. The proteoglycans are noted for their polydisperse nature which not only arises from differences in the number and type of polysaccharide chains (Hardingham & Muir, 1974*a*) but also from variation in amino acid composition of the protein core (Hardingham, Ewins & Muir, 1976). It is not known if the core protein is made up of a single polypeptide unit or of multiple polypeptides arranged either in parallel or series (Serafini-Fracassini & Smith, 1974) linked together by hydrophobic bonds involving specific sequences on non-polar amino acids (Wells & Serafini-Fracassini, 1973). However, the present evidence is in favour of a single covalent structure (Hardingham *et al.*, 1976). The possibility that the core protein has a repeating period (Mathews, 1971) arose from the nature of the fragments produced by trypsin or papain digestion of proteoglycans and the evidence of specific sequences in the linkage regions of the glycosaminoglycan side chains. A revised model of proteoglycan structure (Fig. 29) incorporated the 'doublet' (Anderson *et al.*, 1965; Luscombe & Phelps, 1967) as its structural feature, i.e. glycosaminoglycans attached to the core in pairs, a sequence of about 35 amino acid residues between pairs and the members of a pair lying within 10 residues of each other (Mathews, 1971). However, more recent evidence suggests that glycosaminoglycans are attached in clusters of 1–10 side chains, rather than as regularly spaced doublets (Heinegard & Hascall, 1974*b*).

The amino acid composition of aggregating and non-aggregating proteoglycans show that the two classes have different core proteins (Mayes *et al.*, 1973; Hardingham *et al.*, 1976). Hardingham *et al.* (1976) suggest that this can be accounted for in terms of the proteoglycan structure suggested by Heinegard & Hascall (1974*a*) in which the core protein of aggregating proteoglycans contains two distinct regions:

(i) The hyaluronate-binding region (molecular weight about 90000) is of constant size and has some keratansulphate side chains. It is a globular portion, maintenance of the tertiary structure depending on disulphide linkages, hydrophobic interactions and hydrogen bonding (Hardingham *et al.*, 1976). Basic groups of amino acids such as lysine and arginine may interact directly with the carboxyl groups of the hyaluronate.

(ii) The main polysaccharide-bearing region (molecular weight about

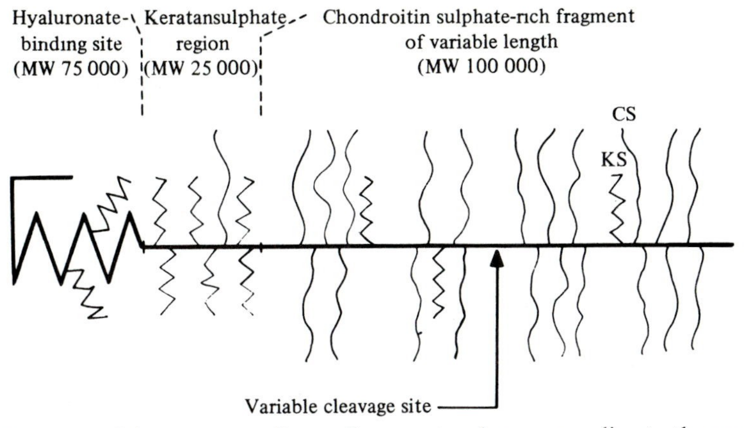

Fig. 34. Diagram of the structure of a cartilage proteoglycan according to the concept of Heinegard & Axensson (1977). The molecular weights refer to the protein core of the molecule. CS = chondroitin sulphate; KS = keratansulphate.

110000) though variable in length contains most of the chondroitin and keratan sulphate side chains; it has no disulphide linkages (Hardingham *et al.*, 1976).

The original structure has been modified (Fig. 34) slightly by Heinegard & Axensson (1977) to include an invariate hyaluronate-binding site (molecular weight 60000–80000) with a few short keratansulphate chains, a chondroitin sulphate-rich fragment (molecular weight 100000) of variable size and an intermediate keratansulphate-rich fragment (molecular weight 20000–25000) lying between them.

Using this type of model, Hardingham *et al.* (1976) hypothesise that cleavage of the protein core at a single point in the polysaccharide-bearing region could account for the polydispersity of aggregating proteoglycans (Fig. 34). Analysis of subfractions of different buoyant density (Hardingham *et al.*, 1976) suggests that with diminishing molecular size the relative concentration of protein and the proportion of the hyaluronate-binding and keratansulphate-rich regions increases in aggregating proteoglycans (Heinegard, 1977). The cleavage mechanism also accounts for the occurrence of the small non-aggregating proteoglycans which have no hyaluronate-binding site and a relatively low content of protein and keratansulphate. Hardingham *et al.* note that electron microscopic observations of monomolecular preparations lend some support to the hypothesis since the length of proteoglycans in intact aggregates from articular cartilage (Rosenberg *et al.*, 1975), though not other forms (Thyberg, Lohmander & Heinegard, 1975), varied over the predicted range of 100–400 nm.

ELECTRON MICROSCOPY OF PROTEOGLYCANS

Proteoglycans extracted from cartilage have been visualised in monomolecular layers using the electron microscope, both as the subunit and in aggregate form with hyaluronate (Rosenberg *et al.*, 1970, 1975). The structures displayed in this manner agree closely with models based on the biochemical properties of these molecules. However, although they can be elegantly depicted in isolation, ultrastructural demonstration of the molecules *in situ* has proved more difficult owing to the presence of other polymers in the tissue, the effect of staining on molecular configuration and probably also molecular size in relation to section thickness. There remain also the usual histochemical problems of incomplete preservation of proteoglycans during preparation of the tissue for electron microscopy (Engfeldt & Hjertquist, 1968; Thyberg *et al.*, 1973) and inadequate specificity of the stains used.

Early studies of the matrix (Robinson & Cameron, 1956; Godman & Porter, 1960; Revel & Hay, 1963) revealed angular and filamentous material forming a meshwork between the fibres and, more especially in mature cartilage, in the pericellular region within the lacunar rim. There are still many uncertainties about this meshwork but there appear to be two morphological components, granules and filaments.

Matrix granules

This term has come into general use following the work of Matukas, Panner & Orbison (1967) who observed granules 20–70 nm in diameter throughout the matrix (Fig. 35*a*), which appeared to be different from fibres or mineral deposits. The granules are stained by colloidal iron, are removed by hyaluronidase or trypsin digestion and have a substructure exhibiting 1.5–4-nm electron-dense areas. On this basis, Matukas *et al.* concluded that the granules consist of one or more proteoglycan molecules compacted by fixation and staining. Corroborative evidence was obtained by Smith and his colleagues who located irregular groups of 3-nm spots stainable with bismuth nitrate in the middle of stellate particles; the minute spots can be removed by hyaluronidase (Smith, Peters & Serafini-Fracassini, 1967; Smith, 1970). They are similar to bismuth-stained spots lying in rows across collagen fibres, earlier identified as collapsed glycosaminoglycan chains (Serafini-Fracassini & Smith, 1966). The identification of matrix granules with fixed proteoglycan has been confirmed by semi-quantitative comparisons of the size and number of granules before and after extraction procedures (Anderson & Sajdera, 1971; Thyberg *et al.*, 1973) and by their positive reaction with ruthenium red and alcian blue.

Alcian blue used in magnesium chloride solution, according to the critical electrolyte concentration (CEC) technique of Scott & Dorling (1965), reveals hyaluronidase-sensitve rods and 'ribbons' (Fig. 35*b–d*) instead of the usual

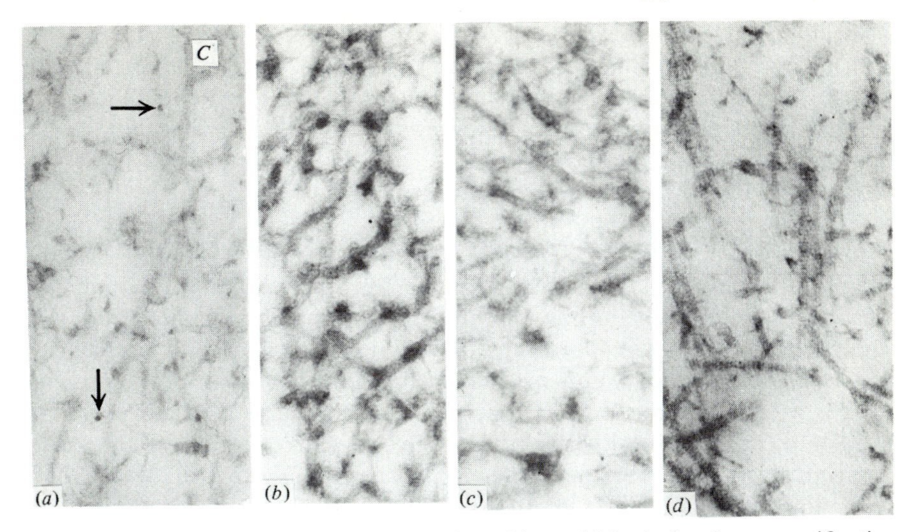

Fig. 35. Rat epiphysial cartilage (two weeks). Glutaraldehyde fixation, magnification × 59000. (*a*) Stained with uranyl acetate and lead citrate showing matrix granules (arrows) and fine filaments, between collagen fibres (*C*). (*b*) Tissue bulk-stained with alcian blue pH 5.8 before embedding, sectioning and staining with uranyl acetate and lead citrate. (*c*) As (*b*) but bulk-stained with alcian blue in 0.4 mol litre^{-1} magnesium chloride. (*d*) As (*b*) but bulk-stained with alcian blue in 0.9 mol litre^{-1} magnesium chloride. Ribbon-like structures in (*b*) and (*c*) replace the matrix granules observed in (*a*). In (*d*) the 'ribbons' are absent and matrix granules re-appear.

40-nm granules (Scott, 1973; Ruggeri, dell'Orbo & Quacci, 1975; Schofield, Williams & Doty, 1975). These strands of material are 20–35 nm wide and up to 500 nm long and in some cases are still stainable at 0.9-M magnesium chloride (Scott, 1973), suggesting that the structures contain both chondroitin sulphate and keratansulphate. Using cinchomeronic phthalocyanine, a low molecular weight analogue of alcian blue (Scott, 1974), the shape and dimensions of the stained molecules more closely resemble the 'ideal' structures depicted by Rosenberg *et al.* (1970) in monolayer.

Filaments

Matrix granules are always associated with projecting filaments 2–5 nm in diameter which also interconnect with collagen fibres, forming a loose meshwork (Fig. 35*a*). The filaments stain with uranyl acetate and lead citrate (Smith, 1970; Ruggeri *et al.*, 1975) and rather weakly with ruthenium red (Myers, Highton & Rayns, 1973; Thyberg *et al.*, 1973) but do not label with ^{35}S-sulphate (Myers, 1976). Though unlikely to contain proteoglycan, their chemical identity is uncertain (Smith, 1970; Meachim, 1972) but may correlate with the large amounts of protein and/or glycoprotein (neither collagenous nor proteoglycan core) which are present in cartilage, particularly in ageing tissue (Szirmai *et al.*, 1967; Muir, Bullough & Maroudas, 1970). The ultrastructural appearance of a network is probably artefactual: as a result of fixation, the protein precipitates and ensheaths the other components of the matrix so producing the 'stellate reticulum' (Smith, 1970; Meachim, 1972; Serafini-Fracassini & Smith, 1974).

COLLAGEN–PROTEOGLYCAN INTERACTION

The maintenance of the integrity of the collagen network and the immobilisation of the large hyaluronate–proteoglycan complexes within it are the basis of the stability of the matrix and of the physico-chemical and mechanical properties of cartilage. A non-specific mechanical entrapment of the proteoglycans within the fibre mesh (Fessler, 1960; Hamerman & Schubert, 1962) is indicated by the ease with which proteoglycans can be freed from the tissue by agents which reduce the proteoglycan particle size as in dissociative extraction (Sajdera & Hascall, 1969). Disruption of the collagen mesh has been implicated as one mechanism by which the matrix proteoglycan may be depleted in fibrillated articular cartilage (Maroudas, 1976). However, following dissociative extraction there still remains a proteoglycan fraction amounting to as much as 20–30% of the total in articular cartilage (Simunek & Muir, 1972) which is very firmly associated with collagen since it can be released only after collagenase digestion of the collagen mesh. Cartilage collagen is much less soluble than other tissue collagens (Miller, van der Korst & Sokoloff, 1969) and this may be due to the association with proteoglycan (Strawich & Nimni, 1971; Deshmukh & Nimni, 1973) as well as collagen cross-linking. Electron histochemical investigation of embryonic and mature cartilage indicates that the polyanionic material is closely related to collagen fibres (Serafini-Fracassini & Smith, 1966; Ruggeri *et al.*, 1975). It seems probable therefore that there may be a specific interaction between the two kinds of macromolecule which would aid both the retention of proteoglycan within the mesh and promote the stability of the network itself.

In polymeric collagen, as in fibrogenesis, direct binding of proteoglycan to collagen could be facilitated by electrostatic interactions (Podrazky *et al.*, 1971) between the acid groups of the polyanions and the basic groups of amino acids in the α-chain sequence. The a_1–a_4 and c_2 bands carry a net positive charge (Doyle *et al.*, 1974*a*, 1975) and histochemical methods using bismuth nitrate and ruthenium red agree in demonstrating anionic material in the region of the *a* bands in cartilage (Serafini-Fracassini & Smith, 1966) and other tissues (Nakao & Bashey, 1972). Other studies also note stainable anionic material distributed periodically along cartilage collagen fibres (Myers *et al.*, 1973; Ruggeri *et al.*, 1977) with some evidence of ^{35}S-sulphate uptake in the same regions (Myers, 1976).

Earlier models (Mathews, 1965) predicted that the protein core and its flexible glycosaminoglycan side chains were aligned parallel to the long axis of the fibre. Serafini-Fracassini & Smith (1966) advocate a modification of this arrangement such that the proteoglycans are coiled transversely around thick fibres; in the case of thin fibres in cartilage (Smith *et al.*, 1967) and cornea (Smith & Frame, 1969) the proteoglycan core may lie tangential to the fibre at the *a* band. Alcian blue staining (Ruggeri *et al.*, 1975) demonstrates filaments 4–12 nm thick (thought to be proteoglycan) lying orthogonal to collagen fibres at 64-nm intervals.

Thus there is some evidence for a direct interaction although it may be noted that fibres formed *in vitro* in the absence of proteoglycan show little affinity for isotopically labelled proteoglycans (Oegama, Laidlaw, Hascall & Dziewiatkowski, 1975). Furthermore, it is difficult to ascertain the precise relationship of the molecules from histochemical studies since cationic dye-binding or staining with heavy metals leads to glycosaminoglycan chain collapse.

Alternatively, proteoglycans might be associated indirectly with collagen via non-collagenous proteins or glycoproteins which could also link fibres together. While the ultrastructural appearance of a filamentous meshwork in the matrix is almost certainly artefactual, it has been suggested that this material might link fibres to proteoglycan (Myers *et al.*, 1973). There is some biochemical evidence for an indirect link between collagen and proteoglycan. For example, the large 600S complex found in articular cartilage contains proteoglycan, collagen and considerable amounts of protein not detectable in the hyaluronate–proteoglycan aggregate. Rosenberg *et al.* (1973) suggest that hydrophobic bonds are involved.

Further support comes from the properties of a non-collagenous glycoprotein extracted from cartilage which has similar staining and structural characteristics to the 'reticulum' filaments in the native tissue (Shipp & Bowness, 1975). In semi-purified fractions, this material shows periodic orthogonal binding to residual collagen fibres although the precise bands to which it is attached cannot be distinguished. The purified glycoprotein contains no proteoglycan, is distinct from the 'link' glycoprotein of the hyaluronate–proteoglycan aggregate and contains a high content of acidic amino acids. Since there is a tenacious association with proteoglycan, this glycoprotein could act as an anchoring material between collagen and proteoglycans. Possibly the alkali-stable bond between keratansulphate and glutamic acid might act as the cross-link between the proteoglycans and glycoproteins, as suggested by Hopwood & Robinson (1974*b*).

MATRIX ORGANISATION

Although the macroscopic appearance of cartilage gives the impression that it is a homogeneous material, several techniques demonstrate that there is a considerable degree of organisation. Neither the chemical components of the matrix nor the cells are randomly disposed, a number of factors including

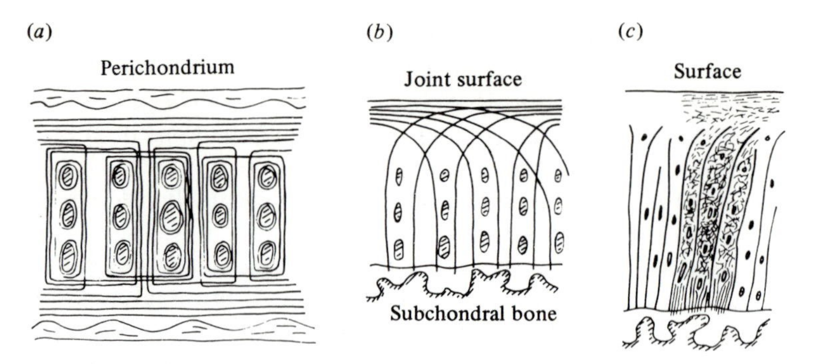

Fig. 36. Schematic diagram of the arrangement of collagen fibres in (*a*) hyaline cartilage, and (*b*) articular cartilage, according to the concept of Benninghoff (1925). In articular cartilage, electron microscopy indicates randomly orientated fibres in the middle zone, organised into 'layers' running radially through the tissue (*c*).

species, age of the organism, and the type, anatomical site and size of the cartilage influencing the pattern of distribution. Within a given cartilage specimen, the matrix varies according to the distance from the tissue boundary and the distance from the cell margin. The structure of the tissue boundaries themselves in most cases are considerably modified compared with the tissue they enclose, on account of mechanical, metabolic and nutritional requirements (see Chapter 5). Except where the boundary is calcified, there is a gradual change in the structure and chemistry of the tissue from the boundary inwards to the typical cartilage. Hence the matrix variation with respect to the cell is influenced by the depth in the tissue at which the cell lies, and description of pericellular variation is simplified by considering only the deeper part of the cartilage.

COLLAGEN FIBRE DISTRIBUTION

Special techniques of light microscopy are required to visualise the fibres of hyaline cartilage matrix, since the polyanions of the ground substance mask the staining sites of the collagen. Using polarised light, Benninghoff (1925) described an arrangement of interlacing continuous arcades which ran radially in the middle of the cartilage, curving at the periphery to run parallel to the tissue boundary before returning into the tissue depths (Fig. 36). In the case of articular cartilage the arcades commenced by running radially from the calcified layer.

In the case of non-articular hyaline cartilage, the internal stresses of the tissue (Gibson & Davis, 1958) are compatible with Benninghoff's ideas. In articular cartilage, which has been investigated most thoroughly, there is no doubt that the fibres in the superficial zone run parallel to the articular surface (Figs. 36, 69) compatible with the tangential tensile stresses occurring in this region (Zarek & Edwards, 1963). There is no evidence, however, that they are continuous with deeper fibres. With respect to the deeper issue, Benninghoff's ideas have been disputed on the biochemical grounds that obliquely orientated fibres would better withstand applied mechanical stress (MacConaill, 1951). Certainly many high-magnification studies demonstrate unequivocally that the middle zone contains randomly orientated oblique or coiled fibres (Little, Pimm & Trueta, 1958; Davies *et al.*, 1962; Weiss *et al.*, 1968; Clarke, 1971*c*), although in the deep zone the fibres may become radial during ageing. Such a meshwork of randomly orientated fibres is fully compatible with the need for entanglement and restraint of proteoglycan molecules (Fessler, 1960) or with the concept of randomly orientated tensile forces in the middle zone set up by compressive load (McCutchen, 1965). On the other hand, investigations at low magnification, particularly with the scanning electron microscope, suggest that the randomly orientated fibres are organised into radially orientated bundles (Fig. 36) or layers (Bullough & Goodfellow, 1968; McCall, 1969; Clarke, 1971*c*; Minns & Steven, 1977). The factors governing bundle or layer formation have not been satisfactorily determined. Thus while individual collagen fibres are neither radially orientated nor follow a continuous arched pathway, there is nevertheless some evidence supporting the general pattern described by Benninghoff.

The arrangement of collagen with respect to individual chondrocytes or cell groups changes during maturation. In fetal cartilage (Robinson & Cameron, 1956; Godman & Porter, 1960; Stockwell, 1971*a*) collagen fibres are of narrow diameter (15–30 nm) in all parts of the matrix, showing no variation in thickness or orientation with respect to the cell with which they may be in contact. However, in growth plate cartilage, fibres in the longitudinal septa between the cell columns tend to be thicker than those lying closer to the cells. In older immature tissue, thicker fibres develop (30–70 nm); fine fibres (15–30 nm) are found next to the cells, giving way quite abruptly to the coarser fibres of the general matrix (Davies *et al.*, 1962; Bonucci, Cuicchio & Dearden, 1974). In mature cartilage, while few thick fibres (150–280 nm) may be found close to degenerate cells (Barnett *et al.*, 1963; Bonucci *et al.*, 1974), the matrix in the immediate vicinity of the normal chondrocytes is characterised by an absence of banded collagen fibres (Fig. 37). This zone forms a corona around the cell, consisting of finely textured matrix (Meachim & Roy, 1967; Weiss *et al.*, 1968). It contains a meshwork of unbanded filaments (5–10 nm in diameter), usually exhibiting an abrupt transition to the surrounding coarsely fibrous matrix. The corona, termed the pericellular matrix by Meachim & Stockwell (1973), varies in breadth (up to 1–3 μm) but may be absent in some parts around the cell, the coarse matrix then approaching close to the cell surface. However, even where the pericellular matrix is a micron or more thick, projecting cell processes may pass through it to enter the outer coarse matrix. The width of the finely textured matrix and the definition of its outer limit increase at first with age but in later life thick collagen fibres may occur in the region (Bonucci *et al.*, 1974; Dearden, Bonucci & Cuicchio, 1974).

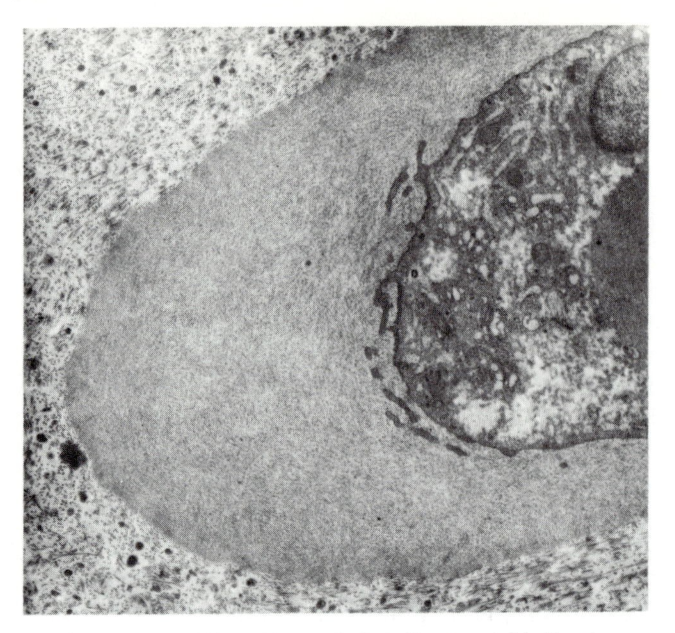

Fig. 37. Human articular cartilage (56 years). Osmium tetroxide fixation, magnification × 8000. Part of a middle zone chondrocyte with a broad pericellular zone of finely textured matrix. Note the well-defined margin (the lacunar rim) at the junction with the coarsely fibrous matrix and the dense bodies (lipidic) near this junction. (From Stockwell, 1978*b*.)

Electron-dense membranous vesicles are often seen just beyond its outer margin (Fig. 37) and may be partly accounted for by detachment of the tips of projecting cell processes (Ghadially *et al.*, 1965). These structures probably account for much of the extracellular lipid found in cartilage. This component of the matrix is discussed fully in Chapters 5 and 8.

The orientation of the collagen fibres in the outer coarse matrix is controversial. Benninghoff (1925) described groups of cells or 'chondrones' enclosed by fibre ellipses, lying between the radially orientated fibre bundles (Fig. 36). In agreement with this concept, Szirmai (1969) found that many chondrones remain intact following high-speed homogenisation of nasal septal cartilage, suggesting that they form mechanical units ensheathed by collagen. Transmission electron microscopy gives some support since many fibres appear to be arranged tangential to the margin of the pericellular matrix (Barnett *et al.*, 1963) and may resemble 'basket-like' enclosures around the cells (Weiss *et al.*, 1968). Scanning electron microscopy can also provide useful information about fibre architecture since specimen fracture permits the study of the matrix around cells on the exposed surface. Mulholland (1974) has studied human articular cartilage by this means, following partial digestion of the fractured surface with papain; he also considers that protective basket-works are commonly present at the margin of the pericellular matrix. On the other hand, Clarke (1974) finds no special orientation or increased density of fibres at the margin of the pericellular matrix. Although he describes and illustrates fibrillar layers arranged concentrically around many cells, he does not accept the idea of 'protective envelopes' around the cells. Nor does microprobe analysis of articular cartilage provide support for the idea of a collagen basket-work since there is no evidence of an increase in the dry mass of the tissue at the margin of the pericellular matrix (Maroudas, 1972).

PROTEOGLYCANS AND GLYCOSAMINOGLYCAN DISTRIBUTION

Unlike collagen distribution, the distribution of proteoglycan and glycosaminoglycans is amenable to light microscope histochemistry, which reveals a marked heterogeneity of the tissue. Early studies using cationic dyes demonstrated that basophilia and metachomasia diminish toward the boundaries of the tissue (Schaffer, 1930; Hirsch, 1944; Collins & McElligott, 1960; Szirmai, 1963). In adult tissues, CEC techniques (Scott & Dorling, 1965) used in association with enzyme extraction of control sections (Stockwell & Scott, 1965; Quintarelli & Dellovo, 1966; Scott & Stockwell, 1967; Mason, 1971) demonstrate the same trend in both chondroitin sulphate and keratansulphate (Fig. 38). In articular cartilage, however, the deep zone near the calcified layer shows stain distribution similar to that of the central regions of non-articular cartilages. Microchemical techniques (Stockwell & Scott, 1967; Szirmai *et al.*, 1967; Maroudas, Muir & Wingham, 1969; Lemperg, Larsson & Hjertquist, 1974) confirm the histochemical distribution of glycosaminoglycans. Hexosamine forms about 12% of the dry weight in the middle of young adult nasal septal cartilage compared with only 2% in the subperichondrial zone; articular cartilage contains less hexosamine and there is not so much variation with distance from the margin of the tissue. All studies show that keratansulphate is maximal in the middle of cartilage specimens or in the case of articular cartilage in the deepest part of the uncalcified tissue (Fig. 39) where it may form more than 50% of the total glycosaminoglycan (Stockwell, 1970).

Little is known of change with depth in the character of the proteoglycans of permanent cartilages. In the growth plate, reduction in glycosaminoglycans and other constituents of the matrix (Campo & Dziewiatkowski, 1963; Hirschman & Dziewiatkowski, 1966; Greer, Janicke & Mankin, 1968) occurs toward the metaphysial side of the growth plate. Similarly the size of proteoglycans (Larsson, Ray & Kuettner, 1973) and the number of aggregates are reduced (Lohmander & Hjerpe, 1975), but these changes relate to impending calcification.

The heterogeneity of the ground substance in relation to the cell was noted by the early microscopists. Variations in basophilia and/or alternating zones

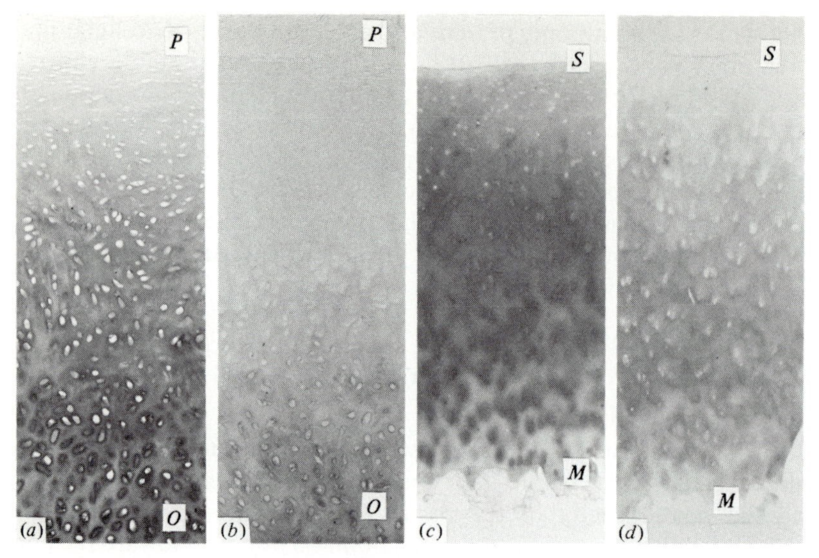

Fig. 38. Histochemical distribution of glycosaminoglycans in cartilage. (*a*), (*b*) Bovine nasal cartilage (young adult). Formalin fixation, magnification ×32. (*c*), (*d*) Human articular cartilage (73 years). Formalin fixation, magnification ×20. Frozen sections were stained with alcian blue in 0.4 mol litre^{-1} magnesium chloride in (*a*) and (*c*), and with alcian blue in 0.9 mol litre^{-1} magnesium chloride in (*b*) and (*d*). Stain density due to both chondroitin sulphate and keratansulphate shows the same distribution in nasal (*a*) and articular cartilage (*c*) diminishing toward the perichondrium (*P*) or the joint surface (*S*), respectively. Staining due to keratansulphate only shows the same distribution in (*b*) and (*d*) and is maximal in the central region (*O*) of nasal cartilage and near the calcified region (*M*) of articular cartilage.

of basophilia and acidophilia around the cell demonstrated differences in glycosaminoglycan staining reaction which seemed to show that the matrix was divided into zones more or less concentric to the individual cell or cell groups (Fig. 40*a*). Interpretation of these staining patterns was made all the more difficult because they altered during maturation and ageing and varied between different parts of the same cartilage specimen. Szirmai (1969) attributes much of the apparent heterogeneity in glycosaminoglycan distribution to the presence of other chemical substances, such as non-collagenous protein. At low pH, cationic dyes must compete with the ionised, positively charged groups of tissue protein lying in the vicinity of the anionic glycosaminoglycans. Thus false negative results may occur, depending on the relative proportions of protein and polysaccharide (Szirmai & Doyle, 1961; Szirmai, 1963).

The effect of protein competition is seen in the central region of aged horse nasal septal cartilage where cationic dye-binding is considerably inhibited at pH 2 but not at pH 6 (Szirmai, 1963). The phenomenon is also relevant to the concentric zones around the cell observed in the same region. Dye-binding

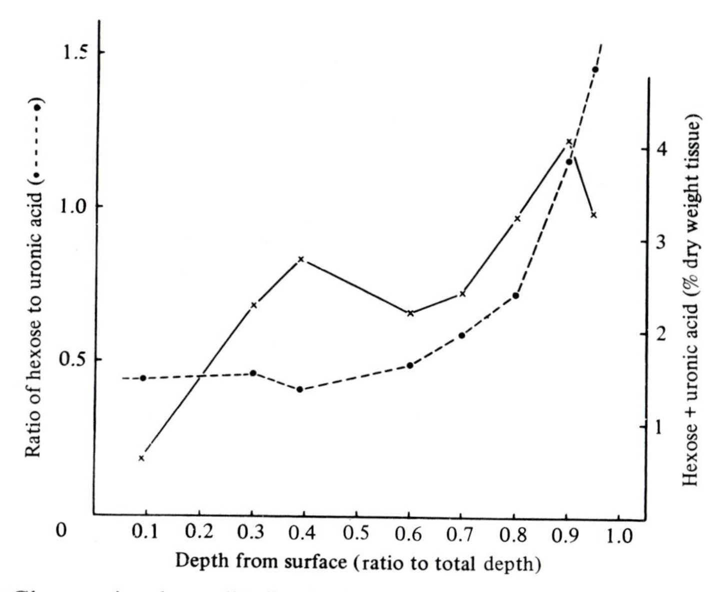

Fig. 39. Glycosaminoglycan distribution in human femoral condylar articular cartilage (32 years). Uronic acid is a component of the repeating disaccharide of chondroitin sulphate, hexose (galactose) of keratansulphate.

at low pH differentiates between a strongly basophilic area around the cells, a narrow unstained 'border zone' (Fig. 40*a*) at the periphery of this area and moderately stained matrix external to the border zone (Galjaard, 1962; Szirmai, 1969). At neutral pH the border zone is as basophilic as the rest of the matrix (Fig. 40*b*). This suggests that there is in fact a fairly homogeneous distribution of glycosaminoglycan. Furthermore, chemical analyses of micro-dissected regions of the matrix detect little difference either in concentration or composition of glycosaminoglycan content of the matrix in relation to distance from the cell (Szirmai, 1963).

The alcian blue–CEC technique, which is not susceptible to the pH-induced phenomenon of protein competition, supports Szirmai and Galjaard's findings in the central part of aged horse nasal cartilage with respect to chondroitin sulphate (Scott, Dorling & Stockwell, 1968); high molecular weight keratan-sulphate is stainable in the matrix near the cells. In the central parts of ageing human costal and bronchial cartilage (and the deep zone of articular cartilage), however, the territorial areas are much more basophilic than the rest of the matrix (Fig. 41), the CEC technique indicating that both chondroitin sulphate and keratansulphate are very largely confined to this area (Stockwell & Scott, 1965; Quintarelli & Dellovo, 1966; Stockwell, 1970; Mason, 1971).

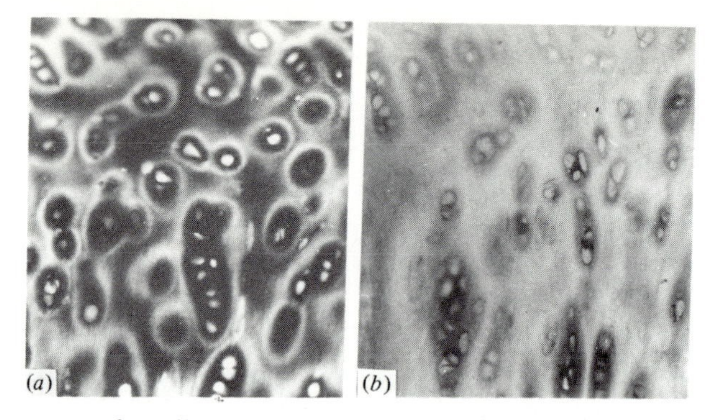

Fig. 40. Human costal cartilage (18 years). Formalin fixation, magnification ×80. (*a*) Stained with azur A at pH2. The appearance of concentric zones of matrix around the cells or cell group is partly due to the unstained ring or 'border zone' separating the densely stained territorial matrix internally from the paler inter-territorial matrix externally. (*b*) Stained with alcian blue in 0.4 mol $litre^{-1}$ magnesium chloride at pH 5.8. The territorial matrix has a higher stain density than the inter-territorial matrix but 'border zones' are absent.

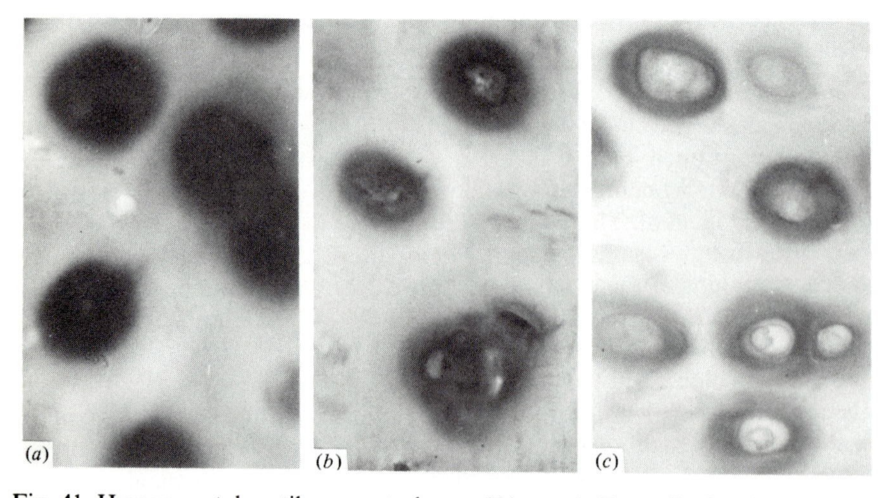

Fig. 41. Human costal cartilage, central zone (64 years). Formalin fixation, magnification ×200. (*a*) Stained with azur A at neutral pH. (*b*) Stained with alcian blue in 0.4 mol $litre^{-1}$ magnesium chloride at pH 5.8. (*c*) Stained with alcian blue in 0.9 mol $litre^{-1}$ magnesium chloride at pH 5.8 after digestion with hyaluronidase, to show the distribution of keratansulphate. In all three techniques, basophilia is largely confined to the territorial matrix.

HISTOLOGICAL SUBDIVISIONS OF THE MATRIX

Whatever the distribution of glycosaminoglycan may be, the multiplicity of zones revealed by early microscopy resulted in a complex and confusing nomenclature. Some of these terms, such as 'balkwerk', 'inter-territorial (amorphous and granular) matrix', 'territorial matrix', 'chondrinbollar', 'cell territory', 'Zellhof', 'capsule', 'lacuna', 'perilacunar rim', 'moat' and so on, are still in contemporary use. Schaffer (1930) in his authoritative review of cartilage distinguishes from within outwards a cell capsule, inner strongly basophilic and outer weakly basophilic parts of a territorial zone and an inter-territorial zone. However, these terms are still apt to confuse. Thus not infrequently in one and the same paper, cell territory, capsule and lacuna are used synonymously *and* to describe different regions of the matrix.

The lacuna

The term lacuna is commonly used but it is not easily defined. Lacunae as true cavities in the tissue, the literal meaning, cannot exist. Early investigators with transmission electron microscopy (Davies *et al.*, 1962) insisted that the chondrocyte completely fills the lacuna which it occupies, implying that the lacuna is the space that would remain if the chondrocyte were removed. Clearly, this does not correspond to common usage in light microscopy, where it is taken to mean the oval region often with a smooth distinct rim enclosing the cell and a small amount of basophilic matrix. Similarly, recent ultrastructural reports, particularly of scanning electron microscopy (Clarke, 1974; Mulholland, 1974), interpret the lacunar margin as corresponding to the outer limit of the finely textured pericellular matrix. Perhaps a useful compromise may be reached if the lacunar margin (or perilacunar rim) is defined as the internal limit of encroachment of the fibrous matrix (Figs. 19, 37, 60). Hence in fetal tissue, the lacuna is indeed the space occupied by the cell since fibres abut on the chondrocyte surface; in mature and ageing cartilage (and in many zones of the growth cartilage) the perilacunar rim encloses the cell 'space' and the finely textured pericellular matrix.

The extra-lacunar matrix

Lacunae should not be confused with the wide basophilic (about 50 μm in diameter) zones surrounding single cells or cell groups. These are usually termed cell territories or territorial matrix by light microscopists, distinct from the outer, less basophilic inter-territorial matrix (Figs. 40, 41). Nevertheless there is little ultrastructural evidence, particularly from scanning electron microscopy, of such zonal subdivisions external to the lacuna. While collagen fibres become coarser and matrix granules less numerous with increasing distance from the cell (Matukas *et al.*, 1967; Campo & Phillips, 1973;

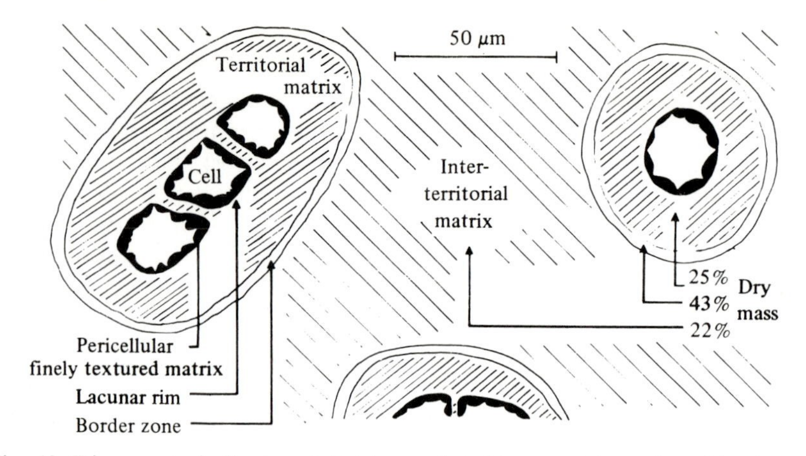

Fig. 42. Diagram to indicate subdivisions of cartilage matrix as found in the central region of ageing horse nasal and some other forms of cartilage. Dry mass data from Galjaard (1962).

Dearden *et al.*, 1974) no special arrangements of collagen fibres have been detected which might demarcate the territorial matrix (Clarke, 1974). The disputed basket-like enclosures are said to be located at the lacunar margin, (Weiss *et al.*, 1968), rather than further out from the cell at the boundary of the cell territory. However, most studies have been made on articular cartilage which normally shows well-formed basophilic territories only in the deepest part of the uncalcified tissue.

The narrow 'border zone' at the margin of the cell territory in horse nasal cartilage is of interest here (Fig. 42). This might have a high protein concentration as suggested by the absence of basophilia at low pH and its dense staining following the periodic acid–Schiff (PAS) reaction. Using interference microscopy, Galjaard (1962) found that this zone has a high dry mass (0.43 g cm^{-3}) compared with the matrix on its inner (0.25 g cm^{-3}) and outer aspect (0.22 g cm^{-3}). Although measurements made by Maroudas (1972) in articular cartilage show no differences in dry mass in various parts of the matrix, it does not follow that the two sets of results are at variance. Maroudas' measurements were made at the perilacunar rim and in the matrix 50 μm outside the rim and therefore might have 'straddled' Galjaard's border zone. Moreover, as regards glycosaminoglycan histochemistry, much of the thickness of articular cartilage resembles the more peripheral parts of non-articular hyaline cartilage: although keratansulphate is stainable only in the inter-territorial matrix, the cell territories are not contained by border zones.

Published scanning electron micrographs of the deep zone of articular cartilage are intriguing in this context since this is the only part of the tissue which exhibits well-defined territories. They appear to show high-density circular ridges corresponding in position to the margins of the cell territories (not the lacunae) seen histochemically. Although these appearances might be artefacts due to the interaction of the electron beam with matrix discontinuities, the authors (Mulholland, 1974, Figs. 12 and 15; Clarke, 1974, Fig. 12*a*) do not comment on them. Further scanning studies of this feature in articular cartilage and in the central region of horse nasal cartilage would be of considerable interest.

Although Szirmai (1969) believes that collagen basket-works are present, he does not consider that collagen itself is responsible for the absence of basophilia at low pH in the border zone. He suggests that this is due to collagen-associated glycoproteins and/or proteins derived from the proteoglycans themselves. In the central parts of adult costal and bronchial cartilage the weak basophilia of the extensive regions of inter-territorial matrix may be partly due to impregnation with protein. The residua of dead chondrocytes may contribute to the protein content since cell density falls during adult life (Stockwell, 1967*a*). At the same time, cell death terminates the synthesis and secretion of proteoglycans and therefore must lead to local depletion of the matrix. Although the territories around dead cells retain their stain density for some time following cell death and the gradual in-filling of the lacunae with fibrous tissue, they eventually lose their basophilia (Mason, 1971; Dearden *et al.*, 1974). The formation of a wide, poorly basophilic interterritorial region, as seen in adult costal cartilage, is thus associated with cell death (Schaffer, 1930; Sames, 1975).

PHYSICO-CHEMICAL PROPERTIES OF THE MATRIX

This very largely concerns the cartilage water and its regulation by the matrix macromolecules. The proteoglycans are the principal agents which control the water, exhibiting ionic effects due to their negative charge and steric effects due to their large size. Although the fibres make a small direct contribution, the main role of the fibrous mesh is to entrap proteoglycan molecules and so retain them within the matrix. In certain situations the function of collagen fibres is more obviously that of resisting highly orientated tensile stress, as at the articular surface. The control of the water and its solutes are relevant to many aspects of chondrocyte function, such as metabolism and nutrition, as well as to some mechanical properties of the matrix. Physico-chemical aspects of cartilage have been discussed by a number of authorities (Laurent, 1968, 1970; Scott, 1968; Ogston, 1970; Maroudas, 1973) and only a short non-mathematical survey is given here.

WATER CONTENT

The water content of cartilage varies. Thus human adult costal cartilage contains as little as 61% while articular cartilage has 75–78% by weight (Lindahl, 1948; Miles & Eichelberger, 1964; Linn & Sokoloff, 1965). Immature cartilage is slightly more hydrated than adult tissue. The superficial zone of articular cartilage (Maroudas *et al.*, 1969) and the subperichondrial region of non-articular cartilage (Szirmai *et al.*, 1967) contain more water than the rest of the cartilage. Local variations in water content within the matrix have been recorded in the central region of aged horse nasal cartilage though similar variation has not been detected in human articular cartilage.

The water is both intra- and extracellular but since the cells occupy only 1–10% of the cartilage volume (depending on the species and cartilage size) the intracellular compartment is relatively small. In articular cartilage there are changes in water content during joint activity and loading. There is also an increase in extracellular water following denervation and immobilisation of joints (Akeson, Eichelberger & Roma, 1958; Eichelberger, Roma & Moulder, 1959) and fibrillated cartilage has a higher water content than normal cartilage (Bollet & Nance, 1966; Maroudas, Evans & Almeida, 1973).

IONIC EFFECTS

When cartilage is immersed in electrolyte, the fixed negative charge of the proteoglycans causes the mobile ions to be distributed between the cartilage and the external solution, in accordance with the Donnan equilibrium. The counterions (cations) accumulate in the cartilage to 'neutralise' the fixed and mobile negative charge; the co-ions are reciprocally diminished in the cartilage but accumulate in the external solution. Ion binding by polyanions is reviewed by Scott (1968). The attraction of the counterions to the proteoglycans depends on the ion activity of the fixed negative charge (associated with the external ionic strength and pH) and the valency of the mobile ion. Thus divalent cations, such as calcium, have a distribution coefficient (concentration in cartilage/concentration in external solution) which is the square of that for monovalent cations such as sodium. Mobile ions show slight differences in affinity to the different polyanions dependent on factors such as the size of the solvation shells. This results in an order of preference which is similar in general for either proteoglycan solutions or cartilage tissue itself. Thus Dunstone (1960) finds the order:

$$K^+ > Na^+ > Mg^{2+} > Ca^{2+} > Sr^{2+} > Be^{2+} > Cu^{2+}$$

but the position of Ca^{2+} is disputed since Maroudas (1973) finds a higher affinity: $K^+ > Ca^{2+} > Na^+$. Preferences also occur in the anions since inorganic sulphate has a higher distribution coefficient than the chloride ion (Maroudas, 1973).

Thus, apart from their electrostatic interaction with collagen the multiple negative charges of the proteoglycans have two effects. First, they partition mobile ions primarily according to the type of charge but also according to the valency and other properties. This regulates the entry of some nutrients into cartilage; hence precursor anions, such as inorganic sulphate, are present in very low concentration (0.003 M) in cartilage. Maroudas (1973, 1975) has stressed the necessity of using the correct intracartilaginous concentration of ionic precursor substances when calculating the rates of turnover and synthesis of glycosaminoglycans. Whatever the precise order of preference of the mobile cations, Ca^{2+} are present in much higher concentration (0.025 equiv. litre^{-1}) than in the external fluid (Maroudas, 1973). This may be relevant to chondrocyte metabolism since Ca^{2+} apparently modify glycosaminoglycan and collagen synthesis. Calcium binding by proteoglycans is also relevant to calcification mechanisms since some proteoglycans may inhibit calcium deposition or perhaps act as carriers between cell and matrix vesicle. These aspects of metabolism and nutrition are discussed elsewhere (Chapters 4 and 5).

The second effect is that the number of mobile ions present in cartilage is enormously increased by the presence of the fixed negative charge. Hence there are many more ions in the cartilage than in the external solution, resulting in a considerable ionic contribution to the osmotic pressure.

STERIC EXCLUSION

The large proteoglycans tend to remain fully expanded by the electrostatic repulsion of their multiple negative charges. In solution they may adopt a spherical conformation (Hascall & Sajdera, 1970), although X-ray diffraction studies suggest that at the high concentrations encountered *in vivo* the glycosaminoglycan side chains may undergo crystalline packing (Atkins, Hardingham, Isaac & Muir, 1974).

As with hyaluronic acid (Laurent, 1970), the space or domain occupied by proteoglycan molecules in solution is very much greater than their anhydrous volume. Hence a considerable solvent volume becomes unavailable to other large solute molecules due to steric exclusion (Laurent, 1968). Ogston & Phelps (1961) showed that if a globular protein is allowed to equilibriate between two phases (Fig. 43), one containing buffer solution and the other hyaluronic acid, the protein concentration is higher in the buffer phase. The distribution coefficient is equal to the available volume fraction (K_{Av}) in the hyaluronic acid phase, and the excluded volume ($V_{excl.}$) is related to the available volume fraction: $V_{excl.} = (1-K_{Av})$/polysaccharide concentration. Thus in a 0.4% solution of hyaluronic acid, 50 ml g^{-1} of polysaccharide are excluded for serum albumin (Laurent, 1964); in a 1% solution of cartilage proteoglycan, the excluded volume for albumin is 55 ml g^{-1} (Gerber & Schubert, 1964).

If the hyaluronic acid is regarded as a rod-like particle (radius r_1) and the globular protein as a sphere (radius r_2), then the centre of the sphere is excluded from a cylindrical volume of solvent (radius $= r_1 + r_2$) around the polysaccharide (Fig. 43). Ogston (1958) showed that in a system of randomly arranged rods (sum of length per cm^3 $= L$) the available volume fraction for

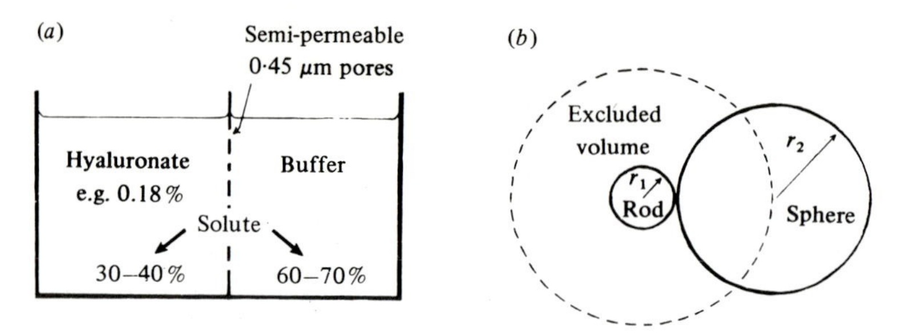

Fig. 43. Steric exclusion effects of macromolecules. (*a*) Use of equilibrium dialysis. The solute, for example a globular protein, can pass freely through the semi-permeable membrane. Data from Ogston & Phelps (1961). (*b*) The centre of the sphere is excluded by the rod from the cylindrical volume of solvent, radius $= r_1 + r_2$.

the spherical particles $K_{Av} = \exp(-\pi L(r_1 + r_2)^2)$. Hence the excluded volume should be sensitive to changes in the concentration (total length) of the rods (i.e. the polysaccharide or proteoglycan concentration) and the spherical particle size. Experimental findings confirm this prediction in cartilage where there is a high concentration of proteoglycans and the domains interpenetrate. Thus the distribution coefficient between cartilage and the external solution diminishes with increasing size of the solute added and with the fixed charge density (i.e. the proteoglycan concentration) of the tissue (Maroudas, 1970, 1976). Large molecules such as haemoglobin and immunoglobulin are almost totally excluded, but even small molecules such as glucose are affected slightly.

Within the cartilage the steric effects of the proteoglycan molecules can be regarded as producing channels and pores (mean radius 2–4 nm) between the domains through which solutes must pass. Hence diffusion rates in the tissue are also affected since the larger the solute molecule the fewer the pores available and presumably the more tortuous the diffusion path. Friction between the solute and the matrix macromolecules also retards the diffusion of large molecules (Maroudas, 1973). Fluid flow through the matrix (hydraulic permeability) is also dependent on the pore size and therefore varies inversely with the proteoglycan concentration. However, permeability is also reduced where the collagen mesh is extremely dense, as in the superficial zone of articular cartilage (Muir *et al.*, 1970).

Thus the principal effect of the large size of the proteoglycans is to produce a molecular 'sieve'. This property is largely responsible for the immunological protection given to the chondrocytes by the matrix. It also affects the turnover of the matrix macromolecules themselves. Thus the passage of both degradative enzymes and newly synthesised macromolecules from the chondrocytes is restricted. There are many unsolved problems regarding these two aspects of matrix turnover but the steric effect tends to confine such activity to within the vicinity of the cell.

WATER RETENTION IN THE MATRIX

The amount of extracellular water in cartilage reflects the balance between the swelling pressure and the restraining forces in the matrix. The swelling pressure is principally due to the properties of the proteoglycans and the restraining forces to the collagen mesh. Small contributions to the swelling pressure are made by the formation of solvation shells around the fixed and mobile ions associated with the proteoglycans; a small amount is associated with the collagen. This bound water is negligible, amounting to only about 1% of the total (Maroudas, 1973). The rest of the water, which is freely exchangeable and available to small solutes (Maroudas, 1970; Maroudas & Venn, 1977), is accounted for by the tissue osmotic pressure and the tendency of the proteoglycans to expand due to electrostatic repulsion.

The osmotic pressure results from the excess of particles in the cartilage compared with the external solution. Thus it has two components: (i) the colloid pressure due to the collagen and the proteoglycans themselves and (ii) the ionic (Donnan) contribution due to the mobile counter- and co-ions. This suggests that the Donnan osmotic pressure should exceed the colloid pressure since the mobile ions far outnumber the colloidal particles (Ogston, 1970). The value of the Donnan osmotic pressure in cartilage (unloaded) has been variously estimated at 1–3 kg cm^{-2} (Ogston, 1970; Maroudas, 1973).

The arrangement of domains and pores is in line with the concept that cartilage ground substance has a colloid-rich phase (the domains) and a water-rich phase corresponding to the pores (Gersh & Catchpole, 1960; Catchpole, Joseph & Engel, 1966). In such a two-phase system, the factors contributing to the swelling pressure of cartilage and Donnan distribution of ions pertain to the colloid-rich phase. A second level of physico-chemical organisation in the matrix may occur in ageing cartilage. If histochemical observations are correct, there is more proteoglycan in the territorial than the inter-territorial regions. Osmotic pressure and exclusion effects would then be more pronounced in the cell territory and preserve mechanical and other types of protection for the cells.

RELATION TO MECHANICAL PROPERTIES

Mechanical properties of cartilage are fully discussed by Kempson (1973). It appears that deformation of cartilage under compression has two components. There is an 'instantaneous' phase occupying a few hundredths of a second, which is independent of fluid flow relative to the cartilage matrix. This is succeeded by a long 'creep' phase due to fluid flow, accounting for most of the change in shape. In physico-chemical terms, water is lost from the region under load because the applied force (together with the internal restraining forces) overcomes the swelling pressure. However, as the matrix particles become more concentrated, the swelling pressure increases and eventually another equilibrium is reached. If the compressive force is removed, the swelling pressure exceeds the restraining force and so the tissue re-imbibes fluid. In articular cartilage, the values of the Donnan osmotic pressure are low compared with the compressive forces (*c.* 30 $kg\ cm^{-2}$) at all stages of joint loading; it is suggested that it is the other components of the swelling pressure which are more important in controlling the equilibrium in prolonged loading (Maroudas, 1973).

The magnitude of the deformation and the stiffness of the cartilage are also related to the rate of fluid expression. This is dependent on the hydraulic permeability of the matrix which in turn is inversely related to proteoglycan concentration. Thus the retention of water under load is doubly dependent on an adequate proteoglycan content.

However, the greater water content of fibrillated, compared with normal, cartilage is somewhat paradoxical, since the proteoglycan content is depleted. Maroudas & Venn (1977) consider that this is due to a diminished restraining force caused by impairment of the collagen mesh, so permitting abnormal cartilage swelling, rather than an increase in the colloid osmotic pressure (Bollet & Nance, 1966).

CELLULARITY

Only 1–10% of the volume of cartilage is occupied by cells and the low cell density is one of the well-known features of the tissue, aiding its histological recognition. It is also relevant when assessing the metabolic activity of cartilage: the low oxygen requirements of chondrocytes together with the low cellularity of the tissue render cartilage a metabolically economical tissue with only minimal blood supply requirements. A price is paid in the poor repair capability of the tissue.

Although cell density is low compared with many other tissues there is a wide range of cellularity according to the type of cartilage, the species, the age of the individual and the dimensions of the cartilage. However, even within a single specimen the cellularity is not homogeneous. There is usually a gradient of cell density from the periphery to the interior whether or not the cartilage is surrounded by a perichondrium; furthermore at any given depth within the tissue the chondrocytes are not distributed at regular intervals.

IMMATURE CARTILAGE

In common with several other tissues, but perhaps somewhat unexpectedly in the case of cartilage, the first sign of differentiation is that of mesenchymal condensation, although the newly established chondrocytes soon become separated by matrix. The cell density at this stage in human embryonic limb buds is $2\text{–}3 \times 10^6$ per mm^3 (see Chapter 6). Thereafter despite mitosis occurring within the tissue the cellularity falls throughout the growth period.

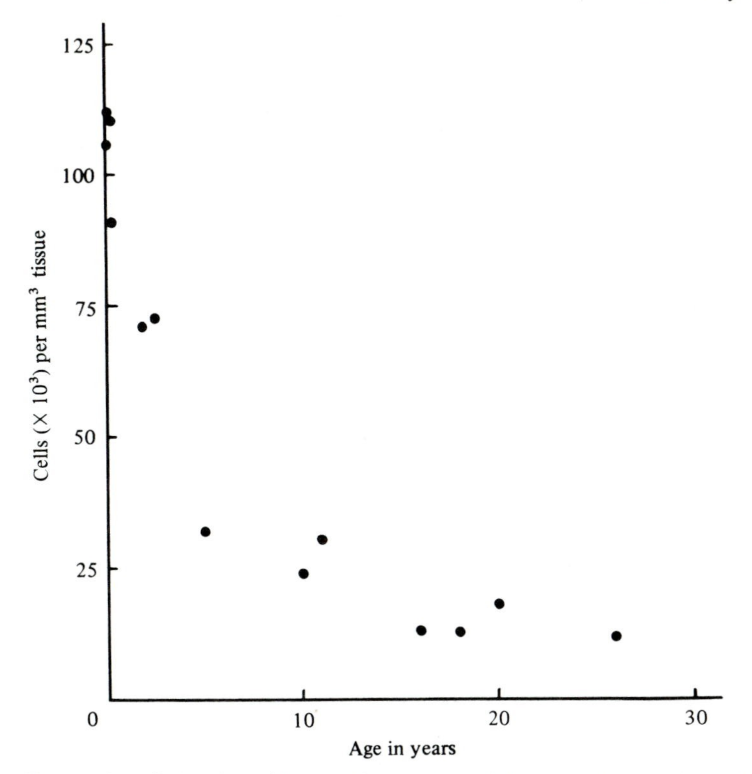

Fig. 44. Change in cell density of human femoral condylar articular cartilage during maturation.

This process carries on in the permanent cartilages until the completion of maturation. Thus in man, the cell density of newborn femoral condylar articular cartilage is about 10^5 cells per mm^3 and this falls to about 15×10^3 cells per mm^3 in the young adult, a sevenfold reduction (Fig. 44). A similar decline occurs in costal (Stockwell, 1967*a*) and tracheal cartilages (Beneke, 1973) and in other species (Bywaters, 1937; Rosenthal, Bowie & Wagoner, 1941; Mankin & Baron, 1965).

The change in cell density is the product of three cellular mechanisms – matrix deposition and cell death, leading to increased intercellular distance, and interstitial mitoses, which tend to reduce intercellular spacing. Cell death occurs from the earliest stages of the pre-cartilage blastema and probably has a morphogenetic role (Glucksmann, 1951) although it is not instrumental in the production of glycosaminoglycan as was once believed. Although both cell death and mitoses are relatively common findings in electron microscopic investigations of fetal epiphysial cartilage, it is difficult to assess quantitatively their effects on the cellularity of the tissue since there are insufficient data. In costal cartilage, the number of cells in a cross-section of the structure declines throughout post-natal life but there is no significant change during the growth period (Stockwell, 1967*a*). However, cell death is more likely to exceed cell reduplication in the central than in the peripheral regions of cartilage specimens since appositional growth predominates in late maturation. Nevertheless it is improbable that cell death occurs on a scale massive enough to account for the fall in cell density of the whole cartilage. Undoubtedly, matrix interposition is responsible for the bulk of the cellularity change during development.

VARIATION WITHIN ADULT CARTILAGE

There is considerable micro-heterogeneity in cell distribution within small volumes of tissue since chondrocytes occur either singly or in small groups. These 'isogenous' cell groups are presumed to be the progeny of a single cell which entered mitosis relatively late in development during the last stages of interstitial growth. The daughter cells remain in close proximity since little interposition of matrix occurs. Such an isogenous cell group is said to occupy the same lacuna: in adult cartilage this is not wholly acceptable since it is rare to find cells which are not separated by coarsely fibrous matrix.

While most groups are probably isogenous in this sense, other explanations are possible. Thus a few cells here and there might remain at the same spacing from fetal to adult life and due to lacunar enlargement could eventually form a group in the adult. Alternatively, an internal remodelling of the tissue might occur involving matrix degradation and cell migration and resulting in a small aggregate of cells. Whatever the explanation for the formation of the cell clusters, single cells and grouped cells must experience environments differing in their nutritional, metabolic and mechanical character.

A second order of variation of cell density within cartilage relates to distance from the tissue boundary. This is partly the result of the juxtaposition and imperceptible merging of the highly cellular perichondrium with the sparsely populated cartilage internally. However, cell density continues to diminish within the cartilage proper, until a low 'plateau' is reached. This is found in adult human costal cartilage, in horse nasal cartilage (Galjaard, 1962, Fig. 18) and in the large avascular mass of the human lumbar intervertebral disc (Maroudas, Stockwell, Nachemson & Urban, 1975). The variation with depth is also present where the cartilage is not limited by a perichondrium as in the intervertebral disc traced from the subchondral bone of the vertebra through the hyaline cartilage end plate to the nucleus pulposus and in articular cartilage traced from the joint surface (Fig. 45). In articular cartilage the fall in cell density is very nearly inversely proportional to distance from the joint surface ($y = x^{-0.95}$) in thin cartilage. In thicker articular cartilage the diminution ceases at varying distances from the joint surface: the thicker the cartilage, the smaller the fraction of its thickness showing a diminishing cell density.

Such variation suggests that the higher cell density near the tissue boundary

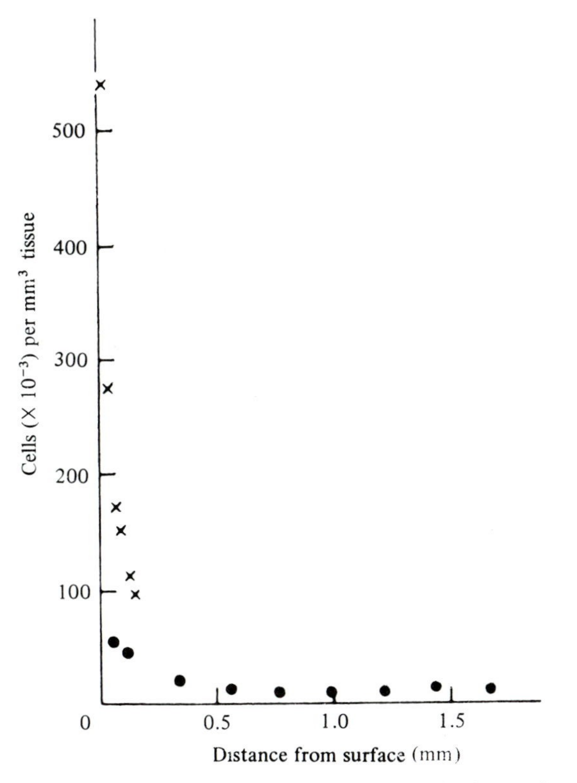

Fig. 45. Change in cell density with distance from the articular surface in adult rabbit (×) and human (●) femoral condylar cartilage. (From Stockwell, 1971*b*.)

is associated with its proximity to a nutritional source. In the case of articular cartilage, distribution is consistent with evidence suggesting that the synovial fluid is the major source of nutrition. The deep zone is of interest since there is evidence in femoral head, though not consistently so in femoral condylar cartilage, that the cell density is slightly greater in the deep than in the middle zone (Vignon, Arlot, Patricot & Vignon, 1976). This may be indicative of a limited nutritional supply from the marrow spaces in the femoral head, as would appear to be the case in the very thick avascular tissue of the intervertebral disc.

The variation of cell density with distance from the boundary may have a slight but significant effect on the limiting thickness of cartilage which can be nourished by the surrounding vascular tissue. Thus it has been calculated that if the cells were distributed *uniformly* within the cartilage, though with the same average cell density for the whole tissue, the limiting distance would be smaller (Maroudas *et al.*, 1975).

TABLE 3

Cartilage thickness and cell density. Data from Stockwell (1971*b*) and J. D. Connolly (*), personal communication

Type of joint and species	Cartilage thickness (mm)	Cell density (cells × 10^{-3} per mm^3)
Knee joint		
Man	2.3±0.5	14.1±3
Cow	1.7±0.1	19.8±4
Sheep	0.84±0.3	52.9±13
Dog	0.67±0.3	44.4±12
Rabbit	0.21±0.07	188±56
Cat	0.33±0.15	108±33
Rat	0.07±0.01	265±57
Mouse	0.06±0.01	334±41
Shoulder joint		
Man	1.6, 1.3	16.8, 12.4
Dog	0.68±0.04	38.5±12
Rat	0.13, 0.17	116, 121
Mouse	0.06±0.01	329±38
Metatarso-phalangeal and inter-phalangeal joints		
Man	0.87±0.2	20.7±2
*Dog	0.24±0.04	106±20
Cat	0.14±0.04	146±13
*Rat	0.044±0.01	303±71
Auditory ossicles		
Man	0.032±0.005	595±109
*Dog	0.022±0.005	444±81
*Rat	0.020±0.006	612±145

SCALE EFFECTS

There are considerable differences in overall cell densities between different adult cartilages even of the same histological variety (Table 3). Thus it is readily apparent that articular cartilage of the mouse knee joint is very much more cellular than that of the human knee joint. In the case of articular cartilage, investigation of a number of species ranging widely in body size demonstrates a similar wide spectrum of cell density, the cellularity of the full thickness of the uncalcified tissue being inversely proportional to body size. This confirms impressions gained from observations on various species (Bywaters, 1937; Rosenthal *et al.*, 1941; Meachim & Collins, 1962; Mankin & Baron, 1965) not always based on analyses of the full thickness of the tissue. However, such variation is not simply a matter of mice and men. If different anatomical joints in each of several species are compared it is found that the

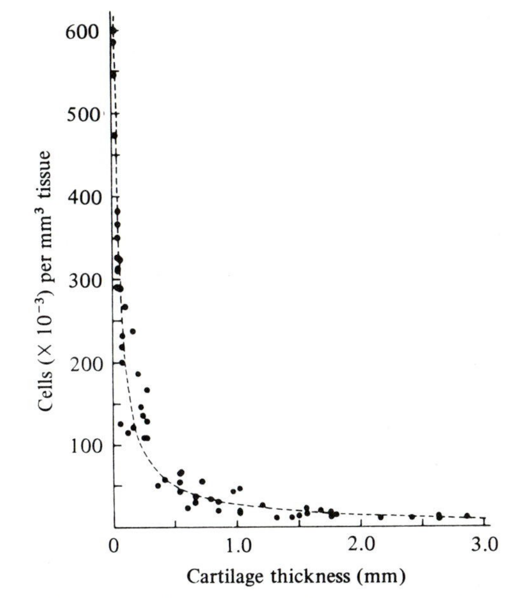

Fig. 46. Relationship of cell density to thickness of articular cartilage ($y = 27900x^{-0.88}$). (From Stockwell, 1971*b*.)

cell density is not species-dependent but rather is inversely proportional to the size of the joint. Thus articular cartilage in the small joints of a large animal may be no less cellular than that in the large joints of a small animal. The cat metatarso-phalangeal joint has the same order of cartilage thickness and cellularity as the rat shoulder joint; a more extreme example concerns the mouse knee and the human incudo-malleolar joint. The key factor appears to be the thickness of the cartilage (Stockwell, 1971*b*); as shown for a large number of adult joints the cell density is inversely related to tissue thickness (Fig. 46). Variations of cartilage thickness over the surface of a joint are associated with corresponding differences in cell density (Vignon *et al.*, 1976).

No systematic comparisons of cell density have been carried out using other hyaline or elastic cartilages, where the imperceptible merging of true cartilage with the perichondrium causes difficulties. However, observations on the central regions of human and equine cartilage suggest also that the thicker the cartilage is, the lower the cell density; this again appears to be the case in the intervertebral disc and in fibrocartilaginous intra-articular discs (R. A. Stockwell, unpublished).

The inverse relationship between cellularity and thickness has the consequence that the total number of cells in the cartilage lying beneath 1 mm^2 of articular surface is approximately the same in joints with cartilage of widely differing thickness and cellularity. The 'cell number' is on average about 25000 deep to 1 mm^2, tending to be lower in very thin and higher in thick cartilage. It is a reasonable hypothesis that the relatively constant cell number is a consequence of the limit to the total number of cells which can be supported by diffusion from a nutritional source located at the boundary of the tissue. Hence nutritional and physico-chemical determinants affect the cell number. The thickness of the cartilage is related to body size and to the degree of incongruity of the joint surfaces, since thicker cartilage deforms more readily and so spreads the load on the underlying bone (Simon, 1970, 1971; Simon, Friedenberg & Richardson, 1973). Since variation in thickness must reflect variation in the amount of matrix, the cell density must also be related to mechanical forces. While there may be a considerable 'physiological reserve' of thickness with respect to the attenuation of local stresses detected in the subchondral bone in simulated hip joints (Freeman & Kempson, 1973), it should not be forgotten that cartilage is a living tissue. The survival of the chondrocytes could depend on the maintenance of more protective matrix between the cells, i.e. a thicker cartilage, than is thought to be required to spare the underlying bone. This may be mediated by 'programmes' inherent in the chondrocyte (perhaps by intercellular communication) and by the effects of microstresses within the tissue, and other extracellular factors regulating chondrocyte metabolism. Such considerations apply to all cartilages at any stage of development.

4

CHONDROCYTE METABOLISM

Chondrocytes, particularly in embryonic cartilage, are useful biochemical tools since they provide a relatively homogeneous cell population which synthesises and secretes proteoglycans and collagen. Much of our knowledge of these important macromolecules has been obtained by studying cartilage. Chondrocytes also have an intrinsic fascination because they can thrive in an avascular environment and at the same time form an organised tissue. The controlling mechanisms are ill understood. Although adult chondrocytes appear to be protodifferentiated with respect to the macromolecules synthesised (Srivastava, Malemud & Sokoloff, 1974), the environment modifies cell activity. More recent investigations have been concerned with the manner in which the interaction with the matrix regulates cell metabolism.

RESPIRATION AND GLYCOLYSIS

Enzyme systems of the glycolytic pathway, the pentose and tricarboxylic acid cycles (Fig. 47) have been demonstrated both biochemically (Lutwak-Mann, 1940; Rosenthal *et al.*, 1942*a*; Kuhlman, 1960; Bernstein, Leboeuf & Cahill, 1961) and histochemically (Balogh, Dudley & Cohen, 1961; Stockwell, 1966; Lenzi *et al.*, 1974; Sprinz & Stockwell, 1976) in various cartilages. Observations suggest that pentose cycle activity is low even in the epiphysial growth plate. As demonstrated histochemically (Fig. 48) the very low level of succinic compared with lactic dehydrogenase activity (Balogh *et al.*, 1961; Stockwell, 1966) reflects the quantitative differences in cellular oxygen uptake and glycolytic rate. Furthermore it has been shown by Tushan, Rodan, Altman & Robin (1969) that the lactic dehydrogenase isoenzyme pattern is consistent with a predominantly anaerobic metabolism since isoenzymes 4 and 5 represent 70% of the total activity.

Cells in the middle of adult cartilage obtain their oxygen and nourishment by diffusion over long distances. Hence the concentrations of these materials are usually lower in the vicinity of the cell than at the source. For example, oxygen tension in the middle of small blocks of cartilage is only about one-third of that in vascular tissues although the concentration of glucose (< 4–8 mM) is diminished to a lesser extent (Silver, 1975). The rates of respiration and glycolysis are in accord with such results (Table 4). Nevertheless, the sluggish metabolism of cartilage on a tissue weight basis is primarily due to its low cellularity. In the adult cartilage, chondrocytes have a glycolysis rate comparable to that in other tissues (Bywaters, 1937) although the rate of

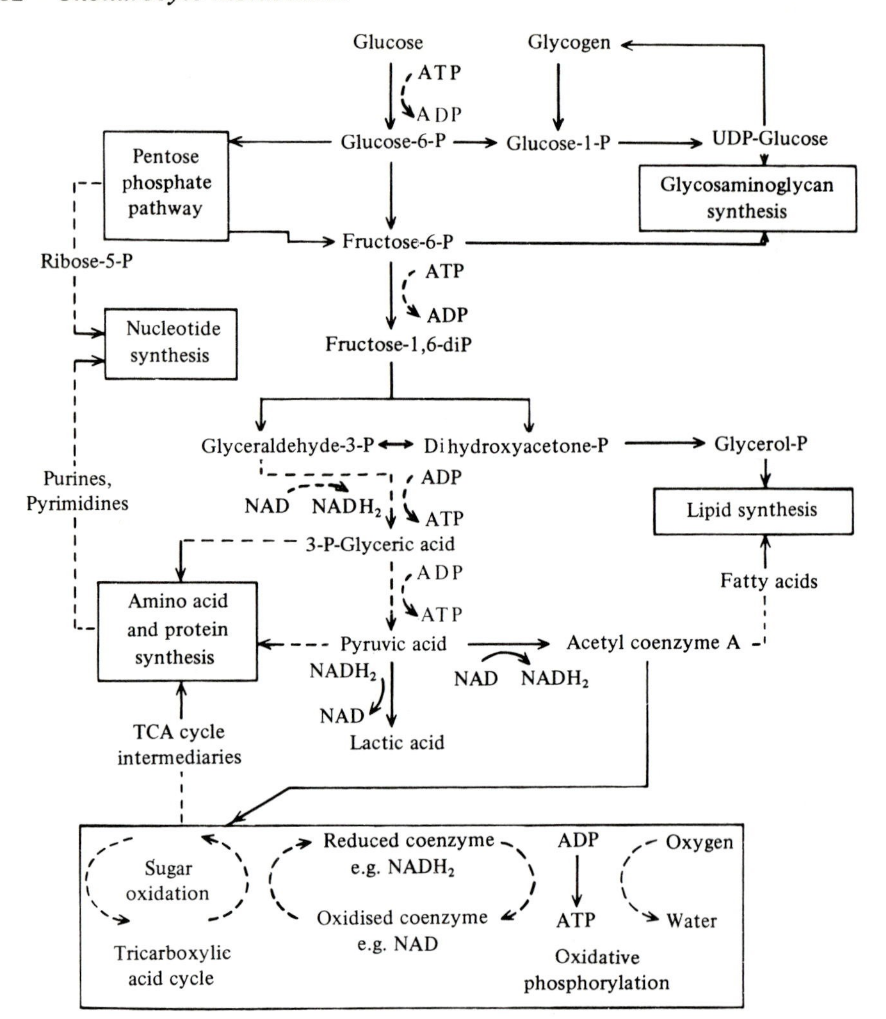

Fig. 47. Glycolysis and some related metabolic pathways. Anaerobic conversion of glucose to lactic acid (glycolysis) produces 2 ATP (3 ATP from glycogen) and no net change in NAD. Oxidative phosphorylation and aerobic conversion of glucose to carbon dioxide and water via pyruvic acid and acetyl coenzyme A produces 36 ATP. Some pathways, such as that of glycosaminoglycan synthesis and the pentose phosphate pathway, produce net amounts of reduced coenzyme while fatty acid synthesis, for example, yields oxidised coenzyme. Incomplete pathways shown in broken arrow. P = phosphate.

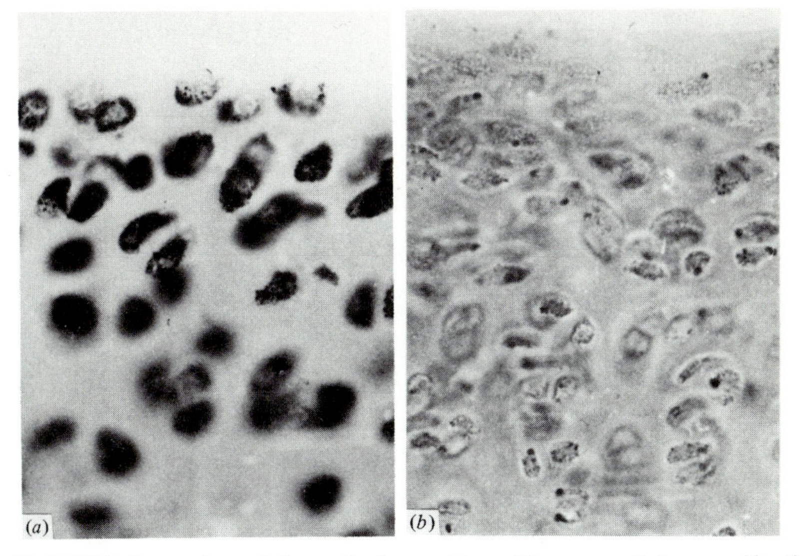

Fig. 48. Rabbit femoral condylar articular cartilage (three months), magnification ×480. Fresh frozen sections incubated to demonstrate activity (dense spots) of (*a*) lactic dehydrogenase and (*b*) succinic dehydrogenase.

oxygen utilisation is very low, only 2–5% of that found elsewhere. The few data suggest that oxygen utilisation is much lower in cartilages from larger than from smaller species. Although oxygen utilisation seems to vary directly with cell density of the cartilage (Fig. 49), this is probably accounted for by differences in basal metabolic rate. It would be interesting, however, to compare high and low cell density cartilages in the same species.

Relatively few reports include rates of glycolysis and oxygen utilisation for different stages of development and ageing. High rates were reported in fetal cartilage (Boström & Mansson, 1953; Boyd & Neuman, 1954); presumably these are accounted for largely by the high cellularity of fetal cartilage. Thus Bywaters (1937) found that the glycolytic rate per unit dry weight of rabbit hyaline cartilage is 10 times higher in the fetus than in the adult, but only twice as much calculated on a per cell basis (Table 4). Similar effects of change in cell density may be inferred for developing rat elastic cartilage (Patnaik, 1967). In post-natal bovine articular cartilage Rosenthal *et al.* (1941) found no significant change in glycolytic rate per cell during maturation or during adult life. However, they observed a significant decline in oxygen utilisation per cell during both these periods. The lower oxygen uptake in aged tissue is associated with a loss of the cytochrome ('oxygen-activating') rather than of the dehydrogenase ('substrate-activating') groups of respiratory enzymes (Rosenthal *et al.*, 1942*b*).

The low oxygen utilisation by chondrocytes is consistent with the low oxygen tension of their avascular environment. On the basis of radiosensitivity, Gray & Scott (1964) calculate that this is about 1.3 kPa in hyaline cartilage

TABLE 4

Respiration and glycolysis of cartilage. The cell density data indicate that while the glycolytic rate per cell is much the same for all (except the fetal) cartilages and that this is similar to the rate in vascular tissues, there are nevertheless wide differences in oxygen utilisation. Note that the cell density values are for the specimens of cartilage used for metabolic studies and are not necessarily representative of the full thickness of the tissue. References: Dickens & Weil-Malherbe (1936); Bywaters (1937); Rosenthal *et al.* (1941); Hills (1940); Patnaik (1967)

Type of cartilage and species	Anaerobic glycolysis (mg lactic acid produced g^{-1} dry weight per hour)	Oxygen uptake (ml g^{-1} dry weight per hour)	Cell density (cells $\times 10^{-7}$ per cm^3)
Human articular	0.60	–	3.1†
Equine carpo-metacarpal articular	0.84	0.005	4.3†
Bovine metatarso- and metacarpo-phalangeal articular			
< 6 months	4.5	0.088	13.3
1–7 years	1.75	0.024	4.7
8–11 years	1.04	0.008	3.4
Rat costal	5.44, 7.40	0.22, 0.68	25.0*
Elastic (auricular)	3.98	0.42	–
Rabbit femoral condylar articular			
Fetal	26.4	–	21.0†
Adult	2.56	0.15	9.0†
Rabbit			
Liver	13.6	7.7	62.0
Kidney	20.8	10.8	110.0

* Author's data.

† Corrected for nuclear height.

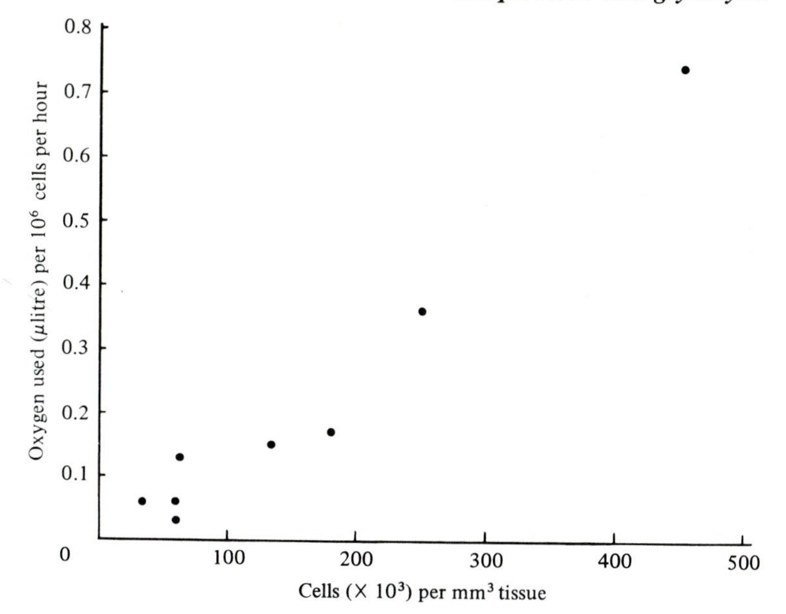

Fig. 49. Oxygen utilisation by chondrocytes. The points on the graph are taken from different species (horse, cow, rabbit, rat and chick) and hence the differences in oxygen uptake probably reflect species differences in basal metabolic rate as much as differences in cell density. Data from Dickens & Weil-Malherbe (1936); Bywaters (1937); Hills (1940); Rosenthal *et al.* (1941); Boyd & Neuman (1954). Cellularities for chick, rat and rabbit from author's data.

and a similar figure has been computed for the dog nucleus pulposus (Urban, Holm, Maroudas & Nachemson, 1978). These estimates are of the same order as the few measurements which have been made. Silver (1975) finds that oxygen gradients in growing cartilage fragments grown in the well-vascularised rabbit ear-chamber are steep compared with fibrous connective tissue: oxygen tensions at the centre of blocks of articular cartilage 0.5 mm thick are 0.7–1 kPa (0.5–1% oxygen). Brighton & Heppenstall (1971), investigating rabbit epiphysial growth plates, find values of 2.7–7.6 kPa (2.5–7.5% oxygen). In both instances the cartilage investigated is immature, of small dimensions and growing very actively. It is probable that oxygen tensions are well below these levels where the diffusion distances are greater, as in adult human costal and articular cartilage. Thus Marcus (1973) calculates that the deeper cells of human articular cartilage may exist in anaerobic conditions, assuming that the cartilage is nourished solely by the synovial fluid in which the oxygen tension is about 5.3–10.6 kPa (Falchak, Goetzl & Kulka, 1970; Lund-Oleson, 1970; Treuhaft & McCarty, 1971).

It is often believed that cartilage thrives under near anaerobic conditions. Thus there is evidence that a low oxygen tension is conducive to cartilage differentiation (Hall, 1970). Also, connective tissue cells cultured under compressive stress in low oxygen atmospheres differentiate into cartilage whereas high oxygen atmospheres produce bone (Bassett & Herrmann, 1961). Similarly, low oxygen tension results in increased synthetic activity by cultured embryonic chondrocytes (Pawalek, 1969; Nevo, Horwitz & Dorfman, 1972). However, although chondrocytes from adult cartilage tolerate low oxygen conditions better than fibroblasts do, nevertheless sulphate incorporation (presumably all synthetic activity) is inhibited when the cells are cultured in atmospheres as low as 1% oxygen (Marcus, 1973; Brighton, Lane & Koh, 1974). Oxygen utilisation by cultured chondrocytes increases linearly with increasing oxygen tension in the flask atmosphere (Lane *et al.*, 1976). The cells have a broad tolerance for oxygen over the range 5–60% (5–40 kPa) but sulphate incorporation is again inhibited at levels of 90–95% oxygen (Brighton *et al.*, 1974).

Although undetected by earlier workers it appears that both in monolayer (Marcus, 1973) and in organ culture (Lane *et al.*, 1976), cartilage displays a Pasteur effect, i.e. an increase of glycolytic rate as oxygen tension is reduced. Either chondrocytes are near-facultative anaerobes or two cell populations are involved, the larger group being obligatory anaerobes and the smaller group consisting of aerobes (Lane *et al.*, 1976). If indeed there are two populations, these could either be randomly distributed in the tissue or might be organised with respect to the tissue boundaries. It may be speculated that *in vivo* the peripherally situated cells could be the aerobes since they are more advantageously placed in relation to the source of oxygen. They might have a higher oxygen utilisation than the deeper cells even if there is only a single population of cells, since chondrocyte oxygen consumption is greater where oxygen tensions are high. Such an hypothesis could explain the results of Rosenthal *et al.* (1941) who found a declining oxygen uptake in ageing articular cartilage: loss of the superficial cells (due to arthritic change) would then account for the age change instead of a general decline in oxygen utilisation of the whole population. There is some difference in function of chondrocytes as regards glycosaminoglycan synthesis rates in different layers of cartilage but no direct evidence as to differences in respiration and glycolysis, however.

COLLAGEN BIOSYNTHESIS

The polypeptide chains of the collagen molecule are synthesised on ribosomes of the granular ER (Fig. 50), each chain requiring a separate monocistronic messenger RNA associated with a 350S polysome (Harwood *et al.*, 1975). Translation begins at the N-terminal end (Vuust & Piez, 1972) at the rate of about 209 amino acid residues per minute (total time six minutes).

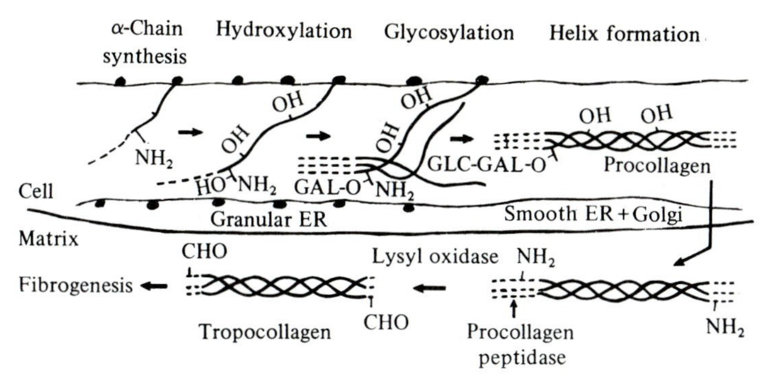

Fig. 50. Schematic diagram to indicate the intra- and extracellular locations where post-translational modifications occur during collagen synthesis and secretion. ER = endoplasmic reticulum.

PROCOLLAGEN

Collagen is synthesised as procollagen which is larger than the tropocollagen molecule, since the α-chains possess random coil extensions at either end of the central helical section. The extension on the N-terminal end of the polypeptide chain is synthesised first, the entire polypeptide being termed a pro-α-chain (Bellamy & Bornstein, 1971). The N-terminal extension has a different amino acid composition to the rest of the chain, resembling an acidic glycoprotein or the polypeptide core of proteoglycan molecules; it is rich in acidic amino acids and contains cysteine and tryptophan (von der Mark & Bornstein, 1973). Estimates of the molecular weight of pro-α-chains vary but the data suggest around 115000 (Bornstein *et al.*, 1972; Vuust & Piez, 1972), compared with 95000 for the α-chain itself. Larger values of 250000 or more probably represent dimers of pro-α-chains (Martin, Byers & Smith, 1973). Although this extension forms about 17% by weight of the whole pro-α-chain, it occupies only 8% (25 nm) of the total length of the SLS form (Lapière, Lenaers & Pierard, 1973). Hence the additional peptide is believed to remain in a random coil state after synthesis, rather than forming part of the triple helix of the procollagen molecule (Bornstein *et al.*, 1972). A non-helical region also exists at the C-terminal end of the α-chain (Rauterberg *et al.*, 1972; Miller, 1972) and pro-α-chain (Tanzer *et al.*, 1974). During synthesis disulphide bonds are formed between the non-helical extensions of adjacent pro-α-chains (Smith, Byers & Martin, 1972). These may align the three chains into the correct positions for triple helix formation, thus acting as 'registration peptides' (Speakman, 1971). Although originally assigned to the N-terminal end (Dehm, Jimenez, Olsen & Prockop, 1972), recent findings indicate that disulphide linkages are situated at the C-terminal non-helical extension of cartilage and tendon procollagens (Olsen, Hoffman & Prockop, 1976).

HYDROXYLATION

Each chain is rich in proline and lysine; hydroxylation (Fig. 50) of certain proline and a number of lysine residues occurs on the nascent polypeptide (Rhoads, Udenfriend & Bernstein, 1971). This is usually completed before the triple helix is formed (Vuust & Piez, 1972). If delayed experimentally, hydroxylation can be effected completely after release of the polypeptide from the ribosome (Prockop *et al.*, 1973; Harwood *et al.*, 1975). Two enzymes are involved, prolyl and lysyl hydroxylases, requiring similar co-factors – oxygen, ferrous ions, ascorbate (or other reducing agent) and α-ketoglutarate which is stoichoimetrically decarboxylated to succinate (Prockop, Kaplan & Udenfriend, 1962; Peterkofsky & Udenfriend, 1965; Rhoads & Udenfriend, 1968). Prolyl hydroxylase is membrane bound in the granular ER (Diegelman, Bernstein & Peterkofsky, 1973; Olsen, Berg, Kishida & Prockop, 1973; Harwood *et al.*, 1975); it is activated by lactate (Prockop *et al.*, 1973) and inhibited by lack of oxygen or by α,α-dipyrridyl, an iron chelator (Juva, Prockop, Cooper & Lash, 1966; Prockop *et al.*, 1973). Lack of ascorbate (less than 10 $\mu g\ ml^{-1}$ medium) results in sub-optimal hydroxylation, proline being more severely effected than lysine (Levene & Bates, 1973).

If hydroxylation is inhibited, protocollagen is formed (Juva *et al.*, 1966) and retained within the cell (Bhatnagar, Kivirikko & Prockop, 1968). This modified form of collagen is largely in the form of a triple helix and contains the same N-terminal extension as does procollagen (Prockop *et al.*, 1973). Hydroxylation appears to be necessary for secretion of collagen from the cell, requiring a certain minimum quantity of trans-4-hydroxyproline (Rosenbloom & Prockop, 1971; Bornstein & Ehrlich, 1973). Curtailed hydroxylation of lysine also results in abnormal secretion although this aspect of hydroxylysine is less important than its role in cross-link formation.

GLYCOSYLATION

The peptide chains of collagen carry a number of short carbohydrate side chains synthesised by transferase enzymes requiring manganese ions (Spiro & Spiro, 1971), and utilising uridine diphosphate (UDP) derivatives of galactose and glucose. A proportion of the hydroxylysine residues are glycosylated. At one time postulated to act as a 'passport' across the cell membrane during secretion, the presence of carbohydrate side chains is no longer considered necessary for this purpose (Winterburn & Phelps, 1972). Thus in a variety of the Ehrles–Danlos syndrome (a heritable disorder of connective tissue featuring abnormally loose skin), skin collagen contains only 5–10% of its normal complement of hydroxylysine but there is no obvious disturbance of secretion, other defects involving collagen accounting for the clinical disorder (Bailey & Robins, 1973). The glycosyl transferases are not located in the plasma membrane (Harwood *et al.*, 1975) as earlier postulated. A higher activity of glucosyl transferase is found in the granular (Fig. 50) than in the smooth membrane fractions (Harwood *et al.*, 1975). This suggests that glycosylation takes place before the triple helix is formed since in chondrocytes procollagen peptides become fully disulphide-bonded in the smooth membranes to which they are transferred from the granular ER. The degree of glycosylation varies in different situations: invertebrate, basement membrane and cartilage collagens have the most carbohydrate and the skin collagens the least; the hexose content and typical fibre diameters are inversely related (Spiro, 1970).

INTRACELLULAR TRANSPORT AND SECRETION

While structural units of collagen are synthesised intracellularly, aggregation into fibres occurs outside the cell (Fig. 50). Although fragments of cross-banded collagen fibres have been observed within cells, these have been attributed either to collagenase-mediated degradation of ingested material during rapid tissue remodelling or to the artefactual appearance of enclosure within the cell, owing to the plane of sectioning (Perez-Tamayo, 1973). It appears that intracellular fibrogenesis does not take place following synthesis of new collagen units because: (i) the procollagen molecule remains soluble under physiological conditions which would permit aggregation of tropocollagen molecules (Layman, McGoodwin & Martin, 1971); (ii) conversion of procollagen to tropocollagen by proteolytic cleavage of most of the non-helical terminal extensions is not completed until after secretion of procollagen from the cell. The enzyme required, procollagen peptidase, is a neutral protease located extracellularly (Layman & Ross, 1973) although a series of cleavage steps may occur which may involve some degree of intracellular modification of procollagen (Bornstein, 1974).

Until recently there has been some controversy about the route of secretion of procollagen from the cell since there has been evidence for at least three pathways through the cell. Procollagen manufactured in the granular ER might:

(i) pass directly into the cytosol and thence to the exterior (Cooper & Prockop, 1968; Reith, 1968; Salpeter, 1968);
(ii) be retained in the ER cisternae, the secretion reaching the exterior either by fusion of the cisterna with the plasma membrane or through the agency of intermediate transport vesicles derived from the cisternae (Ross & Benditt, 1965);
(iii) pass to the Golgi complex, there being packaged into secretory granules as with the normal sequence in secretion from glandular cells (Revel & Hay, 1963).

There is now fresh evidence that the Golgi complex is involved in procollagen secretion. Autoradiography of odontoblasts shows that tritium-labelled proline appears first in the supranuclear granular ER and then passes to Golgi saccules containing filamentous aggregates. The saccules become modified into secretion granules which discharge their contents into the pre-dentine matrix. Exocytosis occurs only from the odontoblast process: since granular ER is absent from this region of the cell, direct transfer to the exterior from the cisternae cannot occur (Weinstock & Leblond, 1974). Furthermore, studies of corneal and tendon fibroblasts using ferritin-labelled antibodies to procollagen locate this molecule in all elements of the Golgi complex as well as in cisternae of the granular ER (Olsen *et al.*, 1975; Nist *et al.*, 1975). Thus there is little doubt that the Golgi complex is involved in procollagen secretion. It has been suggested that microtubules have a role in the intracellular transport of procollagen (Bornstein & Ehrlich, 1973) but results based on the use of microtubule inhibitors are difficult to interpret since these drugs may have other more general effects.

CROSS-LINKAGES

Collagen molecules are subsequently stabilised by covalent cross-linkages (Figs. 20, 23), whose synthesis involves the modification of lysine residues in adjacent α-chains and molecules. Both intra- and intermolecular linkages require the oxidative deamination of lysine (or hydroxylysine) to α-amino-adipic semi-aldehyde (allysine). This is catalysed by lysyl oxidase (Siegel & Martin, 1970), a copper-dependent enzyme. Therefore inhibition occurs in copper deficiency either in 'natural' conditions such as lathyrism or experimentally, using β-aminoproprio nitrile. Hence in lathyrism both intra- and intermolecular cross-linkages of collagen are affected, as are also those of elastin. Lysyl oxidase also requires pyridoxal phosphate (vitamin B6); deficiency during fetal life may lead to weakness in arterial walls by affecting collagen and elastin cross-linking (Levene & Murray, 1977).

Intermolecular linkages also involve hydroxylation of lysine, effected by lysyl hydroxylase which *inter alia* requires vitamin C. Hence in scurvy and experimental scorbutic conditions, the collagen produced by the cells is abnormally soluble. This is due to the effects on the intermolecular cross-links, since the intramolecular linkages are intact (Levene & Bates, 1973). Linkages between molecules are formed during fibrogenesis, following alignment of the monomolecular units within the primary filament.

FIBROGENESIS

Studies of fibrogenesis from tropocollagen solutions (Fig. 51) indicate that there is a nucleation (lag) phase when aggregates of collagen monomer are formed, followed by a growth phase in which there is accretion of collagen particles (Wood, 1960). However, according to Trelstad (1975), nucleation consists of the end-to-end elongation of small collagen aggregates, approximately 200–500 nm long and 5–10 nm wide, into long thin filaments. From

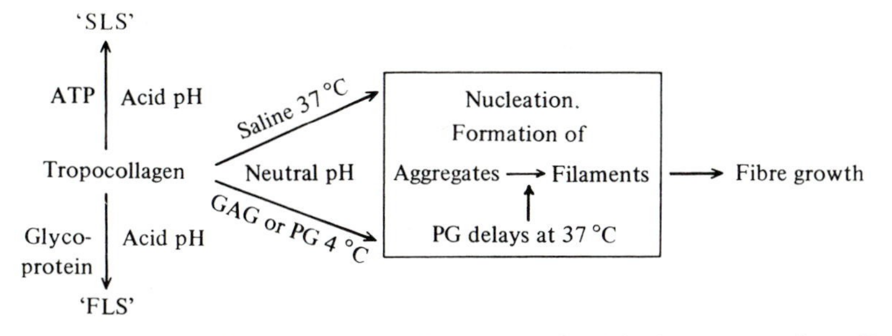

Fig. 51. Factors involved in fibrogenesis. Proteoglycan (PG) has contrasting effects at 4 °C and 37 °C. SLS = 'Segment long spacing' collagen; FLS = 'Fibrous long spacing' collagen.

these dimensions, it seems possible that nucleation involves the formation of the pentameric primary filaments.

Since collagen is synthesised as procollagen, this may be the molecular form involved in fibrogenesis. Thus in the bovine condition of dermatosparaxis, caused by lack of the enzyme procollagen peptidase (Lenaers, Ansay, Nusgens & Lapière, 1971), the procollagen molecule with its intact N-terminal extension can form ribbon-like filaments (Lapière, Nusgens, Pierard & Hermanns, 1975). The length of the procollagen extension (25 nm) would be small enough to fit into the 'gap' zone of the native fibre and hence Lapière *et al.* (1975) postulate that it may have a role in normal fibrogenesis. They suggest that in the nucleation phase, procollagen molecules assemble into fine filaments and that procollagen peptidase then converts the molecules to tropocollagen, permitting 'twisting' of the filaments along their long axis. This hypothesis requires a delay in the action of the enzyme until the molecules have polymerised. However, while the appropriate material for fibrogenesis studies is the collagen molecule as it leaves the cell, most investigations have been performed using tropocollagen.

Influence of proteoglycans

Experimental results indicate that proteoglycans interact with collagen both in fibrogenesis and thereafter in the formed cartilage matrix. Thus proteoglycans added to tropocollagen solutions at 4 °C induce (Toole & Lowther, 1968) and stabilise (Toole, 1969) fibre formation although fibrogenesis from tropocollagen normally requires warming to 37 °C (Fig. 51). An important factor is the pH. If proteoglycans or glycosaminoglycans are mixed with tropocollagen in acid solution, fibres with abnormally long periodicity (280 nm) are produced (Highberger, Gross & Schmitt, 1950) termed 'fibrous long spacing' (FLS) collagen; mixing at physiological, neutral pH yields fibres with the 'native' periodicity (Doyle *et al.*, 1975). In general it appears that under physiological conditions, polyanionic agents which enhance nucleation retard fibre growth (Mathews & Decker, 1968; Scott *et al.*, 1975). Results indicate that an electrostatic interaction with positively charged amino acids in the α-chain sequence is involved and that the tertiary structure of collagen is required (Mathews, 1965). While it has been recorded that chondroitin sulphate, dermatan sulphate and keratansulphate do not react with tropocollagen (Lowther, Toole & Hetherington, 1970), it seems clear that chondroitin sulphate, dermatan sulphate and other polyanions (Table 5) do in fact bind to collagen monomer (Obrink, 1973*a*; Scott, Conochie, Faulk & Bailey, 1975; Greenwald, Schwartz & Cantor, 1975). Interactions are abolished by increasing ionic strength (Mathews, 1965; Obrink, 1973*a*). Since most workers agree that hyaluronic acid and keratansulphate do not bind to collagen, two negative charges per disaccharide may be a minimum requirement (Scott *et al.*, 1975). The steric exclusion effects of hyaluronic acid may be responsible for its observed action in accelerating nucleation and growth (Mathews, 1965; Obrink, 1973*b*). In the tissues and cartilage in particular, fibrogenesis must take place in the presence of proteoglycan subunit or aggregate rather than in the presence of the protein-free glycosaminoglycans. Nevertheless the interaction still appears to be electrostatic in nature (Obrink, 1973*a*); molecular size and glycosaminoglycan chain length are important factors (Mathews, 1965; Mathews & Decker, 1968). A somewhat unexpected finding is the preferential binding of cartilage proteoglycans to Type I rather than Type II collagen (Lee-Own & Anderson, 1976).

While at 4 °C proteoglycans enhance fibre formation from soluble collagen, at 37 °C low concentrations of proteoglycan retard fibrogenesis (Mathews & Decker, 1968; Toole & Lowther, 1968; Oegama *et al.*, 1975), although excess proteoglycan may accelerate fibre growth. Proteoglycan aggregate also binds to tropocollagen and retards fibre growth. This is not dependent on the hyaluronate-binding site or any other part of the core protein (Oegama *et al.*, 1975) although the latter binds to collagen (Greenwald *et al.*, 1975). Nor does the 'link' glycoprotein interact with collagen or affect fibre formation (Lowther & Natargan, 1972; Lee-Own & Anderson, 1975). Oegama *et al.* (1975) find that proteoglycan (although still bound to collagen) does not retard fibre precipitation if added late in the nucleation phase when, according to Trelstad (1975), many fine filaments have already formed. Hence they assign the usual retardation effect of proteoglycans to the stage (Fig. 51) when small collagen aggregates assemble end to end into fine filaments.

In the tissues, the number and diameter of newly formed fibres must be influenced by the type and amount of local proteoglycan (Toole & Lowther, 1968) although other factors such as the degree of glycosylation of different types of collagen are relevant. Since nucleation is accelerated by chondroitin-4-sulphate (the glycosaminoglycan of immature cartilage), fine fibres might be expected and indeed occur in immature cartilage. Chondroitin-6-sulphate and keratansulphate which accumulate during maturation and ageing react to a lesser extent or not at all with collagen and could permit steady fibre growth, eventually producing the thicker fibres found in adult cartilage (Scott *et al.*, 1975).

TABLE 5

Collagen–glycosaminoglycan interaction. Endpoint = dilution of polyanion at which agglutination no longer occurs with sheep red blood cells, tanned and coated with lathyritic chick cartilage collagen. Data from Scott *et al.* (1975)

Type of polyanion	Agglutinating endpoint (μg polyanion ml^{-1})
DNA	0.3
Dermatan sulphate	1.2
Chondroitin-4-sulphate	1.2
Chondroitin-6-sulphate	78.15
Proteoglycan	5.85
Keratansulphate	250.0
Hyaluronic acid	500.0

PROTEOGLYCAN BIOSYNTHESIS

More recent authoritative reviews of this subject include those by Stoolmiller & Dorfman (1969), Muir (1973), Phelps (1973), Roden & Schwartz (1973), Serafini-Fracassini & Smith (1974) and Kennedy (1976). In the cell, synthesis of the protein and carbohydrate portions of the molecule are synchronised. The protein core is assembled first, the polysaccharide side chains subsequently elongating by appositional growth from their attachment sites on the protein; sulphation of the polysaccharide is a late stage in synthesis.

THE CORE PROTEIN

The general mode of formation is thought to be similar to that of other proteins, by translation of the appropriate messenger RNA molecule at the ribosome. As with other proteins, synthesis is inhibited by puromycin and cycloheximide (Haba & Holtzer, 1965; Telser, Robinson & Dorfman, 1965), thereby preventing glycosaminoglycan side chain synthesis.

There is some uncertainty concerning the number of types of core protein synthesised. Thus the number and hydrodynamic size of the major groups of proteoglycan vary in different species, different types of cartilage and at different stages of development. Within a single specimen of cartilage there appear to be at least two major species of extractable proteoglycans, even when there is no possibility of artefactual proteolytic scission during preparation (Simunek & Muir, 1972; Roughley & Mason, 1976; Stanescu, Maroteaux & Sobczak, 1977). Also the distribution of amino acids in the protein core is not random, as indicated by the characteristic differences between the structurally invariate hyaluronate-binding region and the more variable

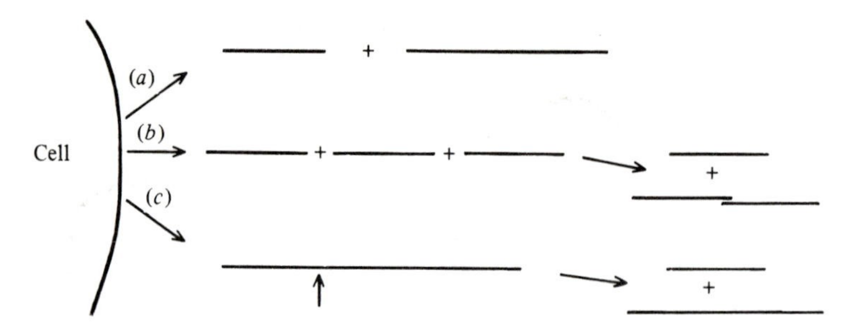

Fig. 52. Schematic diagram showing three possible modes of synthesis and secretion resulting in proteoglycan core proteins of different sizes: (*a*) separate secretion, (*b*) self-assembly, (*c*) secondary cleavage.

chondroitin sulphate-rich fragment (Hardingham *et al.*, 1976). The evidence of a specific amino acid grouping at the attachment point of the chondroitin sulphate chain (Mathews, 1971; Johnson & Baker, 1973; Hopwood & Robinson, 1974*b*) further suggests that there may be short repeating sequences along the chondroitin sulphate-rich portion (Heinegard & Hascall, 1974*b*). There appear to be three possibilities regarding synthesis (Fig. 52):

(i) More than one type of core protein could be formed: Hardingham & Muir (1974*a*) suggest a minimum of two varieties would be required to account for aggregating and non-aggregating proteoglycans, i.e. necessitating messenger RNA molecules of widely different sizes.

(ii) Polydispersity (and possibly any repeating polypeptide units) might result from the synthesis of very small subunits forming a self-associating system (Wells & Serafini-Fracassini, 1973); however, as Hardingham *et al.* (1976) remark there is no precedent for polypeptides being synthesised and subsequently linked together except by disulphide bonds. In the core protein these are restricted to the hyaluronate-binding region.

(iii) Only a single species of core protein may be formed (Hascall & Sajdera, 1970). Physiological post-translational partial cleavage (Fig. 34) might then result in non-aggregating proteoglycans deficient in the hyaluronate-binding region and aggregating proteoglycans with chondroitin sulphate-rich regions of variable size (Hardingham *et al.*, 1976; Heinegard, 1977).

While the current evidence favours the third possibility, progress requires further investigation of the biosynthesis of the protein moiety of the proteoglycans (Heinegard, 1977).

NUCLEOTIDE SUGAR FORMATION

When considering the synthesis of the polysaccharide portion of proteoglycans it is necessary to discuss first the production of the UDP-sugars which provide the 'building blocks' for polymer synthesis (Fig. 30). Hydrolysis of the sugar phosphate bond provides the energy required to synthesise the glycosidic linkage between the liberated sugar and the non-reducing end of the growing polymer. Furthermore, sugars at the nucleotide level are readily interconverted into other forms if required. The greater part of the glycosaminoglycan chain consists of alternating basic and acidic (sometimes neutral) monosaccharides. It is therefore not surprising that there are only two metabolic 'routes' by which nucleotide sugars are formed from glucose or glycolytic intermediates (Fig. 53). The two 'stock' nucleotide sugars – UDP-glucose and UDP-*N*-acetylglucosamine – which are the products of these two metabolic pathways can then be converted where necessary to forms suitable for inclusion in the various glycosaminoglycans synthesised. Synthesis of the nucleotide sugars occurs mainly in the cytosol (Phelps & Stevens, 1975).

UDP-glucose (route 1)

UDP-glucose pyrophosphorylase (Caputto, Barra & Cumar, 1967) catalyses the reaction:

$$\text{UTP} + \text{Glucose-1-P} \longleftrightarrow \text{UDP-glucose} + \text{PPi}.$$

The enzyme is inhibited by AMP and is controlled by the total nucleoside triphosphate/monophosphate (XTP/XMP) ratio in the cytosol (Phelps & Stevens, 1975). UDP-glucose can undergo further reactions to provide neutral or acidic monosaccharides. (*a*) Conversion to UDP-galactose is catalysed by UDP-glucose-4′-epimerase (Leloir, 1951). Since an oxidation reaction is involved in the conversion, NAD is an essential co-factor (Maxwell, 1957) and the rate of conversion is affected by the redox state of the cell and hence the NAD/NADH ratio. UDP-galactose is required for keratansulphate synthesis and the sugar 'bridge' linking chondroitin sulphate chains to core protein (Fig. 30, 31). (*b*) Synthesis of UDP-glucuronic acid is catalysed by UDP-glucose dehydrogenase (Strominger, Maxwell, Axelrod & Kalckar, 1957):

$$\text{UDP-glucose} + 2\text{NAD}^+ + 2\text{H}_2\text{O} \longrightarrow \text{UDP-glucuronic acid} + 2\text{NADH} + 2\text{H}^+$$

Since stoichoimetric amounts of NAD are consumed in the oxidation and the reaction is inhibited by NADH, the NAD/NADH ratio is critical. UDP-glucuronic acid is required for chondroitin-4- and chondroitin-6-sulphates and for hyaluronic acid (Fig. 30). This important reaction was first described in cartilage by Castellani, de Bernard & Zambotti (1957). UDP-glucuronic acid may itself be converted to UDP-iduronic acid (for heparin synthesis) by epimerisation but the decarboxylation reaction (Feingold, Neufeld & Hassid, 1960) is much more important in chondrocytes:

$$\text{UDP-glucuronic acid} \longrightarrow \text{UDP-cylose} + \text{CO}_2.$$

UDP-xylose is required for the protein–polysaccharide linkage. It is also a powerful feedback inhibitor of UDP-glucose dehydrogenase (Neufeld & Hall, 1965).

UDP-*N*-acetylglucosamine (route 2)

This route provides the basic monosaccharides required in glycosaminoglycan synthesis and is more complex than the formation of UDP-glucose. Formation of glucosamine-6-phosphate from fructose-6-phosphate was first observed in cartilage by Castellani & Zambotti, 1956. The reaction is catalysed by an aminotransferase (Ghosh, Blumenthal, Davidson & Roseman, 1960), glutamine donating the amino group. After acetylation and conversion to the 1-phosphate, the end product is again formed by a pyrophosphorylase reaction:

$$\text{UTP} + N\text{-acetylglucosamine} \longleftrightarrow \text{UDP-}N\text{-acetylglucosamine} + \text{PPi}.$$

UDP-*N*-acetylglucosamine is needed for keratansulphate and hyaluronic acid synthesis (Fig. 30). It is also a potent feedback inhibitor (Kornfeld, Kornfeld, Neufeld & O'Brien, 1964) of the reaction:

$$\text{Fructose-6-P} \longleftrightarrow \text{Glucosamine-6-P}.$$

The inhibition is increased by glucose-6-phosphate and AMP but is relieved by UTP (Winterburn & Phelps, 1971).

UDP-*N*-acetylgalactosamine is produced by epimerisation of UDP-*N*-acetylglucosamine (Glaser, 1959). As with UDP-glucose-4′-epimerase, NAD is required in catalytic amounts while NADH is inhibitory. UDP-*N*-acetylgalactosamine is utilised in the synthesis of chondroitin-4- and chondroitin-6-sulphates and in the protein–carbohydrate linkage of skeletal keratansulphate (Figs. 30, 32).

Thus the enzymes which control the diversion of hexose phosphates from the glycolytic pathway into either route are both affected by the XTP/XMP ratio and the supply of glycolytic intermediates (Phelps & Stevens, 1975). Furthermore, the NAD/NADH ratio is an important factor in interconversion. Lastly, both routes are subject to end-product inhibition.

GLYCOSAMINOGLYCAN CHAIN ASSEMBLY

Embryonic chick chondrocytes have been used almost exclusively in the investigation of glycosaminoglycan synthesis. While much is known about the formation of chondroitin sulphate, keratansulphate has been little studied. Biosynthesis of glycosaminoglycan chains may be considered in three phases: (i) chain initiation; (ii) chain propagation; (iii) sulphation.

Chain initiation

Synthesis of the sugar bridge (Fig. 31) at the protein–polysaccharide linkage has been elucidated for chondroitin sulphate. Transfer of each of the three neutral sugars is catalysed by a separate enzyme. Enzyme activity is membrane bound in the microsome fraction (Robinson, Telser & Dorfman, 1966); hence synthesis of glycosaminoglycan begins in the cisternae of the rough ER (Horwitz & Dorfman, 1968). With UDP-xylose as donor, xylose is linked glycosidically to serine residues of the core protein by a xylosyl transferase (Robinson *et al.*, 1966). Only certain serine residues are utilised, presumably in relation to specific sequences of amino acids; free serine is not an acceptor (Baker, Roden & Stoolmiller, 1972).

Each of the two galactose residues is added by a specific galactosyl transferase with UDP-galactose as the donor. The first galactosyl transferase links galactose to xylosyl-serine (Helting & Roden, 1969*a*) in the core protein but will not transfer a second galactose to the oligosaccharide. Since the enzyme can 'recognise' free as well as protein-bound xylose (Helting & Roden, 1969*a*; Helting, 1971), extraneous xyloside can be used to replace the protein core as acceptor (Levitt & Dorfman, 1974; Robinson *et al.*, 1975). This is useful in experimental situations where it is advantageous to dissociate core protein and glycosaminoglycan synthesis, as in studies of chondrogenesis or metabolic control. The second galactosyl transferase requires an acceptor larger than a monosaccharide. It adds galactose to either galactosyl-xylosyl-serine or galactosyl-xylose but not to galactosyl-galactosyl-xylosyl-serine, galactosyl-galactose or free galactose (Helting & Roden, 1969*a*). After the sugar bridge is complete, the first glucuronosyl residue is added by a transferase distinct from that involved in chain elongation (Helting & Roden, 1969*b*).

Chain propagation

This is carried out by addition of *N*-acetylgalactosamine and glucuronic acid to the chain via their UDP derivatives (Perlman, Telser & Dorfman, 1964; Silbert, 1964). The transferases are specific for both the donor and acceptor molecules, and the mono- and not pre-formed disaccharides are added alternately to the growing polymer (Telser *et al.*, 1966). Chain elongation continues as the enlarging proteoglycan and the membrane-bound synthesising enzymes move into the Golgi region (Horwitz & Dorfman, 1968; Olsson, 1972).

Sulphation

Sulphate incorporation is not due to an exchange with pre-existing groups in proteoglycans in the cartilage matrix but is an active process occurring intracellularly (Boström & Manson, 1952, 1953; Campo & Dziewiatkowski, 1962), dependent on biosynthesis of new glycosaminoglycans. As with the transfer of monosaccharide units, sulphation requires an active donor molecule, 3′-phosphoadenosine-5′-phosphosulphate (PAPS). This is formed from inorganic sulphate and ATP; sulphate transfer from PAPS to the carbohydrate polymer is accomplished by a sulphato-transferase (D'Abramo & Lipmann, 1957; Suzuki & Strominger, 1960). Specific enzymes may be required for different glycosaminoglycans, for example chondroitin-4- and chondroitin-6-sulphates, or as in keratansulphate where galactosyl as well as hexosaminyl residues may be sulphated.

Although chain elongation can occur without sulphation taking place (Perlman *et al.*, 1964), the process is usually closely related to chain extension (Derge & Davidson, 1972). It appears that sulphation may be involved in chain termination since glucuronic acid cannot be transferred to a terminal sulphated *N*-acetylgalactosamine residue (Telser *et al.*, 1966).

In cartilage, activation occurs in the cytosol while sulphate transfer is located in the smooth membranes (Horwitz & Dorfman, 1968; Silbert & Deluca, 1969), autoradiography clearly implicating the Golgi complex (Revel & Hay, 1963; Godman & Lane, 1964; Neutra & Leblond, 1966). Since sulphation is a late stage in proteoglycan biosynthesis, sulphate (as ^{35}S-sulphate) incorporation is a convenient means of studying sulphated glycosaminoglycan formation either biochemically or autoradiographically.

GENETIC CONTROL

Chondrocyte differentiation (Chapter 6) itself is a function of the pattern of gene expression which emerges during the development of a cell lineage. In the mature cartilage cell, genetic factors could operate by modifying the type of structural or catalytic protein synthesised. It seems doubtful whether gross changes occur normally, although repeated subculturing *in vitro*, for example, can modify the chondrocyte phenotype into a fibroblast-like form (Mayne *et al.*, 1976*a*). As discussed earlier, the presence of several varieties of proteoglycan with differences in their core proteins (Tsiganos & Muir, 1969; Hardingham & Muir, 1974*a*) and changes in their proportions during maturation and ageing (Brandt & Muir, 1971; Simunek & Muir, 1972) might indicate control via messenger RNA. Variation in the polysaccharide moiety, which is a post-translational modification of the core protein, could be mediated by genetic control of the various enzymic pathways as well as the core protein. Modifications of sulphato-transferase activity during maturation (Robinson & Dorfman, 1969) may provide an example. However, as with the core proteins, changes in the proportions of chondroitin-4-, chondroitin-6-sulphates and keratansulphate during maturation and ageing in whole tissue or even in cultured chondrocytes (Shulman & Meyer, 1968) are difficult to interpret in view of the effects of degradation and differences in turnover rates.

Experimental incorporation of thymidine analogues results in defects of synthesis both of proteoglycan and collagen (Mayne, Schiltz & Holtzer, 1973*b*). However, in pathologically abnormal cells, as in the mucopolysaccharidoses, genetic effects are usually produced by lack of degradative enzyme activity (Silbert, 1973).

ENZYME–SUBSTRATE INTERACTION

Enzyme activity is influenced by the availability of substrates, co-factors and the access of inhibitors. Factors controlling the enzymes involved in UDP-sugar formation and interconversion have already been discussed. In addition, glycosaminoglycan biosynthesis is in competition for UDP-sugar precursors with other biosynthetic pathways, including that of glycogen (an abundant cytoplasmic inclusion of chondrocytes), also formed from UDP-glucose (Fig. 47). It is appropriate therefore that end-product inhibition which effectively controls the 'gate' for entry of glycolytic intermediates (Fig. 53) operates on the second enzyme in route 1 (UDP-glucose dehydrogenase) though on the first enzyme in route 2 (fructose-6-phosphate aminotransferase). Indeed the dehydrogenase reaction: UDP-glucose → UDP-glucuronic acid, appears to be central to the enzymic control of UDP-sugar formation and interconversion and hence glycosaminoglycan synthesis.

UDP-glucose dehydrogenase is inhibited by the end-product of route 1, UDP-xylose, which donates xylose to form the xylosyl-serine linkage (Fig. 31), initiating the chondroitin sulphate chain. It follows that UDP-xylose synchronises core protein and chain synthesis since diminished protein synthesis would result in accumulation of UDP-xylose and hence cessation of UDP-glucuronic acid formation.

The aminotransferase reaction: fructose-6-phosphate → glucosamine-6-phosphate, is inhibited by UDP-*N*-acetylglucosamine, an end-product of route 2 (Fig. 53). This is also subject to indirect control by UDP-glucose dehydrogenase in several ways. Thus in the absence of keratansulphate synthesis, cessation of chondroitin sulphate chain formation due to UDP-xylose inhibition will produce an excess of UDP-amino sugars. In theory, aminotransferase activity will automatically be curtailed. In addition, at known intracellular concentrations of UDP-*N*-acetylglucosamine, the aminotransferase is thought to be permanently bound to its inhibitor and in this state inhibition is increased further by glucose-6-phosphate (Winterburn & Phelps, 1971). A fall in glucose-6-phosphate concentration, caused by its utilisation for UDP-glucose synthesis, for example, permits some enzyme activity, so producing amino sugar precursors for polysaccharide synthesis. However, if UDP-glucose is not utilised in its turn, due to inhibition of UDP-glucose dehydrogenase (or low activity of the epimerase or glycogen transferase) then the glucose-6-phosphate concentration may increase again, leading to inhibition of amino sugar formation.

Within the interconversion system, competition may occur where two

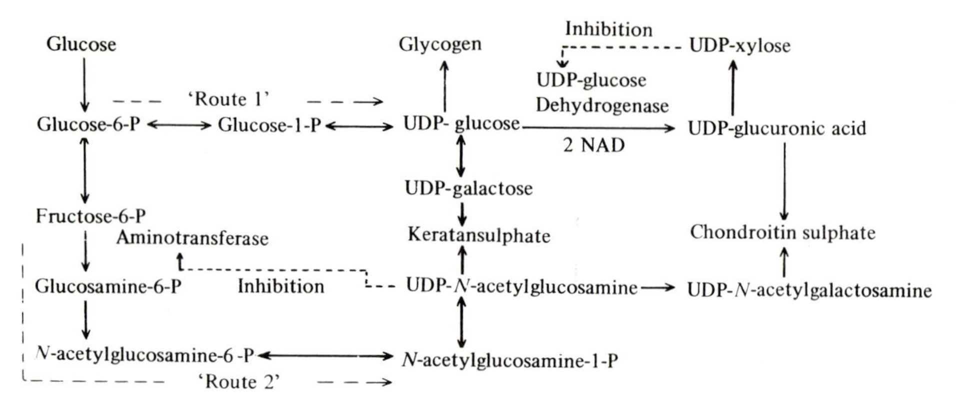

Fig. 53. Pathways involved in the formation and interconversion of the sugar nucleotide precursors of glycosaminoglycan synthesis. P = phosphate.

enzymes use the same substrate. There is a branch point in route 1 (Fig. 53) where either UDP-galactose (via UDP-glucose-4′-epimerase) or UDP-glucuronic acid (via UDP-glucose dehydrogenase) can be formed from UDP-glucose, as precursors of keratansulphate and chondroitin sulphate, respectively. The product formed appears to depend partly on the relative affinity of the enzymes for the substrate (De Luca, Speziale, Balduini & Castellani, 1975). In epiphysial cartilage the relative affinity of the dehydrogenase is much greater and hence UDP-glucuronic acid, the precursor of chondroitin sulphate, is formed in the chondrocytes. The effect of inhibitors is also important. At its physiological concentration in chondrocytes (Handley & Phelps, 1972*a*) UDP-xylose has relatively little inhibitory effect on UDP-glucose dehydrogenase and therefore permits the formation of UDP-glucuronic acid (Balduini *et al.*, 1973). It is of considerable interest that the enzyme characteristics are reversed in cornea where UDP-xylose stimulates keratansulphate synthesis (Balduini, Brovelli & Castellani, 1970). Corneal UDP-glucose dehydrogenase has a much lower affinity than the epimerase for UDP-glucose and is almost completely inhibited by UDP-xylose (De Luca *et al.*, 1975) at the physiological concentrations of substrate and inhibitor found in the cell (Handley & Phelps, 1972*b*). These results are fully compatible with the types of glycosaminoglycan found in cornea (both keratansulphate and chondroitin sulphate) and in epiphysial cartilage (chondroitin sulphate with only a trace of keratansulphate).

Glycosaminoglycan and lipid synthesis

Many of the reactions involving the UDP-sugars are affected by the NAD/NADH ratio. In particular, low levels of NAD could depress UDP-glucose dehydrogenase activity, either causing total cessation of synthesis or favouring keratansulphate production in regions of the tissue remote from a nutritional source (Stockwell & Scott, 1965). It is possible that the large quantities of lipid in chondrocytes (Chapter 2) may be relevant to the maintenance of NAD levels.

Lipid accumulation may be no more than the balance of energy and metabolites not required for polysaccharide synthesis, as this declines with maturation and ageing. However, fat globules are present at an early stage of fetal development (Fell, 1925) when synthesis is very active in chondrocytes and long before true fat cells are distinguishable (Bell, 1909). The amount of fat increases as matrix is produced for growth (Clark & Clark, 1942); if growth is accelerated, fat content also increases (Sacerdotti, 1900). In rabbit articular cartilage, fat droplets increase in size throughout life in the upper middle zone (Stockwell, 1967*b*), yet this zone always shows an active sulphate incorporation (Collins & McElligott, 1960). Other connective tissues also show a parallelism between glycosaminoglycan and fat synthesis, for example polyvinyl sponge granulomata (Bollet, Goodwin & Brown, 1959; Bole, 1963) and the aortic wall.

In aorta, Buddecke, Filipovic, Beckmann & Von Figura (1973) find that there is a fivefold increase in [U-^{14}C]-glucose incorporation into triglyceride lipid under anaerobic conditions, while uptake into glycosaminoglycans is reduced by 50%. The fat content of aorta *in vivo* also rises in animals kept under hypoxic conditions. The authors suggest that fatty acid chain elongation is controlled by the NAD/NADH ratio; utilisation of NADH for fatty acid synthesis (Fig. 47) would keep NAD reduction below 100% and at the same time store acetyl units (Filopovic & Buddecke, 1971).

Similar considerations apply to the chondrocyte living in near-anaerobic conditions. Lipid synthesis may be required as an NADH 'sink' to stabilise the NAD/NADH ratio and prevent it falling to levels which totally inhibit chondroitin sulphate synthesis.

DEGRADATION OF MACROMOLECULES

COLLAGEN

The triple helix of collagen is highly resistant to proteolytic digestion and requires specially potent enzymes usually termed 'collagenases'. Although collagen makes up about 50% of the dry weight of cartilage, detection of collagenase activity has proved to be difficult. There appear to be three possible mechanisms for the degradation of collagen in the tissues (Dingle & Burleigh, 1974; Barrett, 1975).

Collagenases found in the resorbing tadpole tail (Gross & Lapière, 1962) and in the synovial membrane (Werb, 1975) are metallo-proteinases, which act extracellularly. At a neutral pH these enzymes break down collagen fibres by cleavage across the triple helix about three-quarters of the molecular length from the N-terminal end (Fig. 54). The small fragments denature below body temperature and can then be degraded further by other proteolytic enzymes or by the continued action of collagenase. Most collagenases are secreted extracellularly but that of human neutrophil leucocytes is stored intracellularly and is probably distinct from the other enzymes (Barrett, 1975). No collagenase activity has yet been demonstrated in cartilage although very small amounts might go undetected. Type II collagen is in any case much more resistant than other forms of collagen to digestion by the enzyme (Woolley *et al.*, 1973).

Secondly, the elastase found in neutrophil leucocytes (Janoff, 1970) can degrade insoluble collagen as well as elastin and proteoglycan (Barrett, 1975). At neutral pH the enzyme cleaves the non-helical peptides (Fig. 54), removing the cross-linked portion and solubilising the α-chains; the enzyme is much more active against Type II than Type I collagen (Burleigh, 1975). Another neutral proteinase, cathepsin G, also present in leucocytes and spleen (Barrett, 1975), degrades Types I and II equally. However, neither elastase nor cathepsin G has yet been found in cartilage.

The third mechanism involves cathepsin B_1 which is a thiol proteinase hydrolysing arginine-2-naphthylamide and capable of degrading collagen (Burleigh, Barrett & Lazarus, 1974) as well as proteoglycan (Morrison, Barrett, Dingle & Prior, 1973). It initially cleaves the non-helical region of the molecule (Fig. 54), disrupting intermolecular cross-links; it also attacks the triple helix (Dingle & Burleigh, 1974). Cathepsin B_1 has been demonstrated in cartilage and is much more active in young than in adult human articular cartilage. Abnormally high levels are found when the cartilage becomes fibrillated in osteoarthrosis (Ali & Bayliss, 1974). Rabbit articular and ear cartilage are much more active than human tissue (Ali & Evans, 1973*a*). This may account for the collagen turnover found in adult rabbit articular cartilage (Repo & Mitchell, 1971) for which there is little evidence in human cartilage (Libby *et al.*, 1964). However, although cathepsin B_1 is present in cartilage, the enzyme is nearly 10 times more active against Type I than Type II collagen (Burleigh, 1975). Also, the low pH optimum probably restricts its activity to an intracellular lysosomal location unless an extracellular region close to the cell surface was to be rendered acidic by release of lactic acid from the cell (Dingle, 1975).

Collagenase inhibitors

The absence of detectable collagenase activity in cartilage must be considered in the context of tissue and serum inhibitors. In serum, collagenase or almost any active *endo*-peptidase can be inhibited by the α_2-macroglobulin 'trap' (Barrett & Starkey, 1973). The enzyme 'springs' the trap by cleaving a bond in the macroglobulin (molecular weight = *c.* 800000), which enables the two molecules to bind together irreversibly, so sterically inhibiting enzyme activity. Other lower molecular weight inhibitors occur, such as α_1-antitrypsin, which bind to and inactivate granulocyte proteinases reversibly (Ohlsson &

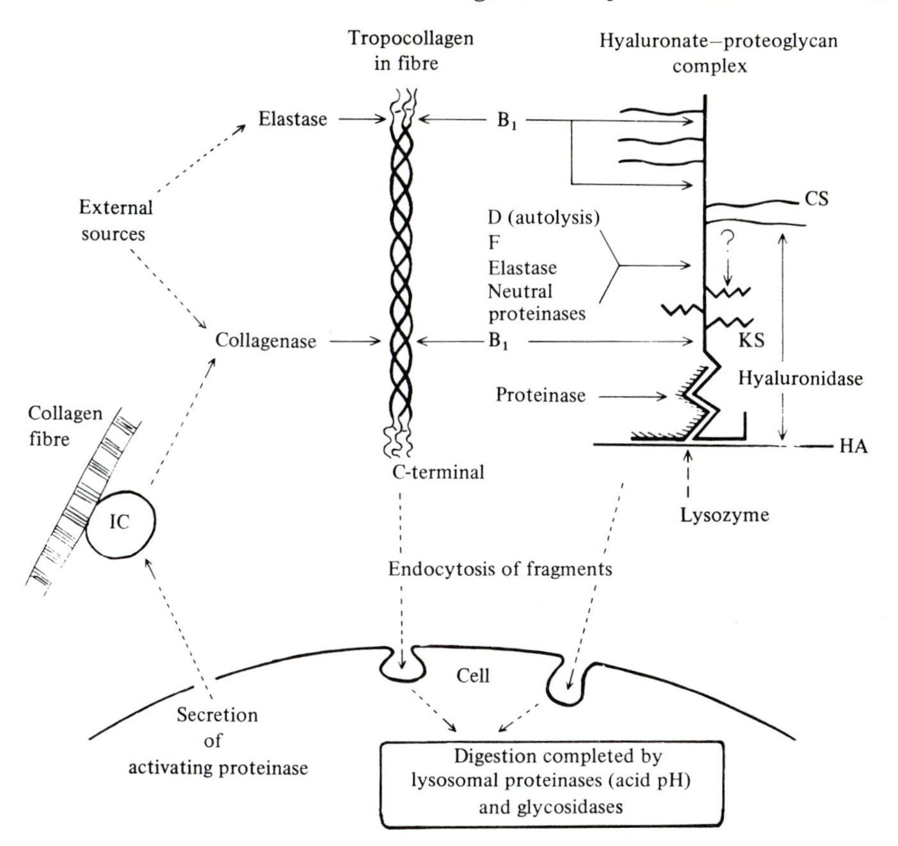

Fig. 54. Enzymes involved in the extra- and intracellular steps of degradation of collagen and proteoglycans. Note the dual role of cathepsin B_1, though not of D and F. According to Reynolds *et al.* (1977), latent collagenase (in bone) may exist as a fibre-associated inhibitor–collagenase (IC) complex which is activated by a proteinase. CS = chondroitin sulphate. KS = keratansulphate; HA = hyaluronic acid.

Delshammar, 1975). A low molecular weight cationic inhibitor (not lysozyme) is extractable from bovine and human cartilage (Kuettner, Hiti, Eisenstein & Harper, 1976). This finding might explain the lack of collagenase activity in cartilage.

Reynolds *et al.* (1977) postulate that all connective tissue cells synthesise inhibitors as a fail-safe device to control local activity of collagenase, the amount of collagen resorption depending on the relative concentration of enzyme and inhibitor. They suggest that the latent collagenase found in tissues such as bone is really an enzyme–inhibitor complex, since activation reduces its molecular weight. A model is proposed in which enzyme–inhibitor complex is bound to the collagen fibre; activation of the complex depends on destruction of the inhibitor by a proteolytic activating enzyme (Fig. 54).

Thus, further work is required before any definite statements can be made

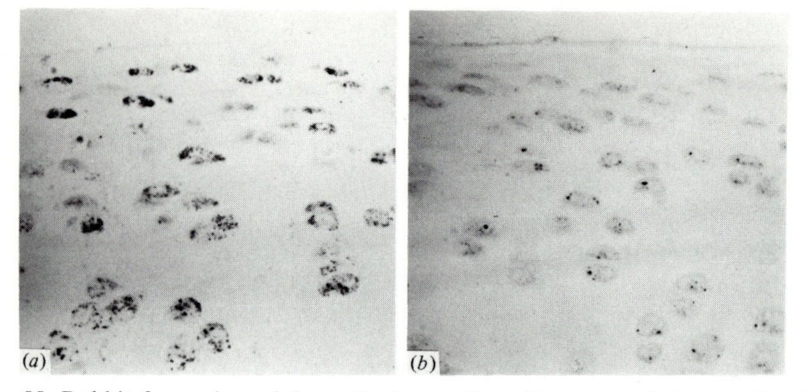

Fig. 55. Rabbit femoral condylar articular cartilage (three months), magnification × 250. Formalin-fixed frozen sections incubated for (*a*) *N*-acetylglucosaminidase; (*b*) acid phosphatase.

about the status of endogenous collagenase activity in cartilage. However, degradation of collagen and other constituents through the action of exogenous collagenase and other proteases occurs not only in pathological conditions such as rheumatoid arthritis (Krane, 1975) but also in experimental cartilage breakdown in the presence of soft tissue (Dingle, Horsfield, Fell & Barratt, 1975) or normal cartilage resorption as in tadpole tail metamorphosis.

PROTEOGLYCANS

The structure of proteoglycans and hyaluronate–proteoglycan aggregates indicates that two major groups of enzymes – proteinases and glycosidases – are needed for degradation (Fig. 54). In theory, hyaluronic acid and the chondroitin sulphate side chains of proteoglycans could be broken down by hyaluronidase (an *endo*-glycosidase) and degraded completely with the aid of *exo*-glycosidases such as *N*-acetyl-β-glucosaminidase and β-glucuronidase. The mode of degradation of keratansulphate chains in mammalian tissue is not known, however (Leaback, 1974). The core protein of the proteoglycans and the 'link' protein of the aggregate should be susceptible to proteinases.

In practice, glucosaminidase (Pugh & Walker, 1961) and glucuronidase (Gubisch & Schlager, 1961) are present in chondrocytes but hyaluronidase activity is not detectable (Bollet, Bonner & Nance, 1963; Bollet & Nance, 1966; Leaback, 1974). In rabbit articular cartilage (Fig. 55), glucosaminidase is active in all chondrocytes although acid phosphatase is less active in superficial than in deeper cells (Sprinz & Stockwell, 1976). Thus chondroitin sulphate degradation must be brought about by glucosaminidase and glucuronidase only; since these enzymes are unlikely to be effective outside the cell (Leaback, 1974) chondroitin sulphate chains must either diffuse out of the tissue or be digested inside the chondrocyte. The apparent lack of hyaluronidase is perhaps more significant with respect to the hyaluronate 'backbone' of the proteoglycan aggregate. Obviously scission of this part of the aggregate is necessary for normal turnover but must also have profound effects on the local physico-chemical properties of the matrix. However, even if hyaluronidase was present in cartilage, it is unlikely that the hyaluronate backbone would be susceptible to the enzyme, owing to steric shielding (Sajdera, 1974; Dingle, 1975). How then is the aggregate disaggregated? There appear to be two general possibilities. Slow chemical breakdown might occur due perhaps to the formation of free radicals in the tissue. Thus ionising radiation will depolymerise hyaluronic acid (Balazs, Davies, Phillips & Young, 1967) while cartilage treated with periodate, which produces free radicals in solution, suffers large losses of proteoglycan from the matrix (Scott, Tigwell & Sajdera, 1972). However, it is more likely that enzymatic cleavage by lysozyme is responsible (Fig. 54). Lysozyme is found in the matrix (Sorgente, Hascall & Kuettner, 1972); in the hypertrophic zone of epiphysial cartilage, disaggregation of hyaluronate–proteoglycan complexes occurs physiologically due to lysozyme, just prior to hydroxylapatite crystallisation (Pita, Muller & Howell, 1975).

Cleavage of the protein core effectively degrades proteoglycans into fragments small enough to diffuse out of the tissue or to be ingested by the chondrocyte (Dingle & Burleigh, 1974). Three lysosomal acid proteinases capable of degrading proteoglycans (Fig. 54), cathepsins D, B_1 and F, have been identified in cartilage (Barrett, 1975). Activities of cathepsins D and B_1 are elevated where matrix degradation is progressing, as in fibrillated cartilage (Ali & Bayliss, 1974). Cathepsin D appears to be mainly responsible for cartilage autolysis, since this can be prevented by specific antisera or inhibitors such as pepstatin. In living tissue where resorption has been stimulated by vitamin A, cathepsin inhibitors have no effect although these molecules may have difficulty in penetrating the matrix (Barrett, 1975; Dingle, 1975).

SITE OF DEGRADATION OF MATRIX

Resorption of matrix is thought to occur both extra- and intracellularly (Fell, 1969; Dingle, 1973). In the first stage, enzymes are released from the chondrocyte which partially degrade the macromolecules into fragments small enough to be readily diffusible. In the second stage, these fragments are endocytosed and digestion completed within lysosomal vacuoles (Fig. 54).

There are biochemical and anatomical problems concerning the first stage. Extracellular localisation of enzyme has been elegantly demonstrated in cultured cartilage by the use of immunofluorescent techniques employing monospecific antisera to cathepsin D (Poole, Hembry & Dingle, 1974), but the same mechanism may not operate *in vivo* (Dingle, 1975). Furthermore the low pH optimum of the proteinases known to be present in cartilage raises problems about their activity *in vivo* in extracellular sites, although fully compatible with digestion in secondary lysosomes in the second stage of resorption.

According to Dingle (1973, 1975) there are three possible mechanisms for the transfer of enzymes into the matrix.

(i) An incompletely fused secondary lysosome might regurgitate its contents by its open end into the extracellular space.

(ii) A primary lysosome might fuse with the plasmalemma and secrete its contents into the matrix.

(iii) Plasmalemma membrane-bound enzymes might be projected into the matrix several microns from the cell by means of long cell processes.

It is possible that acidic conditions prevail in the immediate vicinity of the chondrocyte, due perhaps to the release of lactic acid, particularly in autolysis. This would validate the first two mechanisms with respect to the pH optimum of the cathepsins. However, the few measurements of hydrogen ion concentration in small blocks of cartilage show this to be close to neutrality – pH 6.95–7.4 (Silver, 1975). It is possible, as Dingle (1975) points out that with the third (long cell process) hypothesis, the physico-chemical characteristics of the membrane-bound enzyme could be modified. Nevertheless it is unusual to find cell processes more than 2 μm long.

A theoretical fourth possibility is that chondrocytes (perhaps only a small proportion of the cell population) might change their location in the tissue very slowly by 'burrowing' through the matrix. In fibrillated articular cartilage, for example, chondrocyte cluster formation may be partly accounted for by dissolution of matrix between the cells. Changes in normal cell location would probably require collagen turnover but there is little evidence of this in cartilages with low cell density where cell movement might be most valuable in matrix turnover.

It seems fair to comment that any or all of these suggestions implies that there is decreasing cellular control of the matrix with increasing distance from the cell. Hence there is little doubt that cell density (Chapter 3) must be an important factor affecting degradation, secretion and turnover of matrix macromolecules. Since biochemically there seem to be several different glycosaminoglycan pools with markedly different rates of turnover, these may correlate with the histological heterogeneity of the tissue (Lohmander, Antonopoulos & Friberg, 1973), the half-life being related to the distance from the cell. Indeed there is little evidence that matrix lying remote from the chondrocytes in low cell density cartilage turns over at all.

MATRIX TURNOVER

In immature cartilage, the turnover of matrix macromolecules is a rapid process and for this reason embryonic chondrocytes in particular have been valuable material for the study of biosynthetic and degradative processes. Turnover of matrix constituents in adult cartilage is also of interest in relation to pathological processes. Many of the data apply to small laboratory animals although more recently human cartilage has been examined.

COLLAGEN

Little information is available. However, ^{14}C levels in post-mortem human cartilage, related to variations in atmospheric levels of the isotope during the life of the individual, suggest that there is no collagen turnover in the adult (Libby *et al.*, 1964). Nevertheless in adult rabbit articular cartilage there is a significant rate of synthesis (Repo & Mitchell, 1971), half of the labelled hydroxyproline disappearing from the matrix after 45 days.

In the rabbit, collagen turnover may be related to the basal remodelling thought to be associated with the persistence of alkaline phosphatase activity in the deepest zone of ageing rabbit articular cartilage (Stockwell, 1966). However, autoradiographs of the labelled rabbit cartilage show that cells in all zones are actively incorporating ^{3}H-proline. Possibly the difference between man and rabbit may lie in the very different cellularities of their cartilages (Table 3, Chapter 3) or perhaps in the difference on the absolute time scale at which the two species become mature.

PROTEOGLYCANS

Turnover of proteoglycans has been studied by numerous investigators, usually employing ^{35}S-sulphate as the labelled precursor. The results, briefly reviewed by Maroudas (1975), indicate considerable biochemical heterogeneity within a single sample of cartilage and between different cartilages and species.

Boström (1952) obtained a half-life of 17 days for ^{35}S-sulphate in adult rat costal cartilage. Subsequent investigations (Gross, Mathews & Dorfman, 1960) suggested the presence of more than one metabolic pool possibly due to the much slower turnover rate of keratansulphate than of chondroitin sulphate (Davidson & Small, 1963*a*, *b*). Thus in rabbit nucleus pulposus, keratansulphate has a half-life of 60–120 days and this is very much longer in costal cartilage. Lohmander *et al.* (1973) calculated from data obtained on six-week-old guinea pig cartilage (nucleus pulposus, costal and nasal septa) that there are at least three pools of chondroitin sulphate with half-lives of a few hours, three days and 30–80 days. Keratansulphate also shows moderately fast and slow pools (four and 90 days) in costal cartilage although nucleus pulposus has only a single pool (half-life 30 days).

Appreciable differences are found between different types of proteoglycans. Thus Davidson & Small (1963*b*) found that the soluble fraction has a much shorter half-life (20 days) than the insoluble fraction (120 days) in rabbit costal cartilage. Hardingham & Muir (1972*b*) and Rokosova & Bentley (1973) confirm that the insoluble fractions (released from cartilage by papain only) have the slowest turnover rate and that small proteoglycans (extractable with 0.15-M sodium chloride) have a shorter half-life than larger proteoglycans (extracted with 4-M guanidinium chloride).

Maroudas (1975) finds much slower rates of turnover for ^{35}S-sulphate in adult human (300–800 days) than in adult rabbit cartilage (100 days). These are mean values (Table 6) for all metabolic pools and do not exclude the presence of fast fractions with a half-life of about eight days as found by Mankin & Lippiello (1969*b*) in adult rabbit articular cartilage, now agreed to form less than 5% of the total glycosaminoglycan (Mankin, 1975*b*; Maroudas, 1975).

Lohmander *et al.* (1973) suggest that the different metabolic pools arise from histological heterogeneity of the tissue, i.e. variable distance of the matrix from the tissue surface and from chondrocytes, as well as from biochemical factors such as differences in proteoglycan hydrodynamic size.

TOPOGRAPHICAL FACTORS IN METABOLISM

In general, immature tissue has a more active metabolism than adult tissue, due in large part to differences in cellularity. However, within a single mass of immature cartilage, for example the growth plate, specific zones have different types of activity – cell proliferation in one zone, matrix synthesis in another zone where the chondrocytes are maturing, disaggregation and proteolytic degradation of proteoglycans in the zone of hypertrophy. There also appear to be regional differences in the molecular size of proteoglycan produced in the growth plate, low molecular weight species being synthesised in the hypertrophic zone (Larsson, 1976). Zonal activity is also found in the developing intervertebral disc, where sulphate incorporation is at a maximum in the inner part of the anulus fibrosus (Souter & Taylor, 1970).

ZONAL HETEROGENEITY IN THE ADULT

In the adult there are differences between human costal and articular cartilages (McElligott & Collins, 1960) which may reflect the interrelationship of factors such as size, variation in cellularity and nutritional supply as well as functional differences. Within a single specimen, zonal variations are observed. For example, in costal cartilage, sulphate incorporation is highest in the peripheral zone (Curran & Gibson, 1956), less activity occurring in the larger central cells. The distribution of synthetic enzymes (UDP-glucose dehydrogenase) is similar in rodent cartilage (Balogh & Cohen, 1961). However, most data are available for articular cartilage, in view of its relevance to arthritic processes.

In human articular cartilage it has been known for some time that sulphate

TABLE 6

Turnover of proteoglycans in articular cartilage *in vitro*. Metabolic data from Maroudas (1975); cell density data from Mankin & Baron (1965), Stockwell (1967*a*, 1971*b*)

Specimen	Cell density (cells × 10^{-3} per mm^3)	Mean half-life of sulphated proteoglycans (days)
Rabbit		
Three to five months	240	67
*Two years	180	100
Dog, adult	44	150
Man		
Five weeks	100	130
Adult		
Femoral head	11	800
Femoral condyle	14.5	300
Femoral condyle		
Superficial 0.25 mm	36.5	400
Middle zone	12.5	200
Deep zone	12.5	300

* Measured *in vivo*.

incorporation is much less active in the superficial zone than in the deeper tissue (Collins & McElligott, 1960; McElligott & Collins, 1960; Collins & Meachim, 1961). Maroudas (1975) has further investigated topographical factors in relation to sulphate metabolism (Table 6). The results confirm that immature tissue is more active than adult tissue and that sulphate incorporation is much higher in the middle than in the superficial zone. She also finds that sulphate uptake is greater in the thin articular cartilage of small animals than in the thicker cartilage of man; furthermore, human femoral condylar is more active than femoral head cartilage. Turnover rates follow a similar pattern but with a wider divergence between the hip and the knee than indicated by the rates of sulphate incorporation. This is because the proteoglycan–sulphate concentration is lower in the femoral condyle than in the femoral head cartilage and the half-life equals the proteoglycan–sulphate concentration/2 × 24 × hourly rate of sulphate incorporation. The adult species differences and the higher rates in growing cartilage can be directly related to the cell densities of the specimens. However, Maroudas concludes that differences between the femoral condyle and head cannot be accounted for on this basis since on theoretical grounds the cell densities are probably very similar. She suggests that the concentration of inorganic sulphate is reduced where the cartilage has a high proteoglycan content, in accordance with the Donnan equilibrium: since sulphate uptake is concentration-dependent (Maroudas & Evans, 1974), uptake will be less in the femoral head.

This does not explain the differences between the highly active deeper tissue and the much less active superficial zone, which is more cellular and also has a lower concentration of proteoglycan. The suggestion has been made that oligosaccharides derived from hyaluronic acid in the synovial fluid could inhibit proteoglycan synthesis in the superficial layers both in normal (Mankin, 1975*a*) and in fibrillated cartilage (Freeman, 1975). Similar considerations may apply to the subperichondrial zone of other hyaline cartilages.

PERICELLULAR HETEROGENEITY

The mechanisms resulting in the heterogeneity of the matrix seen histochemically and ultrastructurally (Chapter 3) must also be related to the occurrence of different populations of proteoglycans and glycosaminoglycans. As regards proteoglycan population differences, evidence from localisation studies is based principally on the resistance of lacunar compared with extra-lacunar proteoglycans to extraction with guanidinium chloride (Anderson & Sajdera, 1971). This certainly suggests that they are different from those in the rest of the matrix. Although mechanical enclosure by the collagen mesh around the lacuna is a plausible explanation, a more efficient barrier at the perilacunar rim is probably required than has been observed so far (Chapter 3). Possibly the lacunar proteoglycans are intimately bound to non-fibrous collagen and therefore resist extraction. Immunofluorescent studies also suggest that the lacunar proteoglycans have unique qualities (Loewi, 1965; Barland, Janis & Sandson, 1966).

There is more evidence in the case of the glycosaminoglycans. Most histochemists agree that there are chondroitin sulphate- and keratansulphate-enriched zones in the matrix, that localisation of these areas changes with the passage of time and that different parts of the cartilage specimen have different patterns of localisation. Four mechanisms have been suggested, not mutually exclusive, to account for changes in localisation of glycosaminoglycans. Thus Shulman & Meyer (1968) propose that cultured chondrocytes pass through a 'young' stage (synthesising proteoglycans containing very little keratansulphate) and an 'old' stage (synthesising keratansulphate-rich proteoglycans). A second mechanism involves the turnover rates of the glycosaminoglycans, since keratansulphate is very much slower than chondroitin sulphate (Davidson & Small, 1963*a*, *b*). Thirdly there is evidence that the type of glycosaminoglycan synthesised can be affected by the properties of the synthetic enzymes and the supply of substrates and co-factors (Balduini *et al.*, 1973; Phelps & Stevens, 1975). These properties suggested a nutritional hypothesis formulated in simple terms of the availability of oxidised NAD, whereby increasing distance from the tissue boundary and source of oxygen favours the synthesis and accumulation of keratansulphate (Stockwell &

Scott, 1965). The final mechanism is of course that of cell death (Schaffer, 1930).

In terms of the proteoglycan structure (Fig. 34) proposed by Heinegard (1977), synthesis of molecules containing a high proportion of keratansulphate implies that the variable chondroitin sulphate-rich part of the molecule is smaller relative to the hyaluronate-binding site and the keratansulphate-rich fragment. Similarly an area of matrix rich in keratansulphate contains proteoglycans in which either the variable fragment is reduced or the chondroitin sulphate side chains are very short. The maturation change reported by Shulman & Meyer (1968) might be related either to synthesis of short protein cores or enhanced post-translational scission of the variable fragment. However, the heterogeneity of the matrix cannot be accounted for solely in terms of ageing phenomena in the synthetic processes. Thus in adult human costal cartilage there is no significant difference in the age of cells between the outer central and inner peripheral regions and yet matrix staining patterns are quite different (Stockwell & Scott, 1965). Indeed the deepest part of the uncalcified portion of ageing articular cartilage has the 'oldest' staining pattern (territorial keratansulphate) and yet arguably the youngest cells.

The localisation of high molecular weight keratansulphate in the inter-territorial matrix of the peripheral region of many cartilages (Fig. 38, Chapter 3) may be related to the slow turnover of keratansulphate (Stockwell & Scott, 1965). As a consequence of the presumably maximal turnover conditions operating close to the cell, the juxtacellular molecules are most likely to be maintained (in a dynamic equilibrium) in a chondroitin sulphate-rich state (Fig. 56). However, associated with their slow turnover, keratansulphate-rich fragments of partly degraded proteoglycans might escape further breakdown, diffuse outwards towards the inter-territorial matrix and there accumulate (Mason, 1971).

Results of the CEC technique suggest that in the central regions of cartilage, keratansulphate diminishes from the inter-territorial matrix and accumulates in the cell territory (Fig. 41, Chapter 3). This may be a consequence of glycoprotein accumulation; either collagenous or non-collagenous basic proteins are involved (Quintarelli & Dellovo, 1966). Thus mild proteolysis or hyaluronidase treatment enhances basophilia in the inter-territorial zone, although this is less effective in aged cartilage (Mason, 1971; Sames, 1975). Hence the diminution of keratansulphate basophilia in the inter-territorial matrix of the central regions of cartilage specimens might be due partly to complex formation with basic proteins as well as a real loss associated with cell death. The enhancement of keratansulphate basophilia in the cell territory cannot be so easily explained. Mason (1971) considers that the inter-territorial matrix becomes blocked by entangled cross-linked macromolecules. Hence keratansulphate-rich proteoglycans can no longer diffuse away from the cell territory (Fig. 56).

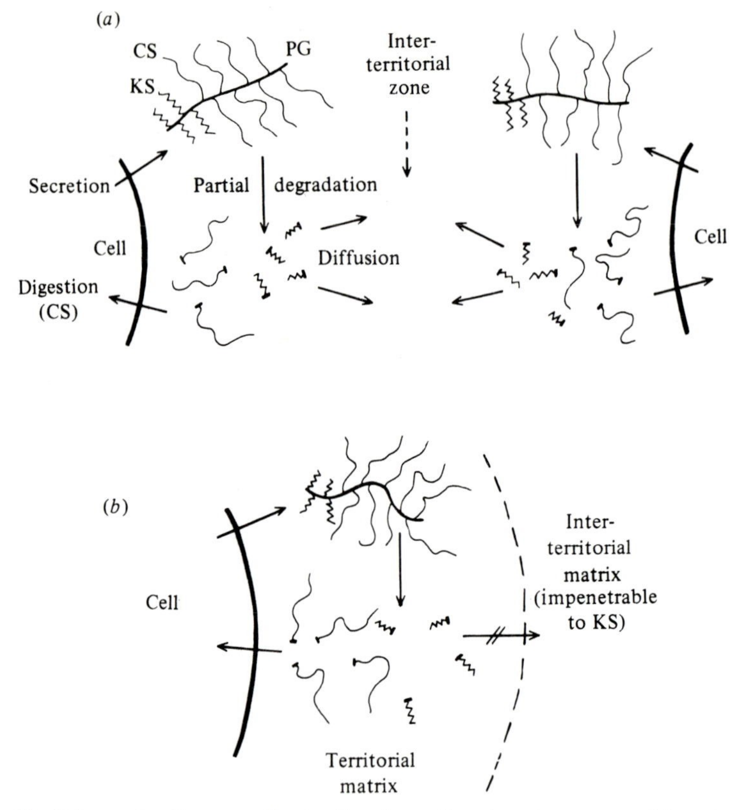

Fig. 56. Schematic diagram illustrating how the slow turnover of keratansulphate might permit it to diffuse into the inter-territorial zone, in (*a*). In (*b*) this zone becomes blocked and keratansulphate is restricted to the territorial zone. PG = proteoglycan, CS = chondroitin sulphate, KS = keratansulphate.

CHONDROCYTE HETEROGENEITY

In addition to differences in chondrocyte activity related to topography there may be a local cell heterogeneity superimposed on the zonal variation.

Histochemical studies of the middle zone of adult canine articular cartilage indicate differences even within an isogenous group. Only about half of the cells contain proteoglycan and the alcian blue–CEC technique demonstrates that some cells of this group do not contain keratansulphate (Kincaid, Van Sickle & Wilsman, 1972). Similar results are obtained with enzyme histochemistry. While the cells are homogeneous with respect to enzymes such as lactic dehydrogenase, both active and inactive cells (1:1) are observed on studying UDP-galactose-4′-epimerase. This enzyme converts UDP-glucose to UDP-galactose (at the branch point in route 1) so affecting chondroitin sulphate synthesis (Wilsman & Van Sickle, 1971). Immunofluorescent techniques also demonstrate heterogeneity in cathepsin D release from chick embryonic chondrocytes (Poole, 1975).

There is no simple subdivision into active and inactive cells, however, Ultrastructural studies of adult rat articular cartilage demonstrate that some cells are rich in glycogen while others have abundant granular ER. Autoradiography using ^{35}S-sulphate and ^{3}H-proline suggests that while all cells produce some proteoglycan, the glycogen-rich cells predominantly synthesise proteoglycans and the reticulum-rich cells produce collagen (Mazhuga & Cherkasov, 1974).

It is probable, therefore, that within a small volume of tissue the chondrocytes are in different states of activity, at least in the case of proteoglycan turnover. This is relevant to the production of hyaluronate–proteoglycan aggregates; if different cells produced hyaluronic acid (or link protein) and the proteoglycan subunits, assembly might then occur at some distance from the cells. Similarly it might be useful if proteoglycan synthesis and degradation were desynchronised.

Heterogeneity could arise either from cyclical changes in function, age-related modifications or the presence of at least two distinct and committed cell populations. While age changes must occur up to and including cell death, it is unlikely that this factor alone would produce a 1:1 distribution of cell types. Until further data are available, no conclusions can be made as to whether the functional heterogeneity of chondrocytes is cyclical or permanent.

CHONDROCYTE–MATRIX INTERACTION

Loss of matrix results in a chondrocyte reaction tending to restore normality. Thus in rabbit ear cartilage depleted of proteoglycan by the action of intravenous papain the chondrocytes show an enhanced ^{35}S-sulphate incorporation (McElligott & Potter, 1960; Sheldon & Robinson, 1960) and the animals' floppy ears eventually become straight again. In deeply fibrillated articular cartilage of the knee joint where there is pathological loss of proteoglycan, the evidence suggests increased sulphate uptake (Collins & McElligott, 1960) although this does not seem to occur in the hip joint (Byers *et al.*, 1977), possibly due to differences in the assay procedure. The situation in degenerating cartilage is complicated by factors such as the degree of cell multiplication (Chapter 7) and proteoglycan degradation; cell clusters where formed show a local increase in sulphate uptake. In different regions of the epiphysial plate the rate of synthesis is inversely related to the glycosaminoglycan concentration (Greer *et al.*, 1968), although this relationship does not apply in different zones of articular cartilage (Maroudas, 1975). In cartilage fragments grown in organ culture and depleted by papain (Bosmann, 1968), hyaluronidase (Fitton Jackson, 1970) or trypsin (Millroy & Poole, 1974), the matrix is replenished rapidly during post-incubation in normal media.

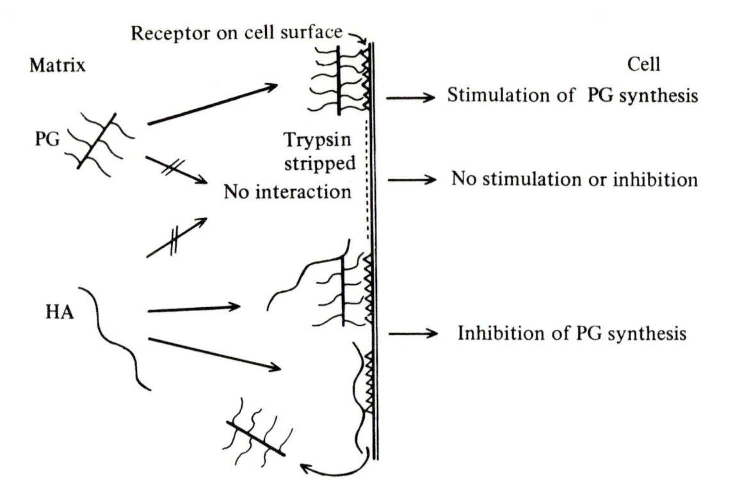

Fig. 57. Schematic diagram illustrating possible interactions of proteoglycan (PG) and hyaluronic acid (HA) with receptors on the chondrocyte surface.

Similarly isolated adult chondrocytes stripped of all matrix by enzyme digestion form proteoglycan and collagen *in vitro* and *in vivo*, eventually producing small cartilage nodules, although there is an initial period of inhibition of synthesis (Huang, 1977).

INTERACTION WITH MACROMOLECULES

There are various ways in which such changes in the matrix might affect the chondrocyte but recent investigations have shown that where considerable depletion has occurred the matrix macromolecules may interact with the cell directly (Fig. 57). Using suspension cultures of isolated embryonic chondrocytes, Nevo & Dorfman (1972) have shown that addition of cartilage proteoglycans, glycosaminoglycans (though not hyaluronic acid) and other substances such as dextran sulphate and agar to the medium causes increased synthesis of proteoglycans. The stimulation is dose-dependent up to 2 mg ml^{-1} and is specific for proteoglycan synthesis since there is no effect on total protein or collagen. Similar findings have been obtained in studies of chondrogenesis. The macromolecules do not enter the cell and since mild trypsin digestion of the cells temporarily prevents the response, Nevo & Dorfman suggest that a cell surface interaction is involved. Similarly, cultured chondrocytes treated with phospholipase C fail to synthesise metachromatic matrix. The inhibition is associated with the liberation of proteins from the cell surface (Nameroff, Trotter, Keller & Muner, 1973).

The action of hyaluronic acid has also proved to be of considerable significance (Fig. 57). Studying chondrogenesis, Toole, Jackson & Gross (1972) noted that hyaluronate at very low concentrations (1 ng ml^{-1})

prevents cytodifferentiation of dissociated chick limb bud and somite cells. Subsequently Wiebkin & Muir (1973, 1975), using suspensions of adult pig laryngeal chondrocytes, found that inclusion of hyaluronic acid (0.1 μg ml^{-1}) in the medium reduces sulphate incorporation by 50%. The effect is specific for chondrocytes and as with proteoglycan stimulation, previous treatment of the cells with trypsin reversibly abolishes the inhibitory effect. Solursh, Vaerewyck & Reiter (1974) have also observed hyaluronate inhibition on chick embryo sternal chondrocytes in confluent cell culture but find that higher concentrations of hyaluronate (20–500 μg ml^{-1}) are needed to obtain 25% inhibition of sulphate incorporation. The marked differences in hyaluronate concentration in the three systems – prechondrocytes, chondrocytes in suspension and in confluent culture – may be related to differences in the amounts of pericellular matrix already *in situ*.

The action of hyaluronate seems to be specific for chondrocytes (Wiebkin & Muir, 1973) and for proteoglycan synthesis since there is no depression of growth, collagen or total protein synthesis (Solursh *et al.*, 1974). While sulphate incorporation into macromolecules is reduced, the mechanism of sulphate uptake into the cell is not affected. Handley & Lowther (1976) confirm that total protein synthesis is unaffected and have shown that hyaluronate does not inhibit glycosaminoglycan chain formation provided that exogenous xyloside is provided as acceptor. This suggests that hyaluronic acid affects chain initiation, either by decreasing the availability of the core protein (by changes in its synthesis) or repressing the activity of xylosyl transferase.

The inhibitory action of hyaluronic acid might be mediated in various ways. Hyaluronate is taken up at or near the cell surface but does not enter the cell (Wiebkin, Hardingham & Muir, 1975). Its action might be purely physical, binding proteoglycan aggregates at the cell surface and so preventing their deposition in the matrix (Solursh *et al.*, 1974). However, decasaccharides derived from hyaluronate also have an inhibitory effect yet the smallest oligosaccharide which can bind more than one proteoglycan unit is much larger than a decasaccharide. Hence an interaction of a more specific nature is indicated. At the same time, Wiebkin & Muir (1975) believe that hyaluronic acid interacts with proteoglycans on the cell surface since (i) the characteristics of hyaluronate–proteoglycan aggregation are similar to those of hyaluronate cell inhibition and (ii) mild trypsinisation of the cells abolishes the inhibitory action of hyaluronate and removes cell-associated ^{14}C-hyaluronate. Hyaluronate might act either by detaching proteoglycan molecules from receptors on the cell membrane (Huang, 1977) or by itself attaching to membrane receptors (Fig. 57). Whatever the nature of the interactions of proteoglycans and hyaluronic acid at the cell surface, they may in turn cause perturbations of the cell membrane or changes in pericellular ionic structure, e.g. calcium concentration, which affect synthesis perhaps via intracellular agents.

The biological significance of the stimulatory action of proteoglycan on cartilage *in vivo* is not fully understood. Maximum stimulation of synthesis by proteoglycan, for example, occurs at concentrations (2 mg ml^{-1} = 0.2% wet weight) far below those found in normal cartilage (Table 1), so that the chondrocyte is normally fully stimulated. In the case of hyaluronic acid, the amounts present in normal hyaline cartilage, about 0.5–1% of the total glycosaminoglycan (Hardingham & Muir, 1974*a*), are of the same order as the highest concentrations required for inhibition. Hyaluronate probably functions as a negative feedback mechanism, inhibiting proteoglycan synthesis when matrix quantities are adequate. During growth, factors such as somatomedin would be necessary to override the inhibition: a number of hormones such as growth hormone, thyroxin and other components present in serum (Toole, 1973; Solursh *et al.*, 1975) antagonise the action of hyaluronate. However, since some proteoglycans do not interact with hyaluronic acid, their synthesis may not be affected (Wiebkin & Muir, 1975).

CATIONS

The ionic environment also affects the metabolism of the chondrocyte, and may do so by direct action on the cell. Sulphate incorporation by cultured fragments of embryonic chick vertebral cartilage is stimulated by physiological concentrations of ionised calcium (0.5–1.5 mM); Mg^{2+} have a similar but less powerful action. Puromycin blocks the stimulatory effect, suggesting that core protein synthesis is involved. Shulman & Opler (1974) postulate that stimulation is mediated by affecting the negative charge density of the pericellular glycosaminoglycan.

Collagen synthesis is also affected by Ca^{2+} concentration. Deshmukh, Kline & Sawyer (1976) note that rabbit articular chondrocytes grown to confluency in monolayer culture and then transferred to suspension culture produce Type II collagen if no ionised calcium is present in the medium. If Ca^{2+} are present, Type I collagen is produced. Since the cells also produce Type I collagen if dibutyryl adenosine-3′:5′-cyclic phosphate (dibutyryl cyclic AMP) is present in the calcium-free suspension culture medium, or if they are pre-treated with a calcium ionophore (which increases calcium uptake into the cell and elevates the intracellular cyclic AMP), Deshmukh and his colleagues consider that intracellular cyclic AMP is acting as a second messenger. Pre-treatment with calcitonin and parathyroid hormone (Deshmukh *et al.*, 1977) also causes influx of calcium into the cell, cyclic AMP elevation and the production of Type I collagen in calcium-free suspension culture. However, the action of Ca^{2+} seems to vary in different situations. Thus Norton, Rodan & Bourrett (1977) claim that calcium influx causes inhibition of adenyl cyclase activity (resulting in a lower level of cyclic AMP) in epiphysial proliferative cells but has no significant effect on the same enzyme in hypertrophic cells. Thus, the precise action of Ca^{2+} *in vivo* in intact cartilage is not immediately apparent from these *in vitro* studies.

Monovalent cations also affect synthesis *in vitro*. Very high concentrations of K^+ (60–70 mM) cause a considerable increase in glycosaminoglycan synthesis; however, although minimal degradation of the synthesised products occurs, little or no matrix is found in the culture (Daniel, Kosher, Hamos & Lash, 1974). This could be related to a recent finding that intracellular glycosaminoglycans, which are unswollen, exist as potassium salts but extracellularly become calcium or magnesium salts (Hunt & Oakes, 1977). If the secreted proteoglycans remained unexpanded (due to the high potassium environment) they would be more diffusible and less likely to be trapped in the pericellular matrix. The lack of formed matrix around the cells might in turn account for the stimulation of synthesis.

MECHANICAL FACTORS

Cartilage matrix is concerned with resistance to compressive, tensile and shearing forces. Yet surprisingly little is known about the effects of mechanical stress on metabolism of the cell, which produces and maintains the matrix. Studies of the interrelationship of cell density, cartilage thickness and species size (Stockwell, 1971*b*) suggest that the amount of matrix per cell varies according to the stresses experienced by the block of cartilage tissue; this also applies to weight-bearing and non-weight-bearing areas within a synovial joint (Vignon *et al.*, 1976). The maintenance of the appropriate volume of matrix may be mediated by microstresses affecting the cell (Stockwell & Meachim, 1973).

A number of studies have investigated the effects of immobilisation and altered mechanical stress on joints. Following pathological loss of joint function, articular cartilage becomes thin, probably accounted for by loss of matrix since numerous chondrocytes are found (Collins, 1949). A decreased content of chondroitin sulphate is found in paralysis due to poliomyelitis (Eichelberger, Miles & Roma, 1952) or experimental denervation where the cartilage may become thinner (Akeson *et al.*, 1958). Experimental immobilisation by external or internal joint fixation results in a lower hexosamine concentration (Akeson *et al.*, 1973); the cartilage becomes thinner but is more cellular (Sood, 1971). Hence both the volume of matrix per cell and the concentration of glycosaminoglycans in the remaining matrix are reduced. Immobilisation involving compression of a joint (Salter & Field, 1960: Trias, 1961; Ginsberg, Eyring & Curtiss, 1969; Thompson & Bassett, 1970) also causes loss of metachromasia and cartilage thickness but this is accompanied by chondrocyte degeneration and cartilage fibrillation and erosion. Although it may appear that mechanical factors are solely and directly responsible, in the case of compression–immobilisation the changes are almost certainly because the compressed cartilage is cut off from its nutritional supply from the synovial fluid. Similar arguments apply to immobilisation without compression since the lack of joint movement will again result in a diminished nutrient supply (Maroudas *et al.*, 1968). However, alterations in mechanical forces may act in addition to the nutritional changes.

Where additional stress is placed on a joint, there are changes the reverse of those seen after disuse. Thus following partial amputation or immobilisation of the contralateral limb there is an increase in glycosaminoglycan content in the cartilage of the loaded limb (Kostenszky & Olah, 1972; Lowther & Caterson, 1974). Similarly, following extrasynovial severance of the cruciate ligaments, knee joint cartilage shows an increase in thickness and water content. This happens before there are any signs of an arthritic disruption of the articular surface (McDevitt & Muir, 1976); the proteoglycans synthesised contain more chondroitin sulphate than normal, relative to keratansulphate.

More direct studies of chondrocytes can be made in tissue culture. There is no doubt that connective tissue cells *in vitro* vary in their response to different conditions of mechanical stress. Thus fibrous tissue is deposited along lines of tensile stress; this is due to the orientation of connective tissue cells, secondary to their interaction with ground substance colloids aligned along lines of tension and fluid flow (Weiss, 1933). Sokoloff (1976) regards the anchorage of isolated chondrocytes to the vessel wall (in monolayer culture) as predisposing the tissue against the chondrogenic expression found in suspension cultures where no such surface interaction occurs. Glucksmann (1939) showed that after an initial phase of flattening due to compression, chick embryonic perichondrial or periosteal cells respond to pressure by forming pericellular capsules of ground substance.

Mechanical stress interacts with other factors. Thus Bassett & Herrmann (1961) found that embryonic chick cells in different environments produce either fibrous tissue (high oxygen tension plus tensile stress), bone (high oxygen tension plus compressive stress) or cartilage (low oxygen tension plus compressive stress). To some extent the form and metabolism of chondrocytes in the native tissue relates to local stresses. Thus the superficial zone of articular cartilage is subject to tensile forces (Zarek & Edwards, 1963), the cells are flattened and discoidal with a low rate of glycosaminoglycan synthesis (Maroudas & Evans, 1974); the deeper cells are subject to compressive stress, are rounded and are much more active in synthesis. Similar considerations apply to the cells of other forms of hyaline cartilage, for example the subperichondrial region also has flattened cells and is subject to 'interlocked' tensile stress (Fry & Robertson, 1967). One effect of cartilage compression is to cause fluid flow relative to matrix macromolecules and to the cells. This may set up 'streaming potentials' which might stimulate cell synthesis (Bassett, 1971).

It is well known that considerable compressive force (28–31 g mm^{-2}) must be applied to the epiphysial growth plate to retard bone growth (Sissons, 1971). More recently it has been demonstrated that small changes in hydrostatic pressure (60 g cm^{-2}) applied to tissue fragments and isolated

TABLE 7
Changes in cyclic AMP content of 16-day chick embryonic tibial epiphyses in response to hydrostatic pressure (60 g cm^{-2}) and calcium uptake. Values calculated from Rodan *et al.* (1975) and Bourrett & Rodan (1976*a*, *b*)

Specimen and treatment	Cyclic AMP content (experiment: control)
Pressure:	
Whole epiphysis	0.72
Isolated cells	
Proliferative zone	0.79
Hypertrophic zone	1.07
Calcium ionophore:	
Isolated cells	
Proliferative zone	0.85
Hypertrophic zone	1.46

cells from chick embryonic growth plates stimulate ^{3}H-thymidine uptake (Rodan, Mensi & Harvey, 1975). Bourrett & Rodan (1976*a*, *b*) have further shown that pressure causes a decrease of cyclic AMP in proliferative cells, due to adenyl cyclase inhibition; they consider that the pressure effect is brought about by an increased penetration of Ca^{2+} into the cell. The cyclic AMP level in hypertrophic cells is not changed significantly although calcium enters the cell, possibly because cell maturation involves membrane alterations affecting the responsiveness of adenyl cyclase (Table 7). It is not known whether the pressure acts directly by cell deformation or by the generation of electrical signals (Bassett, 1971). Norton *et al.* (1977) find that whole epiphyses placed with their long axis parallel to electric fields (900 volts across 1.5 cm) show elevated cyclic AMP levels, while pressure causes a decrease in cyclic AMP. With isolated cells, both mechanical and electrical forces reduce the level of cyclic AMP. Norton *et al.* comment that the matrix and/or the arrangement of the cells is important.

HORMONAL CONTROL

Certain hormones have profound effects on chondrocyte metabolism. Although these are more marked in immature tissue during the regulation of growth, adult cells are also influenced by endocrine factors. In addition to their role in the control of normal adult metabolism, hormones may be of pathological significance. For example, excessive amounts of growth hormone (in acromegaly) may cause degenerative changes in joints, and the cortisone group are potentially harmful to cartilage in high concentration (Shaw & Lacey, 1973).

HORMONES PROMOTING SYNTHESIS

In vivo growth hormone or somatotrophin promotes the growth of long bone by indirectly stimulating cell proliferation and matrix production in epiphysial cartilage; similar effects are produced in adult articular cartilage (Silberberg & Hasler, 1971). In mammals, growth hormone acts on chondrocytes via an intermediary substance in the plasma, 'sulphation factor' (Salmon & Daughaday, 1957), now designated 'somatomedin' (Daughaday *et al.*, 1972). Although growth hormone itself has no direct action on costal chondrocytes, it appears to depress synthesis of all glycosaminoglycans in articular chondrocytes (Smith, Duckworth, Bergenholtz & Lemperg, 1975). Synthesis in avian chondrocyte monolayer culture is stimulated directly by mammalian growth hormone, possibly associated with the fact that birds have no growth hormone (Sokoloff, 1976).

Thyroxin primarily affects maturation. Sulphate incorporation and chondroitin sulphate synthesis are depressed in the cartilage of thyroidectomised animals (Dziewiatkowski, 1964) and thyroxin stimulates sulphate incorporation in chondrocyte cell culture (Pawalek, 1969) containing serum. The hormone does not have a stimulatory effect *in vitro* on costal cartilage from hypophysectomised rats (Salmon & Daughaday, 1957). Similarly in skin, thyroxin restores sulphate incorporation in the hypothyroid rat but not in the hypothyroid and hypophysectomised animal (Schiller, 1966). In isolated epiphysial proliferative cells, thyroxin antagonises somatomedin-stimulated DNA synthesis though it does not depress normal levels (Ash & Francis, 1975). Hence thyroxin works synergistically with growth hormone.

Insulin acts synergistically with amino acids to promote sulphate incorporation in cartilage from hypophysectomised rats (Salmon & Daughaday, 1957; Dziewiatkowski, 1964); chondroitin sulphate synthesis is depressed in diabetic skin (Schiller & Dorfman, 1963). However, insulin itself does not alter either DNA or proteoglycan synthesis in chondrocyte monolayer culture (Sokoloff, 1976).

Serum growth factors

The growth hormone intermediary, somatomedin, is produced in the liver (McConaghey, 1972; Sledge, 1973) and stimulates DNA, proteoglycan and collagen synthesis in rat costal cartilage *in vitro*. As regards proteoglycans, the primary effect of somatomedin is thought to be the stimulation of core protein synthesis (Salmon, 1972). The chondrocyte response to somatomedin declines with age in costal cartilage (Heins, Garland & Daughaday, 1970) and with maturation and cell hypertrophy in the cell columns of the growth cartilage itself (Ash & Francis, 1975).

Three forms of somatomedin are found in human plasma all of which have insulin-like effects on target tissues (Hall, Takano, Fryklund & Sievertsson,

1975). Somatomedin A is a neutral, B an acidic and C a basic polypeptide, each with a molecular weight in the range 6000–8000 although in the plasma each is attached to carrier molecules of high molecular weight (*c.* 50000). Somatomedin A acts primarily on chick cartilage but binds to rat cartilage, while somatomedin C stimulates DNA synthesis and sulphate incorporation in rat cartilage. Other chondrocyte growth factors, apparently not identifiable with the somatomedins (since they depress proteoglycan synthesis), have been detected in pituitary extracts (Jones & Addison, 1975; Sokoloff, 1976). It is possible, however, that growth hormone-dependent factors with 'non-suppressible insulin-like activity' (NSILA) are closely related to the somatomedins (Hall *et al.*, 1975).

In various cell types, including chondrocytes, Tell, Cuatrecasas, Van Wyk & Hintz (1973) find that somatomedin (like insulin) reduces adenyl cyclase activity, presumably depressing intracellular cyclic AMP levels. However, the role of cyclic AMP in chondrocytes is not clear.

Cyclic AMP

Cyclic AMP is produced by the action of adenyl cyclase, an enzyme in the cell membrane, and destroyed by phosphodiesterase activity in the cytosol. In the cell, cyclic AMP changes the activity of already existing enzymes (Pastan & Perlman, 1971). Many hormones, for example thyroxin, act by increasing the level of intracellular cyclic AMP. Others, such as insulin, cause a fall in cyclic AMP in liver, muscle and adipose cells, resulting in storage of glycogen and fat. Since somatomedin has the opposite effect (Tell *et al.*, 1973) to the catabolic (adenyl cyclase-stimulating) hormones (Hall *et al.*, 1975) reduced levels of cyclic AMP in chondrocytes could cause stimulation of DNA and proteoglycan synthesis. Several findings are compatible with this view. Thus Kosher (1976) finds that butyrylated cyclic AMP derivatives (which readily enter the cell) inhibit cartilage expression in late chick somites. Norton *et al.* (1977) find that raised hydrostatic pressure causes diminution in intracellular cyclic AMP levels, accompanied by a rise in DNA synthesis in the proliferative zone of growth cartilage. A number of anti-inflammatory drugs, including indomethacin, which inhibit glycosaminoglycan synthesis in cartilage (McKenzie, Horsburgh, Ghosh & Taylor, 1976) also inhibit phosphodiesterase activity (Newcombe, Thanassi & Ciosek, 1974), which should permit cyclic AMP levels to rise.

However, Drezner, Eisenbarth, Neelon & Lebovitz (1975) believe that growth hormone-dependent factors in the serum act by raising intracellular cyclic AMP. Thus although in cultured cartilage fragments 0.5-mM cyclic AMP causes a 30% inhibition of sulphate, leucine and uridine incorporation into proteoglycan, total protein and RNA, butyrylated cyclic AMP derivatives stimulate incorporation by 50–100% (Drezner, Neelon & Lebovitz, 1976). They consider that exogenous cyclic AMP (but not the butyrylated derivative) is degraded extracellularly to adenosine which can be shown to inhibit precursor incorporation. In support of this view, Thanassi & Newcombe (1974) find that high concentrations of thyroxin cause inhibition of phosphodiesterase in epiphysial cartilage. This suggests that the action of thyroxin in stimulating glycosaminoglycan synthesis by the chondrocyte is mediated by a rise in cyclic AMP as in other tissues. Toole (1973) also finds that hormones normally producing an increase in cyclic AMP antagonise the hyaluronate inhibition of cartilage expression. This agrees with Drezner *et al.*'s (1975, 1976) concept of the action of cyclic AMP in the chondrocyte, although a number of substances non-specific with respect to cyclic AMP also reversed hyaluronate inhibition.

No firm statement can be made, therefore, concerning the action of cyclic AMP in controlling proteoglycan synthesis by the chondrocyte. However, the hormonal control of chondrocyte proliferation and synthetic activity involves complex relationships. Thus Froesch *et al.* (1976) suggest that three factors are required for sulphation in chick embryonic cartilage: (i) NSILA (growth hormone-dependent); (ii) an unidentified factor (not under endocrine control); (iii) thyroxin. The separate stages of growth (i.e. cell division and DNA synthesis) and matrix synthesis (i.e. cell maturation/sulphate incorporation) may be controlled by the appropriate balance between these stimuli as well as the receptivity of the chondrocyte.

SEX HORMONES

Testosterone stimulates growth because of its effect on nitrogen retention (Silberberg & Silberberg, 1971) and has no direct action on sulphate utilisation (Salmon, Bower & Thompson, 1963).

Oestrogens appear to work synergistically with growth hormone in the regulation of body growth but have peripheral actions antagonising the effects of growth hormone. They inhibit linear bone growth and accelerate skeletal maturation. Glycosaminoglycan synthesis is depressed in chondrocytes (Priest & Koplitz, 1962; Dziewiatkowski, 1964). Oestradiol inhibits DNA synthesis both in cell culture (Sokoloff, 1976) and in the intact animal, so antagonising the stimulatory effect of growth hormone (Strickland & Sprinz, 1973; Gustafsson, Kasstrom, Lindberg & Olsson, 1975).

CORTISONE

Cortisol and related steroids have been widely investigated on account of their side effects when used as therapeutic agents. It has long been known that cortisone inhibits synthesis of chondroitin sulphate (Layton, 1951; Boström & Odeblad, 1953). It also reduces the rate of degradation by stabilising the lysosomal membrane (Weiss & Dingle, 1964; Dingle, Fell & Lucey, 1966) and so diminishing cathepsin release. In addition, it may retard the synthesis and ageing of collagen (Silberberg & Silberberg, 1971). These basic effects are consistent with the corticosteroid-induced changes in glycosaminoglycan content of cartilage (Kaplan & Fisher, 1964; Barrett, Sledge & Dingle, 1966; Mankin, Zarins & Jaffe, 1972) and of other tissues (Schiller, 1966).

At low doses (0.1–1 mg kg^{-1} body weight day^{-1}) for short periods, corticosteroids cause little change in total cartilage glycosaminoglycans provided the functional demand on the tissue is small, as in mature costal cartilage (Kaplan & Fisher, 1964), although there may be sufficient loss for superficial fibrillation to develop in articular cartilage (Shaw & Lacey, 1973). However, if the functional demand is large, for example in embryonic tissue normally growing very actively (Barrett *et al.*, 1966) or in articular cartilage subjected to increased mechanical loading (Olah & Kostenszky, 1976), there is a more marked loss of glycosaminoglycan. If the matrix is depleted by administration of papain prior to treatment, then low doses of cortisone prevent recovery by inhibition of synthesis (Shaw & Lacey, 1973). With high doses given over a prolonged period, the loss of glycosaminoglycan is accentuated (Mankin *et al.*, 1972).

Some evidence suggests that cortisone depresses chondroitin sulphate rather than keratansulphate content in articular cartilage (Olah & Kostenszky, 1976); in costal cartilage, keratansulphate constituents may actually increase (Kaplan & Fisher, 1964). This may be consistent with the slow rate of turnover of keratansulphate and the apparent lack of any clear indication whether or how it is degraded (Leaback, 1974).

5

CHONDROCYTE NUTRITION, CARTILAGE BOUNDARIES AND PERMEABILITY

Any cell or organism is adequately nourished when its nutritional requirements are satisfied by the supply of nutrients. A full discussion of nutrition involves both aspects but there is little information as to the precise requirements of the chondrocyte. Rates of utilisation of oxygen and glucose (Bywaters, 1937) and sulphate (Maroudas & Evans, 1974) have been measured, but such *in vitro* studies may not accurately reflect metabolic activity *in situ* and, for example, it is not known how long the cells can survive in complete anaerobic conditions. Even less is understood as regards other nutrient materials. The macromolecular products synthesised and secreted by the chondrocyte indicate the types of material needed. For example, studies of cultured chondrocytes producing collagen in chemically defined media confirm that ascorbic acid is required (Levenson, 1969) and radioactive isotope studies *in vivo* demonstrate that amino acids, glucose and of course inorganic sulphate are incorporated by the cells. Further information may be gained from the study of various deficiency diseases.

However, as McKibbin (1973) points out, there is little direct information as to the precise form in which these metabolites, including sulphate, actually reach the cell in the tissues. Furthermore the nutrient 'pool' is derived from local degradation and turnover of the pericellular matrix as well as from nutrients entering the cartilage from external sources. Although the concentration of the intracartilaginous pool of sulphate, for example, may be calculated from physico-chemical data and must be taken into account when assessing precursor incorporation rates (Maroudas, 1973), we know little about the relative contributions from endogenous and exogenous sources or of local variation in the tissue. In the immature animal, of course, growth requirements demand a plentiful supply of exogenous nutrient. It is probable in the adult that most of the exogenous contribution is required for energy production to drive synthetic reactions for cell maintenance and matrix turnover, although losses by diffusion from the tissue as a whole must also be taken into account. Hence it is practicable to restrict discussion of chondrocyte nutrition to the sources of exogenous nutrients only and their access to the chondrocyte.

In this context, therefore, chondrocytes obtain all of their nutrients by diffusion from blood vessels, like most other tissues in the body. Cartilage is peculiar in that its blood vessels lie at the periphery of the tissue or completely outside it. In contrast with vascular tissues, such as muscle, nerve or even bone, where the maximum distance between blood vessel and cell is of the order of 0.1 mm, the chondrocyte survives in blocks of tissue where the maximum diffusion distance can be several millimetres or even centimetres. The low respiratory activity of the chondrocyte and low cellularity of cartilage are related to these conditions, which indeed are produced by the cells themselves during growth. In order to understand the mechanism by which even this small number of relatively inactive cells are sustained it is necessary to consider first the boundaries of the tissue since they contain the nutritional sources and often are of considerable intrinsic interest. Secondly, the characteristics of diffusion within the cartilage regulate the types of molecule which can reach the chondrocyte; this is reviewed in the context of articular cartilage nutrition on which much of the relevant investigation has been done.

TYPES OF BOUNDARY

The structural and functional characteristics of the cartilage tissue and its boundaries must satisfy mechanical as well as nutritional requirements at all stages of development and during adult life. In different hyaline cartilages at least, the mechanical role and properties and hence the structural characteristics of the tissue itself are broadly similar but there can be quite marked differences between the types of boundary. It is common knowledge that cartilage is either surrounded by perichondrium or has a calcified surface or possesses a smooth articular margin. These boundaries are essentially interfaces where cartilage abuts on other solid tissues, i.e. loose connective tissue, bone or cartilage itself, a predominance of dense fibrous tissue, hydroxylapatite mineral or fluid occurring either in the cartilage or between it and the apposing tissue (Fig. 58). The type of boundary may be important biomechanically but from the point of view of nutrition the significant aspects to note are the degree of vascularity and the nature and thickness of the tissues intervening between the blood vessels and the cartilage, i.e. the blood–cartilage barrier. Different surfaces of a cartilaginous mass may exhibit quite dissimilar characteristics and it is then important to know the relative contribution which each makes to the nutritional supply of the chondrocytes.

FIBROUS BOUNDARIES

THE PERICHONDRIUM

This is the most well-known boundary of cartilage and arguably that which has been investigated least, in respect of its nutritional properties. The perichondrium is the only boundary of many cartilages of the upper respiratory tract and of the ear. Conventionally, two strata are described – an outer fibrous and an inner cellular chondrogenic layer. The outer layer contains many elastic as well as collagenous fibres (Amprino & Bairati, 1933)

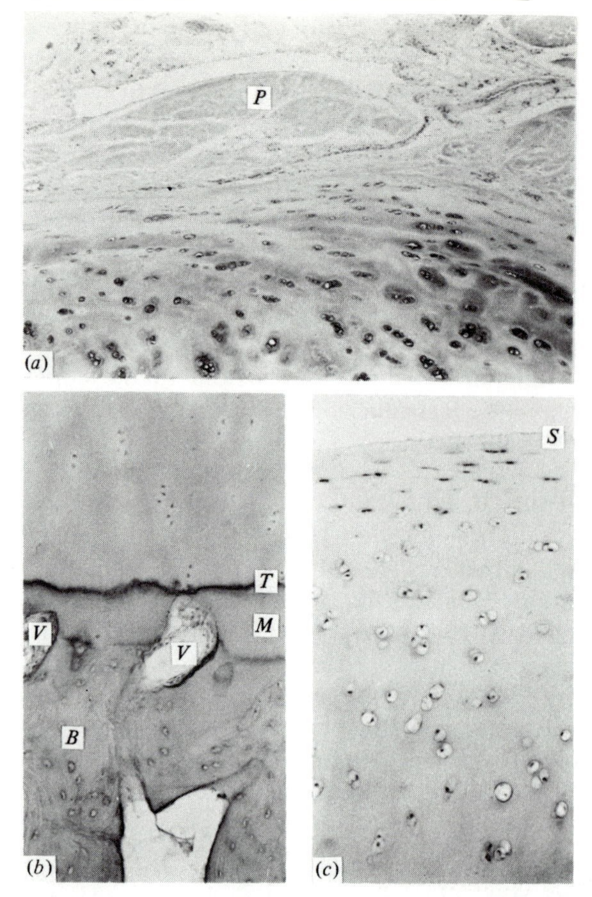

Fig. 58. Human cartilage, formalin fixation. (*a*) Fibrous boundary formed by the perichondrium (*P*) of costal cartilage (32 years). Note the blood vessels in the perichondrium, magnification ×32. (*b*) Osseochondral boundary at the junction of articular cartilage (28 years) with the subchondral bone (*B*). The tidemark (*T*) lies at the interface of calcified (*M*) and uncalcified cartilage. Note the vascular spaces (*V*) in the subchondral bone and calcified cartilage, magnification ×80. (*c*) Articular surface (*S*) where hyaline cartilage is bathed in synovial fluid (20 years), magnification ×100.

and blends with the surrounding connective tissue. The inner layer merges imperceptibly with the subperichondrial cartilage. The perichondrium varies in thickness and in immature tissue the chondrogenic layer is thicker. In the adult a plane of cleavage is found between the cartilage and the fibrous layer, which can often be stripped off. The tensile stresses which tend to cause distortion of cartilage grafts taken from costal and nasal cartilage are resident in the subperichondrial cartilage and not in the perichondrium (Gibson & Davis, 1958; Fry & Robertson, 1967).

The perichondrium is acidophilic and the subperichondrial region is less basophilic than the rest of the cartilage (Fig. 38): both stain well using the PAS reaction (Galjaard, 1962; Stockwell & Scott, 1965), indicating glycoprotein material. Chemical analysis of horse nasal septal cartilage (Szirmai *et al.*, 1967) confirms that total hexosamine (about 2% of the dry weight), particularly as chondroitin sulphate, is lowest in the perichondrium and increases on passing through the subperichondrial zone toward the centre of the tissue. Collagen and non-collagenous proteins are a maximum in the perichondrium, where each forms 45–50% of the dry weight in the aged animal although there is less non-collagenous protein in the young adult. The perichondrium also has a higher water content (77–79%) than the cartilage (65%) itself.

The chondrocytes of the subperichondrial zone become flattened on approaching the perichondrium, lying tangential to the margin of the cartilage. In the perichondrium, the cells are fibroblastic, lying within dense collagenous tissue. In adult and late immature cartilage, the cell density increases on passing from the middle of the cartilage toward the perichondrium (Galjaard, 1962) and rises to about 22000 cells per mm^3 in the subperichondrial zone of human costal cartilage. This cell distribution is in accord with presumptive nutrient concentration gradients declining from the perichondrium to the centre of the cartilage.

The perichondrium is vascular, blood vessels lying principally in the outer layer although occasionally in contact with the subperichondrial region (Fig. 58*a*). Since only fibrous tissue with a high water content lies between the vessels and the cartilage, the perichondrium may be regarded as a container for nutrient fluid bathing the subperichondrial cartilage. The main variable determining the supply of nutrients is likely to be the degree of vascularity, although blood flow rate and the density of the fibrous tissue may also be significant. There seem to be no systematic descriptions of the arrangement of blood vessels in the perichondrium or data concerning variations of vascularity in different cartilages, species or during maturation and ageing. Harris (1933) attributes the lack of calcification in nasal septal cartilage to the proximity of the highly vascularised mucoperiosteum. Probably considerable differences in blood supply occur between the cartilages of the body but these have yet to be assessed and correlated with variables such as cartilage thickness and cell density.

Cell density and dimensional data are available for human costal cartilage (Stockwell, 1966). There are about 10^4 nuclei in a complete 10 μm thick transverse section of the adult cartilage of mean radius 5.3 mm. It may be calculated (allowing for nuclear height/cell count correction) that the number of cartilage cells per mm^2 of perichondrium surface area is about 27000; this may be the number of chondrocytes in the block of tissue which can be adequately nourished by blood vessels in that portion of the perichondrium over the diffusion distances operating in adult costal cartilage. However, in

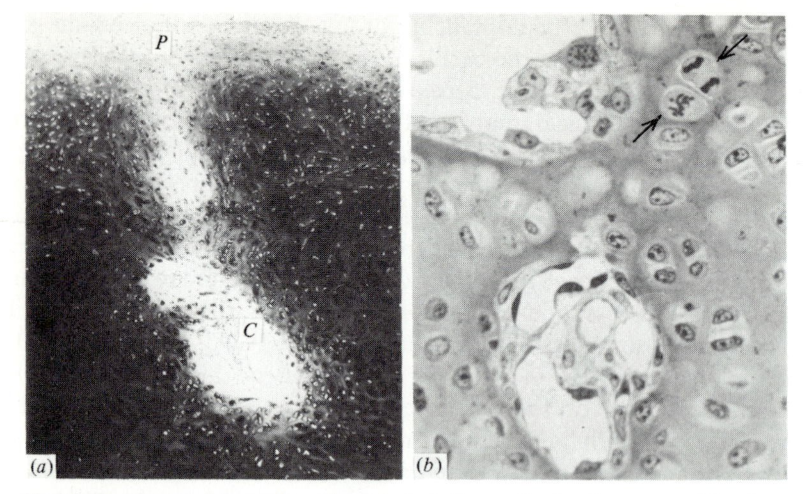

Fig. 59. (*a*) Human costal cartilage (six weeks). Formalin fixation, magnification $\times$ 32. Cartilage canals (*C*) near the site of continuity with the perichondrium (*P*). Each contains several blood vessels and there is diminished basophilia at the canal margin. (*b*) Sheep fetal cartilage, glutaraldehyde fixation, 1 μm araldite section, stained with toluidine blue and pyronin B, magnification $\times$ 410. Mitoses (arrows) near cartilage canals.

infant costal cartilage, the number of cells per mm^2 of perichondrium is much greater (72000) since although the cartilage is much smaller (mean radius 2 mm), the number of cells in a complete transverse section is not very different from that of the adult.

In some situations in immature cartilage, there may be no barrier whatsoever between cartilage matrix and blood vessels lined only by endothelium. The blood vessels concerned lie in cartilage canals which are derivatives of the perichondrium. Therefore in a nutritional context it is appropriate to describe these structures as 'fibrous boundaries' although they lie wholly within the cartilage.

CARTILAGE CANALS

Any large chondro-epiphysis will bleed from its cut surface if sliced open prior to ossification, though it is entirely cartilaginous. Vascular channels, known as cartilage canals, course through the hyaline cartilage of epiphyses and small short bones prior to ossification and are found in permanent cartilages such as costal and laryngeal cartilages into adult life (Fig. 59). They are present in mammals, birds, certain lizards and possibly amphibia (Haines, 1942), although the tiny blocks of cartilage found in small mammals such as rats and mice may be avascular. The fact that fetal cartilage is vascularised has long been known – Haines (1974) assigns priority to Nesbitt (1736) – yet several authorities have chosen to ignore the cartilage canals or to discount their relevance to the physiology of cartilage. Even so, although the canals have stimulated considerable interest, their role and mode of formation in growing cartilage still remain more a matter of conjecture than of proof.

The number of canals in the cartilage models of epiphyses and small bones varies; in the proximal tibia more of them are present in larger than in smaller animals (Levene, 1964). There are species-specific patterns of canal entry and ramification within a single epiphysis and it is of interest that the arrangement of cartilage canals corresponds to, and is the forerunner of, the anatomical distribution of blood vessels in the adult epiphysis (Brookes, 1971). All the canals enter from a surface covered by perichondrium, usually from grooves and sulci, for example the ossification groove in an epiphysis. They do not usually enter from areas of attachment of ligaments and tendons (Hurrell, 1934; Haines, 1937) although Gray & Gardner (1950) describe entry via the cruciate ligaments in the knee: these are shorter than canals from other origins. The articular surface is not penetrated by canals.

Within an epiphysis the lengths of canals vary considerably; anastomoses between canals rarely occur although they have been described in the goat proximal tibia between the members of a closely related group (Levene, 1964). Except in the mandibular condyle, where canals make close approaches to the surface of the articular fibrocartilage (Blackwood, 1966), all observers are agreed that the canals do not at any time enter the region of the presumptive articular cartilage (Hurrell, 1934). Many of them turn down toward the resting zone of the growth cartilage and 'communicating canals' may be found running from the cartilaginous epiphysis into the metaphysis: a sheath of hyaline cartilage persists around the canal as it goes through the growth plate and the primary subepiphysial trabecula (Haines, 1933; Brookes, 1958). Such canals are attributed to extension of shaft ossification into the region of cartilage containing the canal, so effecting continuity with the shaft marrow spaces (Hurrell, 1934). Haines (1937) describes 'tunnel canals' running from one perichondrial surface to another.

Each canal contains a leash of capillaries at its blind end, the tip often expanding into a 'glomerulus' (Wilsman & Van Sickle, 1972). The capillaries are supplied and drained by an arteriole and venule which run through the centre of the trunk of the canal (Fig. 59) from the perichondrium; capillaries lie around the artery and vein along the course of the canal. Occasional lymphatics (Wilsman & Van Sickle, 1972) and unmyelinated nerves are also present. In general, ultrastructural observations of canal structure confirm reports from light microscopy. In sheep fetal epiphyses (Stockwell, 1971*a*), the capillary endothelium is thin (0.04 μm) and contains pores 60 nm wide covered by diaphragms; occasional larger discontinuities are found. A variable number of structures lie at the interface of blood vessels and surrounding cartilage. At some sites along the whole course of the canal but especially at the tips, capillary endothelium abuts directly on hyaline cartilage matrix, normal chondrocytes lying within 5 μm of the endothelium (Fig. 60*a*):

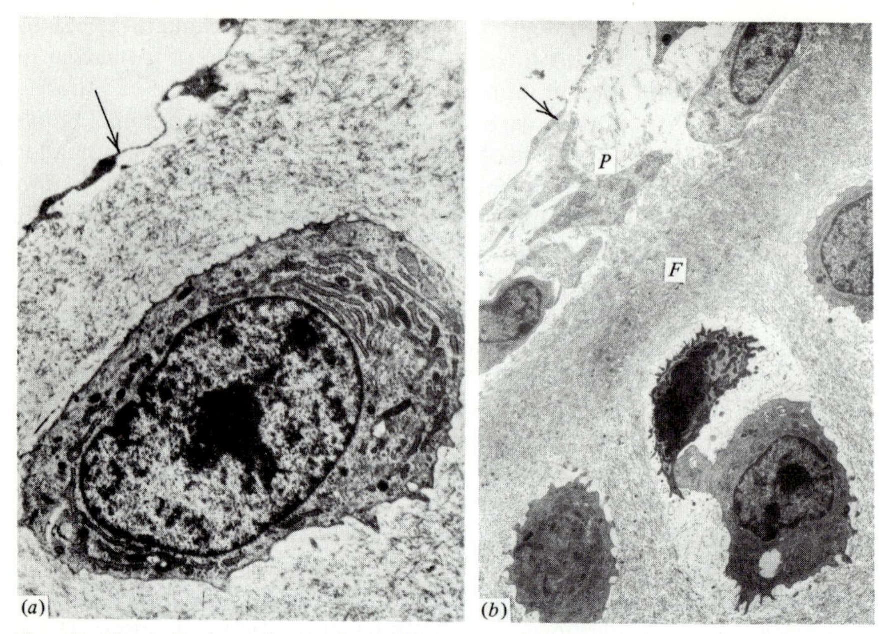

Fig. 60. Sheep fetal cartilage, glutaraldehyde fixation. Ultrastructure of boundaries between cartilage and blood vessel lumina in cartilage canals. (*a*) Endothelium only. A normal chondrocyte lies within 5 μm of the endothelium (arrow); note its relatively smooth contour on the side nearest the canal and that the fibrous matrix abuts on the cell margin, magnification ×6000. (*b*) Cartilage cells and matrix separated from the blood vessel lumen by endothelium (arrow), pleomorphic cells (*P*) and dense fibrous tissue (*F*), magnification ×3000. (From Stockwell, 1971*a*.)

these cells show a smooth contour with few processes on the side facing the endothelium. At other sites, an internal layer of polymorphic cells and an external sheath of collagen fibres (25–50 nm in diameter) are interposed between endothelium and hyaline cartilage forming a barrier 5–10 μm thick (Fig. 60*b*). Degenerate cells or chondrocytes with extremely dilated ER cisternae may be found just outside this type of junction.

Origin of the cartilage canals

There are two quite separate beliefs concerning the mode of formation of the vascular canals, both dependent almost exclusively on morphological evidence. Haines (1933, 1937) proposes the inclusion hypothesis: at various points around the periphery of the cartilage, blood vessels and connective tissue of the perichondrium become passively enclosed by cartilage as the girth of the cartilaginous mass increases by subperichondrial appositional growth (Fig. 61). This suggestion has numerous opponents who favour the hypothesis of active ingrowth whereby vascular channels invade cartilage tissue already

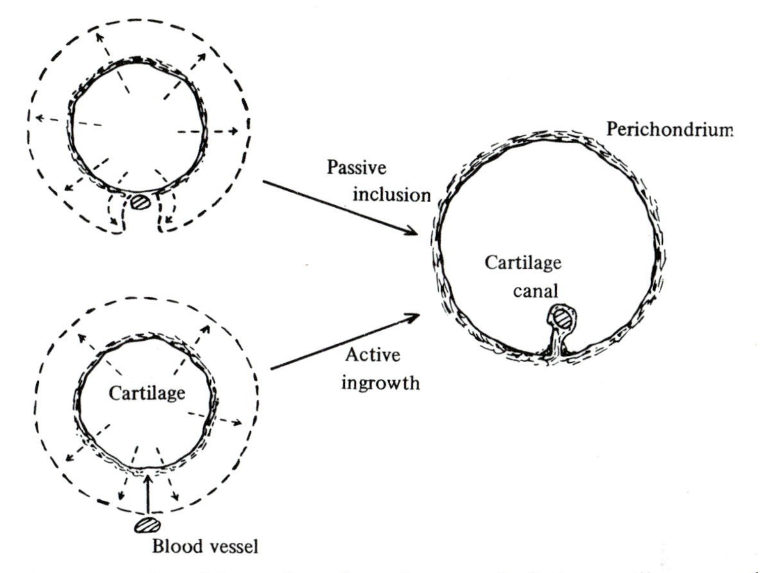

Fig. 61. Possible modes of formation of cartilage canals during cartilage growth.

in situ. Those who accept the passive inclusion hypothesis have to explain the patterns of growth and branching of the canals formed during development, while the hypothesis of active ingrowth has to define the mechanisms whereby invasion occurs.

It must be stated that the first hypothesis does not seem to account completely for the geometry and growth of the canal pattern: some authorities would consider the published evidence sufficient grounds to refute Haines' theory. Nevertheless in many situations, the acquisition of complexly branched systems of canals or even the formation of tunnel canals can be accounted for by appositional cartilage growth, by a certain amount of cartilage 'flow' internally or by 'injection' of new cartilage into old channels (Haines, 1937). However, during the growth of the proximal tibia, epiphysial width increases more than the length increment of the peripheral cartilage canals and much of the extra width is vascularised by elongation of the branches of central canals (Levene, 1964). It is difficult to see how this can be accounted for except by invasive growth of the central branches. Similarly in double ('back to back') growth cartilages, such as the spheno-occipital synchondrosis (Moss-Salentijn, 1975), canals form and elongate during pre-natal life although there is no enlargement of the cartilage as a whole. However, we need to learn more about the distribution of cell proliferation and patterns of cartilage growth in pre-ossification stages, before such examples can be regarded as critical tests of the inclusion hypothesis, especially as there is no means of following the changes in a single epiphysis throughout development.

On the other hand, the evidence relating to mechanisms of active invasion is not really satisfactory. Haines (1933) believed active ingrowth to be unlikely. He considered that cartilage could be removed only in certain situations, for example either after matrix calcification and cell degeneration, as in the growth plate, or as the result of leucocyte action, as in the resorption of normal hyaline cartilage seen in the metamorphosis of the tadpole tail. Pathologically in rheumatoid arthritis, pannus tissue erodes the superficial aspect of articular cartilage apparently through leucocyte activity and a similar mechanism is cited in the normal removal of cartilage in the trochlear notch of the human ulna (Haines, 1976). While some have described the cartilage canals as entering only degenerate regions of cartilage, in the normal course of events the canals come to occupy normal hyaline cartilage. If indeed the canals actively invade the tissue, then liquefaction of cartilage must occur at the site of the ingrowth. There are three possible agents which might facilitate this process. These are: (i) the vascular endothelium; (ii) the pleomorphic cells at the periphery of the canal; (iii) the chondrocytes themselves.

Vascular endothelium was early considered as a candidate (Hurrell, 1934). Thus, capillaries on the metaphysial side of the growth plate appear to erode cartilage, including some uncalcified tissue (Brookes & Landon, 1964; Schenk, Spiro & Wiener, 1967), although this is preceded by changes in at least the proteoglycan moiety of the matrix. Endothelium is known to contain fibrinolysin activators (Astrup, 1969). However, although endothelium at the tips of the canals is exposed to the hyaline matrix, there are no marked ultrastructural changes either in the matrix or in nearby chondrocytes (Stockwell, 1971*a*). Histiocytic action by cells in close contact with the matrix at the front and lateral walls of the canal is postulated by Anderson & Matthiessen (1966) but their only illustration shows histiocyte enzyme activity in the middle of the canal. Stump (1925) proposed that chondrocyte activity resulted in dissolution of chondromucin, the freed chondrocytes merging with the invading tissue. Watermann (1961) also believes that cartilage cells lyse themselves free of their matrix in the path of the incoming canal. At the level of the light microscope several authors have noted round or elongated cells at the margin of the canal which appear to be partly in the canal wall proper and partly in the surrounding cartilage. These have been variously interpreted as liberated chondrocytes becoming incorporated into the mesenchyme around the canal or as the differentiation of mesenchymal cells into chondrocytes (Hurrell, 1934; Haines, 1937; Andersen, 1962; Lufti, 1970*a*; Moss-Salentijn, 1975). Such transitions are less likely where a fibrous sheath is present external to the cellular element of the canal wall. Further evidence is required before any of these mechanisms can be accepted.

The problem of the mode of formation of the cartilage canals may appear to be of an academic and perhaps trivial nature. However, the relationship of cartilage to blood vessels is relevant to the control of tumour vascularisation. Recent investigations of normal tissues have shown that hyaline, though not calcified, cartilage resists penetration by vascular mesenchyme (Eisenstein *et al.*, 1973). Tumours induce capillary proliferation and ingrowth by releasing a diffusible 'tumour angiogenic factor' which is mitogenic to capillary endothelium; if cartilage fragments are interposed between tumour and vascular tissue, no vascularisation of the tumour occurs (Brem & Folkman, 1975). It has been demonstrated that the cartilage fragments produce a factor which inhibits capillary proliferation directly; the inhibitory factor is not proteoglycan or collagen (Brem & Folkman, 1975; Eisenstein *et al.*, 1975). Boiled cartilage is ineffective and viable cartilage produces the factor continuously (Brem & Folkman, 1975) suggesting that it is a product of cell activity. Similarly isolated chondrocytes, separated from their matrix by enzymatic digestion, inhibit lymphocyte-induced angiogenesis and this effect is specific for viable chondrocytes (Kaminski, Kaminski, Jakobisiak & Brzezinski, 1977). The inhibitory factor has a molecular weight of 14000–18000 (Langer *et al.*, 1976) and it is postulated that it is a protease inhibitor (Eisenstein *et al.*, 1975).

Any hypothesis of cartilage canal formation must take into account the results of these studies on vascular inhibition, which provide some support for the passive inclusion theory. If active ingrowth occurs, a first step must be the local elimination of the inhibitory factor, either by chondrocyte or canal wall mesenchyme activity. It seems unlikely that the endothelium could be the primary tissue involved.

Nutritive function

Whatever the mode of their formation, the morphology of the canals suggests that they are essentially extensions of the perichondrium into the interior of a mass of cartilage. Hence it is logical that the canals should have a nutritive and a chondrogenic (see Chapter 7) function. Most authors consider that the cartilage canals have a nutritive role, although there is only morphological evidence for this view. Haines (1933, 1937) believes that this is their main function, since the presence and number of canals in a cartilaginous mass is in general related to its size.

Although small structures such as the sesamoids also contain cartilage canals in man, it cannot be concluded from this that the presence of the canals is not a reflection of the necessity of nutrition of large blocks of cartilage (Gray, Gardner & O'Rahilly, 1957; Gardner, Gray & O'Rahilly, 1959). It is true that the overall size of a cartilaginous mass is unlikely to be the only factor involved. However, measurement of published illustrations indicates that the width of the sesamoids (0.5 mm) at initial vascularisation is of the same order as in other cartilages in man, though at the lower end of the range (0.5–2.0 mm); a similar range occurs in other species. Nevertheless other factors must be taken into consideration and these do not appear to have been investigated. Thus variation in the cell density and in the degree of vascularity of the surface of different cartilages may account for the range of size at the commencement of vascularisation.

There is little quantitative work correlating canal density with the nutritional requirements of the surrounding cartilage. However, in most of the cartilage of the puppy proximal humerus, Wilsman & Van Sickle (1972) find that the mean distance between canals remains constant at about 1.4 mm during the first week of post-natal life. It is noteworthy that this distance falls within the range of diameters of cartilages found at the onset of vascularisation. Hence the tissue which each cartilage canal can nourish adequately probably lies within 0.7 mm of the canal. Wilsman & Van Sickle find a similar interval (0.65 mm) between the articular surface and the most superficial canal glomerulus at birth, increasing during the first week to 1.04 mm. They attribute the increase to an improved nutritive flow via the articular surface, as a result of increasing mobility of the joint and circulation of synovial fluid after birth. Haines (1937) finds a similar distance (0.5–0.6 mm) between the human patellar articular surface and the tips of the subjacent cartilage canals pre-natally: this remains constant for a long period but increases to about 1.0 mm at birth. In the puppy humerus near the growth plate, where a higher metabolic demand might be expected, the interval between canals is shorter (1.15 mm) than elsewhere. The canals may also provide stem cells for growth at this location (Lutfi, 1970*a*, *b*; see Chapter 7).

It is possible to use Wilsman & Van Sickle's (1972) data to estimate the approximate number of cells which can be nourished by the canals. Consider a canal 0.2 mm in diameter (from illustrations) forming the axis of a cylinder of tissue of radius 0.7 mm.

Volume of tissue per mm length of cylinder = 1.54 mm^3
Surface area of canal per mm length = 0.63 mm^2
Volume of tissue per mm^2 canal surface area = 2.44 mm^3

Since cell density (D. C. Van Sickle, personal communication) is approximately 120000 cells per mm^3, about 0.30×10^6 cells are nourished from 1 mm^2 of canal surface. Measurements of cell density, canal diameter and density in sheep fetal epiphyses indicate a similar value (0.28×10^6 cells per mm^2 of canal surface); although the cell density is much higher in the sheep fetus than the newborn puppy the canal profiles are closer together (R. A. Stockwell, unpublished).

Relation to ossification

The establishment of the ossification centre was early believed to be the sole function of the canals. However, although they eventually furnish the vascular element for such centres there is no evidence that this is their primary function. Thus in the distal skeletal elements of human limbs, cartilage canals appear in a definite sequence (Gray *et al.*, 1957; Gardner *et al.*, 1959) but this is not related to the order of chondrification or ossification. In any case the canals

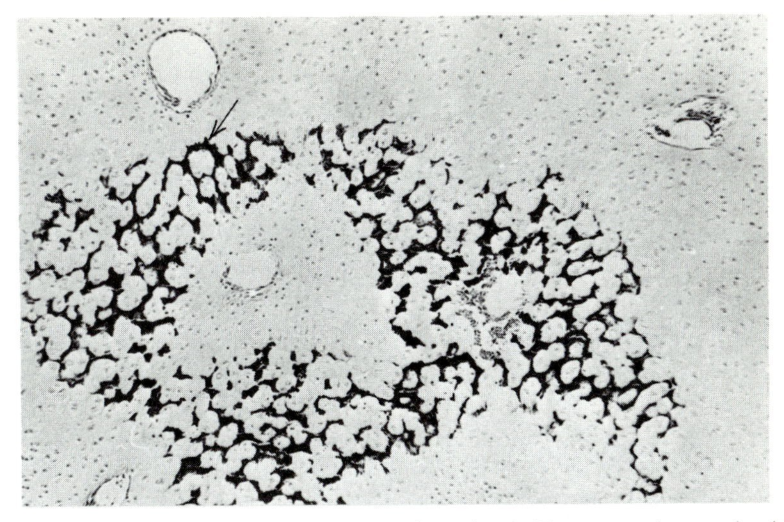

Fig. 62. Canine cartilage, upper humerus (four days). Frozen section, stained Von Kossa, magnification ×37. Location of early matrix calcification (arrow) in relation to cartilage canals. (From Wilsman & Van Sickle (1970).)

can exist in blocks of cartilage for many years before ossification commences, a good example being the trochlea of the humerus (Haraldsson, 1962). Furthermore, calcification prior to ossification can occur where cartilage canals are not present, as in certain rat epiphyses (Haines, 1933); in man, cartilage canals appear in both ends of the metacarpals and phalanges but one end is usually ossified from the shaft.

Although not primarily concerned with ossification, it has been suggested that the distribution of canals in an epiphysis may determine the position of the ossification centre (Haines, 1933, 1974). It has long been argued that areas of calcification found prior to ossification occur in poorly nourished regions (Parsons, 1905). Haines considers that an avascular lamina of cartilage is established between groups of canals and this becomes the initial site of calcification; the subsequent ossification centre is supplied by the canals on either side of the lamina. However, many authors find no evidence of large avascular laminae, the secondary centre occurring instead within vascularised regions (Hintzche, 1928; Waugh, 1958; Haraldsson, 1962; Lutfi, 1970*a*). In the sheep proximal tibia Levene (1964) finds two regions where groups of canals meet, but contrary to Haines' argument only one ossification centre develops, among the branches of the canals. In the puppy humeral head, Wilsman & Van Sickle (1970) find that multiple foci of calcification constitute the first evidence of ossification (Fig. 62), each focus lying immediately adjacent to the end of a canal.

From this evidence it seems unlikely that cartilage canals determine the site of the ossification centre. Nevertheless, the cartilage changes leading to

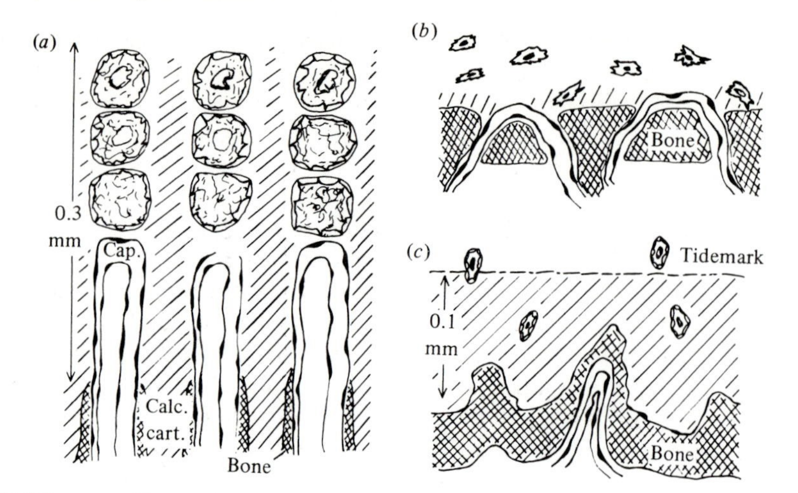

Fig. 63. Diagrams illustrating three types of osseochondral junction. (*a*) A thick lamina of calcified cartilage penetrated by numerous vascular spaces or capillaries (Cap.) apposed to columns of uncalcified cartilage, as at the metaphysial side of the growth plate. (*b*) A thin lamina of calcified cartilage penetrated by numerous blood vessels, as at the epiphysial side of the growth plate. (*c*) A thick lamina of calcified cartilage where vascular penetration is sparse or absent as at the deep surface of articular cartilage.

ossification may be located initially where nutritional conditions are poorest. Although an early stage in the process of endochondral ossification, calcification is a late stage relative to the life of the chondrocyte. Except at certain sites where calcification is very slow, mineralisation of the matrix is preceded by cell hypertrophy, matrix vesicles being formed even earlier (Chapter 8). Hence cell hypertrophy, rather than calcification itself, may be the 'marker' which is more relevant to the problem of the canals and ossification. Hurrell (1934) notes in the human carpus and tarsus that zones of swollen cells are separated from the canal margin by 20–30 cell diameters, i.e. not less than 0.25 mm, and although Lutfi (1970*a*) states that foci of hypertrophic cells appear close to the canals, his illustrations demonstrate a distance of 0.3–0.4 mm, midway between canal profiles.

It is probable that the canals then invade the degenerate and calcifying regions. Thus although the centre is in general formed within the vascular region of the cartilage, changes prior to the formation of the centre within that region appear to be in the parts most remote from the canal.

OSSEOCHONDRAL BOUNDARIES

Junctions of cartilage and bone are found in a number of situations in the body and normally a lamina of calcified cartilage separates the uncalcified cartilage from the bone. The calcified lamina is generally accepted as having the mechanical function of providing uncalcified cartilage with a firm attachment to the underlying bone. However, it also constitutes an impenetrable barrier to diffusion unless perforated by vascular channels. Collins (1949) believed that the stability of cartilage as a tissue could only be served if a continuous imperforate calcified stratum lay between cartilage and the vascular bone.

There is, however, considerable variation in the degree of vascularity, of this type of boundary and in the thickness of the calcified lamina (Fig. 63). In the immature animal, rapid endochondral ossification can occur at this type of junction; the boundary is then extremely vascular but has a thick lamina of calcified cartilage, for example at the metaphysial side of the growth plate. Where endochondral ossification is quiescent and the primary requirement appears to be nutrition, the boundary is still vascular but the calcified lamina is quite thin, as in the epiphysial side of the growth plate or in the hyaline cartilage endplates of the intervertebral discs. Where bone growth and the need for a nutritional source are small or absent, there are few blood vessels and a moderately thick calcified lamina, as in the deep zone of articular cartilage.

THE GROWTH PLATE

The epiphysial disc or cartilaginous growth plate lies between the epiphysis and the shaft of a long bone. The chondrocytes are arranged in columns parallel to the long axis of the bone and five zones may be recognised (Fig. 64) in them. Passing from the epiphysial to the diaphysial side of the plate, these are:

(i) resting or reserve zone, containing stem cells;
(ii) proliferative zone, with stacks of flattened cells;
(iii) hypertrophic zone, with enlarged chondrocytes;
(iv) degenerative zone, with calcified matrix;
(v) zone of erosion, with capillary loops.

Unless a communicating cartilage canal passes through the growth plate, it is avascular. However, it is richly supplied with blood vessels at its boundaries. There are three surfaces of the growth plate and at each a bone–cartilage junction occurs. These are:

(i) the epiphysial surface;
(ii) the metaphysial surface;
(iii) the circumferential surface enclosed by the ossification groove (Fig. 65).

Epiphysial surface

A thin plate of bone six to eight lamellae thick (Trueta & Morgan, 1960) abuts on the cartilage of the growth plate, connected by trabecula to the subarticular

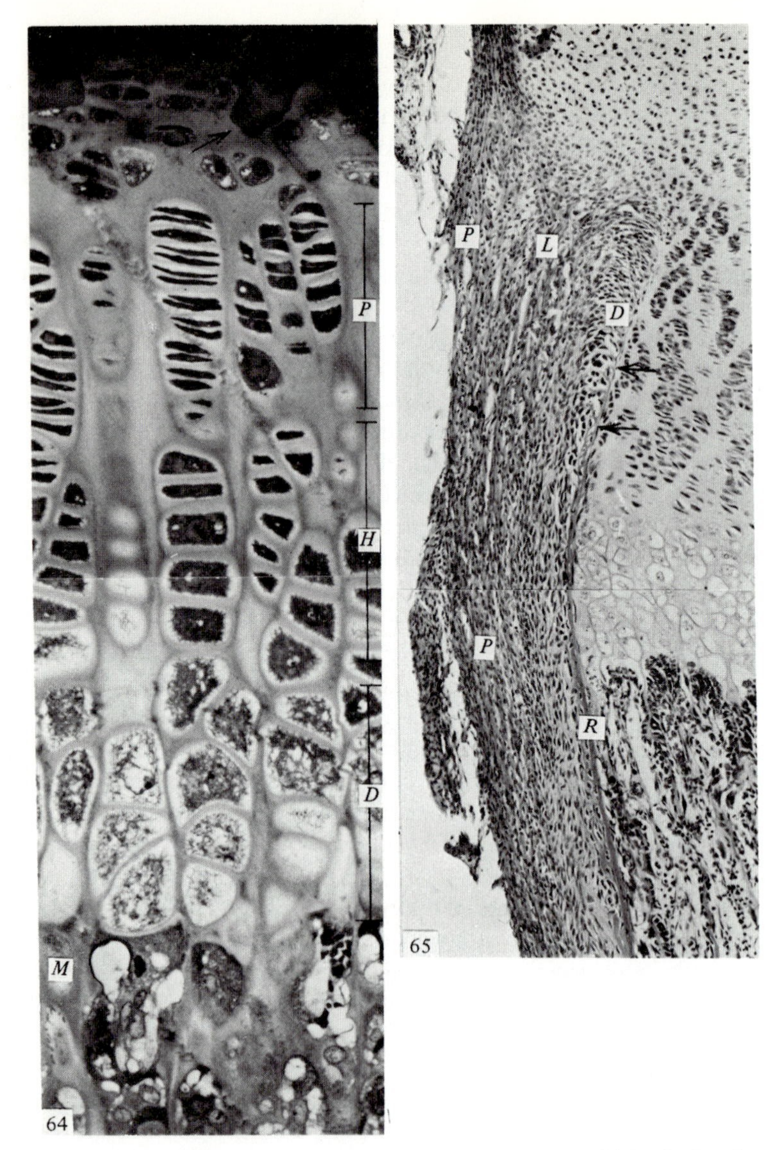

Fig. 64. Rat upper tibial growth plate (four weeks). Glutaraldehyde fixation, 1-μm araldite section stained with toluidine blue and pyronin B, magnification ×320. A zone of calcification (arrow) at the epiphysial side of the growth plate encroaches on the reserve zone lying next to the proliferative (*P*) region of the cell columns. Chondrocytes enlarge in the zone of hypertrophy (*H*); further increase in cell and lacunar height occurs in the zone of degeneration (*D*). Capillary profiles (*M*) are seen in the zone of erosion at the metaphysial side of the plate.

Fig. 65. Rabbit lower femur (newborn). Formalin fixation, magnification ×80. Circumferential boundary of the growth plate. A thin shell of bone (*R*) extends as far as the level of the epiphysial limit of the hypertrophic zone. Beyond this, the margin of the growth plate is limited from within outwards by a collagen lamina (arrows), a zone of densely packed cells (*D*) and a zone of loosely packed cells (*L*) containing blood vessels. The perichondrium-periosteum (*P*), firmly attached to the chondro-epiphysis, ensheathes the whole of the ossification groove.

bone and the rest of the bony epiphysis. There is a thin lamina of calcified cartilage next to the bone plate (Trueta & Little, 1960). However, the junction is densely perforated by epiphysial blood vessels which pass through the calcified layer as far as the reserve zone of the plate (Fig. 64). The terminal expansions of the vessels cover an area corresponding to 4–10 cell columns, forming an almost continuous ceiling for the growth plate (Trueta & Morgan, 1960). There is no difference in vascularity between the central and peripheral parts of the growth plate and casual estimates show that as much as 50% of the interface may be occupied by vascular contacts in young animals. As development proceeds, the degree of vascularity diminishes (Kember, 1973) together with the thickness of the whole plate and the rate of cell proliferation.

Metaphysial surface

The zone of provisional calcification extends three to four cell diameters from the last transverse septum in the cell column on the metaphysial aspect (Trueta & Morgan, 1960) but foci of apatite crystals may be seen with the electron microscope at higher levels in the columns (Bonucci, 1970). The zone of provisional calcification is not completely mineralised, however. Mineralisation is observed predominantly in the central portions of the matrix between the cell columns and only two-thirds of these longitudinal septa are mineralised (Schenk *et al.*, 1967). The pericellular matrix, including the transverse septa and the peripheral parts of the longitudinal septa, are not affected although spurs of calcification may project into the margins of the transverse septa. Metaphysial calcified matrix is more radio dense than immature cortical bone (Owen, Jowsey & Vaughan, 1955).

The last transverse septum of the cell columns forms the limit of vascularisation. The capillary loops have a narrow ascending and a wide descending limb (Irving, 1964). In the fetus, these may consist of clumps of sinusoids (which erode some of the longitudinal septa) but in older growth plates are probably single loops (Brookes, 1971). At the apex of the loop the endothelium apparently ruptures (Fig. 66) and a microhaemorrhage occurs into the newly opened lacuna of the dead chondrocyte (Schenk *et al.*, 1968), the endothelium subsequently re-forming. Erosion of uncalcified cartilage is associated with the capillary endothelium and removal of calcified cartilage is facilitated by chondroclasts (Schenk *et al.*, 1967); the transverse septa are removed first, followed by the rest of the uncalcified cartilage. This leaves longitudinal bars of calcified cartilage projecting towards the shaft of the bone (Fig. 63*a*). Formation of bone commences at an appreciable distance from the diaphysial aspect of the last transverse septum; Serafini-Fracassini & Smith (1974) give a distance of 150 μm. This is equivalent to nearly one day's growth in the rat proximal tibia (Sissons, 1953), and the precise level probably varies with different growth rates.

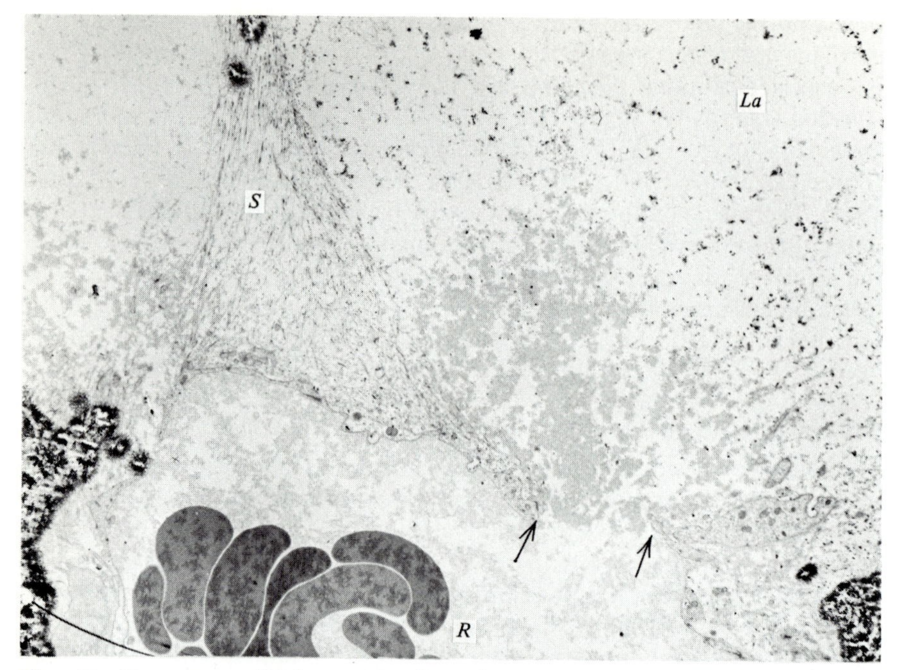

Fig. 66. Rat upper tibial growth plate (four weeks). Glutaraldehyde fixation, magnification ×3000. Rupture of metaphysial capillary endothelium (arrows) as the last transverse septum (uncalcified cartilage) of the cell column is eroded and plasma erupts into the lacuna (*La*). Note the mineralisation confined to the longitudinal septa (*S*). *R* = red blood corpuscles.

Circumference of the growth plate

The growth plate is surrounded by fibrocellular tissue in the 'ossification groove' (Ranvier, 1875) on the deep aspect of the vascular perichondrium-periosteum, itself firmly attached to the expanded epiphysis. However, the osseous 'perichondrial ring' (Lacroix, 1951) of the ossification groove is interposed, lying in immediate contact with the periphery of the cartilage (Fig. 65). This is the continuation of the periosteal tube of bone which keeps abreast of the zone of hypertrophy during growth in length of the diaphysis. Thus the perichondrial ring extends as far towards the epiphysis as the level of the zone of hypertrophy in the growth plate, but is cut off from the rest of the bony diaphysial wall by the resorption process at the metaphysis, although separation does not always occur in human long bones (Gray & Gardner, 1969). Thus the structures forming the circumferential boundary vary at different levels of the growth plate.

The most constant and continuous feature of the perichondrial ring is that the matrix contains coarse collagen fibres lying parallel to the long axis of the cell columns (Weidenreich, 1930; Pratt, 1957). These fibres extend toward the epiphysial side as far as the reserve zone, gradually diminishing in number

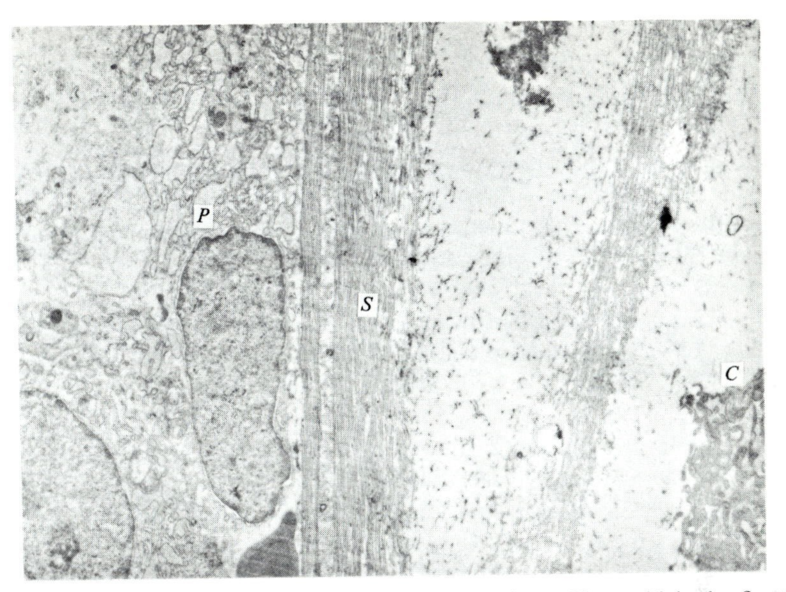

Fig. 67. Rat upper tibial growth plate (four weeks). Glutaraldehyde fixation, magnification × 5800. The collagen lamina (*S*) at the circumferential margin of the growth plate, beyond the epiphysial limit of the bony perichondrial ring. Part of a chondrocyte (*C*) lies in a lacuna on the inner aspect of the collagen lamina while densely packed cells (*P*) lie on its outer aspect.

and finally losing their orientation. At the epiphysial end irregularly shaped perichondrial cells lie external to the collagen lamina (Fig. 67); these become fragmented and degenerate on passing toward the metaphysis (Luxembourger, Malkani & Rebel, 1974). Mineralisation occurs first between cell fragments on the external aspect of the collagen lamina, eventually engulfing the perichondrial cell fragments near the metaphysis. The calcified layer is thickest at the level of the zone of erosion and is covered externally by osteoblasts and a varying amount of bone. Calcification in the perichondrial ring extends further toward the epiphysial end of the columns than it does in the longitudinal septa of the growth plate; small masses of crystals are seen at the level of the proliferative zone and matrix vesicles beyond this.

The chondrocytes at all levels on the internal aspect of the collagen lamina are elongated parallel to the long axis of the cell columns. Luxembourger *et al.* (1974) claim that few of these cells degenerate on passing toward the metaphysial side of the plate, and believe that they have a role in the organisation of the collagen lamina.

There is no doubt that the vessels running circumferentially (Shapiro, Holtrop & Glimcher, 1977) in the ossification groove nourish the periphery of the growth plate and the subperichondrial tissue beyond the epiphysial limit of the perichondrial ring. This is unlikely below the level of the ring since its calcified and bony part shows no evidence of perforation.

Nutrition of the growth plate

Of all the cartilages in the body, the definitive growth plate is most plentifully supplied with blood vessels. This is consistent with a high metabolic rate associated with its rapid cell proliferation and matrix synthesis. Yet despite the structural evidence of rich vascularisation on both epiphysial and diaphysial surfaces, experiments show that the metaphysial vessels have a negligible role in growth plate nutrition, even though, as discussed below, there seems to be no structural barrier to diffusion.

The discrepancy between structure and function was given prominence by the results of vascular ablation experiments carried out by Trueta & Amato (1960) which confirmed views held for a considerable period (Haas, 1917; Kistler, 1936; Foster, Kelly & Watts, 1951). If the epiphysial vessels are destroyed the growth plate dies, though peripheral tissue can survive presumably owing to the vessels in the ossification groove; bone bridges may be formed between epiphysis and diaphysis. By contrast, if the metaphysial vessels are destroyed, the growth plate does not die but increases in thickness several fold; erosion and endochondral ossification cease but are resumed if the surface becomes revascularised, the growth plate reverting to its normal thickness. Trueta & Amato conclude that the whole thickness of the plate is nourished by the epiphysial vessels and that the metaphysial vessels are concerned only with the process of endochondral ossification. Brashear (1963) and Fyfe (1964) have made similar observations supporting these conclusions.

Few doubt the importance of the epiphysial vessels in the maintenance of the proliferative zone of the growth plate. Although, as Brookes (1971) points out, epiphysial cartilage in the early fetal period grows quite satisfactorily before the epiphysial circulation is established, this is at a time when the chondro-epiphysis is very small and diffusion from the perichondrium is presumably adequate. It is possible to follow the parallelism between increase in size of the chondro-epiphysis and the changes in nutritional source: initially from the perichondrial vessels alone, later becoming supplemented by cartilage canals and finally vascularisation of the epiphysial side of the definitive growth plate by branches of vessels derived from the cartilage canals.

The problem really concerns the status of the metaphysial vessels and whether they nourish any part at all of the plate. It is obvious that they cannot nourish the whole growth plate since otherwise the proliferative zone would survive following ablation of the epiphysial vessels. However, even if the hypertrophic and calcified zone alone received their nourishment from the metaphysial vessels, loss of the proliferative zone following epiphysial ablation would still lead to degeneration of the whole plate since the hypertrophic zone would not be replenished as it became used up in endochondral ossification. Nevertheless, as Trueta & Amato (1960) claim, the formation of the abnormally long columns of cells following ablation of the metaphysial vessels

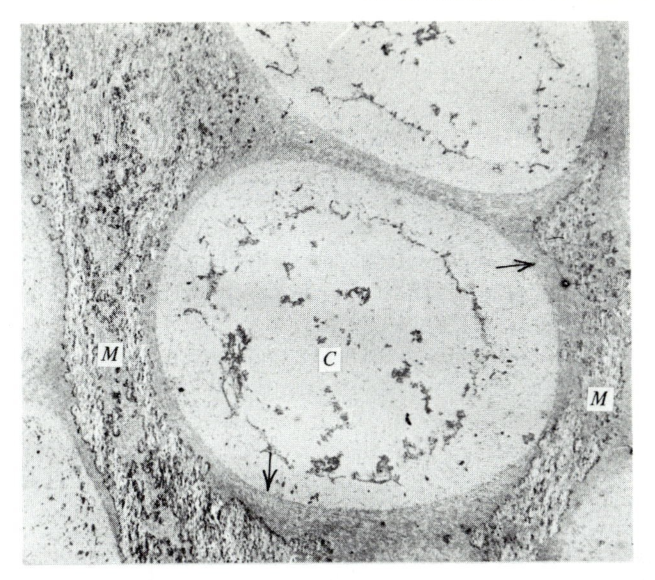

Fig. 68. Rat upper tibial growth plate (four weeks). Glutaraldehyde fixation, magnification ×1970. Lacunae in the degenerative zone. There is a narrow layer (arrows) of uncalcified cartilage between the lacuna containing the degenerate chondrocyte (*C*) and the mineral (*M*) in the longitudinal septum.

appears to discount the possibility that these vessels might have any useful nutritive role while at the same time demonstrating that the epiphysial vessels alone are capable of maintaining the vitality of the whole thickness of the normal plate.

However, it is fair to state that there are observations which are not altogether in harmony with this view. First there appears to be no structural barrier to diffusion. Consider the metaphysial surface in the context of nutrition instead of in terms of cartilage erosion and bone deposition. It consists of subchondral bone with a very high density of vascular perforations occupying most of the available interfacial area. The calcified cartilage zone is quite thick, about 0.3 mm, but does not form a continuous lamina. Cylinders of uncalcified cartilage containing hypertrophied and degenerate chondrocytes extend down from the main cartilage mass into the upper half of the calcified lamina. Subchondral vessels penetrate into the lower half of the calcified lamina and abut on the cylinders of uncalcified cartilage (Fig. 63*a*). Thus in reality there is little calcified cartilage to act as a diffusion barrier between the blood vessels and the uncalcified cartilage of the growth plate. In each column the diffusion pathway consists of (i) a central broad channel by way of the transverse septa and the degenerate and hypertrophic chondrocytes in their lacunae and (ii) a peripheral narrow path (about one-tenth of the total cross-sectional area) through the uncalcified part of the longitudinal septa (Fig. 68). If there is a 'barrier' to diffusion at the metaphysial side of the growth plate, the hypertrophic and degenerate cells themselves may be involved.

Experimental evidence also leaves no doubt that solutes can diffuse from the metaphysial vessels into the growth plate. Brodin (1955*a*) observes that fluorescent dye passes into the cartilage mainly from the metaphysial side. Others have used more physiological materials. After destruction or occlusion of the epiphysial vessels, Prives, Funshein, Shcherban & Shishova (1959) find that ^{32}P-phosphate passes across the growth plate from metaphysis to epiphysis; Fyfe (1964) notes that ^{35}S-sulphate penetrates as far as the middle of the hypertrophic zone. More recently, it has been shown that ^{3}H-adenine, possibly reaching the diaphysis as ATP, penetrates into the growth plate as far as the epiphysial limit of calcification (Goodlad, Stuart & Smith, 1972, quoted by Serafini-Fracassini & Smith, 1974).

The situation regarding nutrient sources is also quite different around the secondary ossification centre in the epiphysis. Here the eroding vessels undoubtedly have a nutritive role: infarction leads to degeneration of the overlying cartilage and cessation of the chondrocyte proliferation which maintains growth of the epiphysis (McKibbin & Holdsworth, 1966). Illustrations (Fyfe, 1964; Goodlad *et al.*, 1972, quoted by Serafini-Fracassini & Smith, 1974) showing penetration of sulphate and adenine into the inner aspect of the epiphysial cartilage around the ossification centre lend support to McKibbin & Holdsworth's results.

Since diffusion can occur from the eroding vessels both into the cartilage around the growing epiphysial centre and the metaphysial side of the growth plate, the question arises why the metaphysial vessels have no nutritive function. McKibbin & Holdsworth (1966) consider that the shorter columns and more substantial bars of matrix permit diffusion adequate for nourishment around the epiphysial centre, but this reasoning in no way negates the possibility that the metaphysial vessels *partially* nourish the growth plate. Irving (1964) suggests that the mature and 'calcifying' chondrocytes cannot metabolise the nutrient products which diffuse from the metaphysial vessels. While this may be true of these degenerate chondrocytes immediately adjacent to the erosion front, the evidence is rather to the contrary as regards the rest of the calcifying and hypertrophic zone. It is true that the hypertrophic cells do not proliferate in response to growth hormone and somatomedins (Ash & Francis, 1975), but oxidative enzyme activity is maximal in this zone (Castellani & Pedrini, 1956; Whitehead & Weidman, 1959) even though its cells lie farthest from nutrient flow via the epiphysial vessels. In addition, while hexosamine content and proteoglycan aggregate size (Greer *et al.*, 1968; Pita *et al.*, 1975) in the matrix diminish toward the calcifying zone, ^{35}S-sulphate incorporation and synthesis of low molecular weight glycosaminoglycans increase in this region (Larsson, 1976).

Although Trueta & Amato's (1960) evidence is persuasive, observations cited above suggest that the diaphysial vessels might act as a 'booster' source for the hypertrophic cells. While diaphysial ablation certainly shows that the epiphysial vessels can support a very long column of chondrocytes for at least a short period, it is not known whether their level of metabolism is normal or reduced although Trueta & Amato noted loss of alkaline phosphatase activity. It would be interesting to investigate the viability of the hypertrophic zone, following epiphysial ablation, under conditions where no vascular erosion could occur.

BASAL SURFACE OF ARTICULAR CARTILAGE

The calcified cartilage lamina forms the basal zone of articular cartilage (Figs. 58*b*, 63*c*). Despite considerable local variation it has in general a thickness of about 0.1 mm which is remarkably constant in different joints and species (Meachim & Stockwell, 1973) unless the articular cartilage is less than 0.1 mm thick. In human patellae, Green, Martin, Eanes & Sokoloff (1970) find a mean thickness of 0.134 mm which remains unchanged during maturation, ageing and in osteoarthrosis. The subchondral bone is often considered to be a well-defined cortical shell but in the mature rabbit Lemperg (1971*a*) finds wide variation in its thickness. Some areas show osteones next to the cartilage, with tangentially arranged lamellae on their deep aspect, while others possess only a few lamellae next to the junction. The superficial contour of the bone is highly irregular, following all the cell indentations (Balmain-Oligo, Moscofian & Plachot, 1973) and is demarcated by a cement line (Green *et al.*, 1970). By contrast the junction of the calcified with the uncalcified cartilage is relatively smooth and exhibits a 'blue line' when stained with haematoxylin, aptly dubbed the 'tidemark' by Fawns & Landells (1953).

The cement line

When stained with Bodian's copper-protargol the cement line is argyrophilic (Green *et al.*, 1970). The nature of its collagen content is uncertain although collagen fibres enter it from the calcified cartilage and do not continue across the osseochondral junction (Sokoloff, 1973). Although it reacts faintly to haematoxylin and the PAS technique, it is not metachromatic or alcian blue positive and hence the amount of mucopolysaccharide present must be small (Balmain-Oligo *et al.*, 1973; Sokoloff, 1973). No neutral lipids or phospholipids are present. A firm statement cannot be made regarding the degree of mineralisation, since some state that the cement line is hypercalcified (Balmain-Oligo *et al.*, 1973), while others note a double layer on the border of the bone consisting of a stain-free (toluidine blue or PAS) line and a granular layer of polysaccharide (Fawns & Landells, 1953). Green *et al.* (1970) describe a mineralised cement line with a hypercalcified lamina on its articular aspect. There are obvious difficulties in comparing the results of different techniques carried out in different sections where the zone examined is so narrow. Similar uncertainty applies to the nature of interosteonal cement lines.

According to Sokoloff (1973) the subchondral cement line shows irregularities of the order of a micron superimposed on the primary contour of the osseochondral junction. This presumably improves the adhesion of the bone to the calcified cartilage.

The calcified cartilage

This contains viable chondrocytes with much larger lacunae than those of the subchondral osteocytes; as observed in the patella, however, their number is reduced in middle-aged as compared with immature tissue and many lacunae are empty (Green *et al.*, 1970). The tissue is mineralised with hydroxylapatite as in bone (Davies *et al.*, 1962) although the radio density of the cartilage compared with the bone varies in different species and in different anatomical sites (Green *et al.*, 1970; Balmain-Oligo *et al.*, 1973). No change is said to occur with respect to age or osteoarthrosis although in the rabbit demineralisation of the calcified cartilage and the subchondral bone is found beneath experimental partial thickness defects, followed after two weeks by a massive re-mineralisation (Lemperg, 1971*b*).

Mineralisation is not uniform. Green *et al.* (1970) describe two heavily mineralised zones: a 2-μm thick layer on the articular aspect of the cement line and a slightly broader lamina of enhanced radio opacity at the superficial edge of the calcified layer deep to and independent of the tidemark. In some cases, parallel lamellae more heavily mineralised than surrounding tissue occur in the intermediate part of the calcified layer due to arrest of calcification during remodelling of the bone cartilage interface (Balmain-Oligo *et al.*, 1973; Lemperg, 1971*a*).

The calcified cartilage is birefringent and contains close-packed collagenous fibres perpendicular to but not crossing the osseochondral junction (Ohnsorge, Schutt & Holm, 1970). The ground substance is heavily basophilic and PAS-positive even after decalcification although it is less metachromatic than uncalcified cartilage (Fawns & Landells, 1953). Like the deep part of the uncalcified cartilage it is less basophilic in adult than in immature tissue (Green *et al.*, 1970). Sharply defined metachromatic 'haloes' of polysaccharide are seen around the chondrocytes (Fawns & Landells, 1953).

It is suggested by Haines & Mohuiddin (1968) that the calcified layer is a form of metaplastic bone since it forms a part of the permanent skeleton and is said to stain like bone. True metaplasia implies the transformation of one specialised or differentiated tissue into another specialised tissue. There is no evidence that this has occurred. As Balmain-Oligo *et al.* (1973) remark, the staining properties of the calcified layer declare its provenance and in no way deny its continuing cartilaginous character. It is basophilic, metachromatic and PAS-positive like the uncalcified cartilage; its lacunae are much larger than those of osteocytes and metachromatic material surrounds the chondrocytes. Collagen fibres do not cross from the calcified layer into the subchondral bone but are continuous into the uncalcified cartilage. Finally, inclusion in the permanent skeleton indicates little more than that the tissue is densely mineralised. Hence, there is little justification for Haines & Mohuiddin's proposal.

The tidemark

This varies in width from 2–5 μm (Redler, Mow, Zimny & Mansell, 1975). It lies between the calcified and uncalcified cartilage and is not itself calcified (Fawns & Landells, 1953). It contains no neutral fat or phospholipid (Green *et al.*, 1970). Although it stains blue with haematoxylin, results with toluidine blue and alcian blue suggest that it contains little mucopolysaccharide (Balmain-Oligo *et al.*, 1973). Hence the cause of the basophilia is not known; it is PAS-positive, however, suggesting that neutral glycoproteins are present.

It is not argyrophilic with Bodian's protargol. As described by Benninghoff (1925), collagen fibres are thought to be 'anchored' in the calcified zone and to run radially through the tidemark; scanning electron microscopic studies support this concept (Redler *et al.*, 1975; Minns & Steven, 1977). However, other patterns occur in the tidemark, the fibres either showing a random orientation or an arrangement parallel to the undulating surface of the calcified cartilage (Redler *et al.*, 1975). Minns & Steven (1977) describe bundle formation (about 55 μm in diameter) in the calcified as well as the deep parts of the uncalcified cartilage (McCall, 1969).

The tidemark and the adjacent deep uncalcified cartilage form a plane of weakness in articular cartilage where horizontal splitting can occur (Fawns & Landells, 1953), presumably due to the abrupt change in elastic moduli on passing from uncalcified to calcified layers (Sokoloff, 1973). Load bearing at the articular surface gives rise to shear stresses in the deep cartilage (Zarek & Edwards, 1963; McCutchen, 1965), producing tensile forces on the radial collagen fibres in the tidemark. Redler *et al.* (1975) suggest that the random and tangential arrangements of other fibres in the tidemark may help to diffuse the stress over a larger area. Splitting does not take place at the osseochondral junction: the greater strength of this junction must be attributed to the properties of the cement line and the 'key' afforded by its highly irregular contour, since no collagen fibres cross from the cartilage into the bone.

Vascular contacts

At random points along the osseochondral junction, vascular tissue from the subchondral marrow spaces penetrates through the bone into the calcified cartilage (Fig. 58*b*) and may occasionally break through the tidemark.

Holmdahl & Ingelmark (1950) describe 'ampullary' (30–50 μm in diameter) and 'canal-like' (11–18 μm in diameter) channels, while Trueta & Harrison (1953) observe single broad capillary loops passing through the calcified cartilage. Scanning electron microscopy reveals 'canaliculi' and 'tubules' within the calcified cartilage as well as interruptions in the tidemark (Mital & Millington, 1970; Redler *et al.*, 1975). Millington & Clarke (1973) describe channels 50–60 μm wide narrowing to 5–10 μm in the superficial part of the calcified zone; they contain pores 3–5 μm in diameter in their walls.

The degree of vascular contact is difficult to assess in vertical sections of articular cartilage, since, *inter alia*, the amount of calcified cartilage separating the vessels from the uncalcified cartilage varies. Nevertheless it seems that vascularity of the deep surface of articular cartilage varies in different parts of a joint, between different joints and at different ages (Holmdahl & Ingelmark, 1950; Woods, Greenwald & Haynes, 1970).

ARTICULAR SURFACES

This type of boundary consists of a bare cartilaginous surface exposed to synovial fluid; the surfaces of intra-articular fibrocartilages are similar. Articular surfaces have been investigated extensively because of their biomechanical and pathological significance. Although lacking both a perichondrium and a calcified layer, the tissue lying near an articular surface (Fig. 58*c*) nevertheless differs in some respects from the rest of the cartilage.

THE SUPERFICIAL MATRIX

All cartilage matrix consists largely of water, collagen and proteoglycan; the superficial matrix is no exception but in many cases also contains unusually large amounts of lipid. It must also be remembered that by virtue of its position, the superficial tissue is particularly susceptible to abnormalities of the synovial fluid. Pathologically, materials such as immune complex, the pigment of ochronosis and other unidentified substances may accumulate and theoretically could affect the physico-chemical properties of the matrix.

Water

Water content is highest next to the articular surface, where in normal post-mortem tissue, it forms about 75% of the wet weight (Maroudas *et al.*, 1969). Although in man the water content of the tissue as a whole may vary between joints – thus on average the femoral head cartilage contains about 10% less than the condylar cartilage (Maroudas *et al.*, 1973) – that of the superficial cartilage appears to be much the same in different joints. However, developmental changes are seen. In newborn cartilage, water content is higher than in the adult, forming about 80–90% of the wet weight (Lindahl, 1948; Linn & Sokoloff, 1965; Maroudas *et al.*, 1973). In fibrillated specimens the

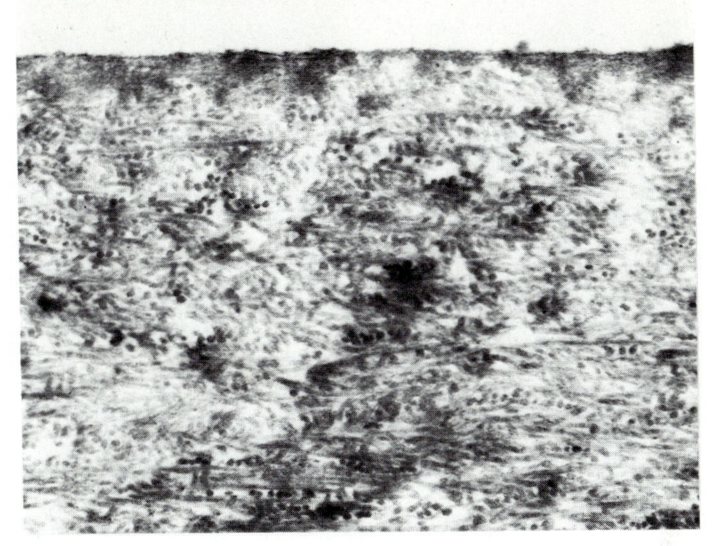

Fig. 69. Human femoral condylar articular cartilage (56 years). Glutaraldehyde fixation, magnification ×26000. Closely packed collagen fibres lying tangential to the articular surface.

tissue becomes more hydrated in all zones, particularly in the middle zone (Venn & Maroudas, 1977). In experimental arthritis, water content increases prior to macroscopic changes in the cartilage (McDevitt & Muir, 1976).

Collagen

Collagen fibres in the superficial matrix are of narrower diameter (about 30 nm) than elsewhere though finer fibres may be present. The fibres are close packed and small bundles may be formed (Davies *et al.*, 1962; Weiss *et al.*, 1968; Meachim & Roy, 1969). Studies with the light microscope (Benninghoff, 1925; MacConaill, 1951) and the transmission electron microscope (Davies *et al.*, 1962; Weiss *et al.*, 1968) show that the fibres run parallel to the articular surface (Fig. 69), an arrangement already present at birth (Cameron & Robinson, 1958; Little *et al.*, 1958). At any given point on the joint surface the fibres have a predominant orientation within the plane tangential to the articular surface. The 'split lines' formed by puncturing the articular surface with a pin (Benninghoff, 1925; Bullough & Goodfellow, 1968) correlate with the principal fibre alignment (Meachim, Denham, Emery & Wilkinson, 1974).

Millington & Clarke (1973) consider that three fibre-containing layers may be distinguished within 10 μm or so of the articular surface. A superficial layer about 50 nm thick contains 'unbridged' parallel fibres; an intermediate layer 3 μm thick contains 'fronded' parallel fibres and a deeper network 4 μm or more thick overlies the random meshwork present in the middle layer of the

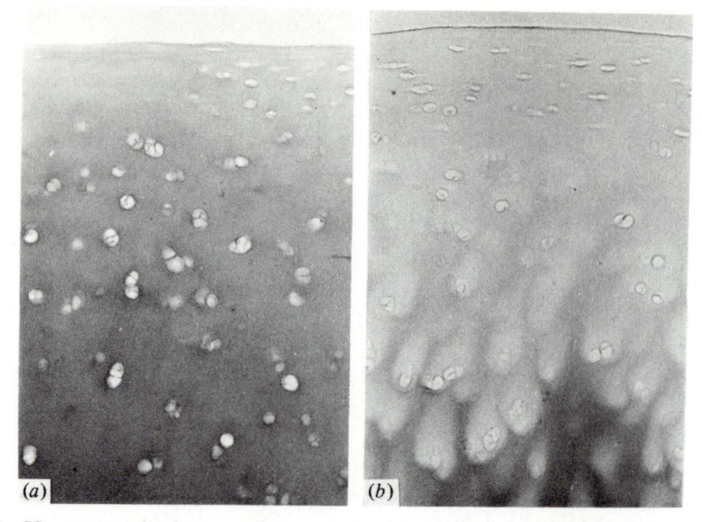

Fig. 70. Human articular cartilage, capitulum of humerus (20 years). Formalin fixation, magnification ×100. Glycosaminoglycan stain density near the articular surface. (*a*) Stained with alcian blue in 0.4 mol litre^{-1} magnesium chloride, for chondroitin sulphate and keratansulphate. (*b*) Stained with alcian blue in 0.9 mol litre^{-1} magnesium chloride, for high molecular weight keratansulphate.

cartilage. Others apportion a considerably greater thickness to the tangential collagen layer, particularly near the periphery of the articular area where it merges with the marginal transitional zone (Weiss *et al.*, 1968).

Chemical analysis of the human femoral condylar cartilage shows that the collagen content is highest in the superficial cartilage where it forms about 60–90% of the dry weight (Bullough, Maroudas & Muir, 1970). It falls sharply with distance from the articular surface. However, in the rat, stereological analysis demonstrates that the proportion of fibres to clear matrix is highest in the middle zone of the cartilage (Palfrey, 1975). Hence there may be species differences as regards collagen content.

Ground substance

Histological techniques show that the glycosaminoglycan content of normal adult articular cartilage declines toward the joint surface (Schaffer, 1930; Hirsch, 1944; Collins & McElligott, 1960). The alcian blue–CEC technique demonstrates that both chondroitin sulphate and keratansulphate diminish toward the joint surface (Fig. 70), keratansulphate becoming almost unstainable (Stockwell & Scott, 1965). Chemical analysis confirms that chondroitin sulphate and keratansulphate concentrations are lowest (Fig. 39) in the superficial zone (Stockwell & Scott, 1967). Determinations of fixed charge density (FCD) of the cartilage glycosaminoglycans, using streaming potential methods, agree with results of chemical analyses (Maroudas *et al.*, 1969; Venn & Maroudas, 1977).

In the human femoral condyle, the superficial layer 0.25 mm thick contains about 0.8% dry weight as uronic acid and 0.33% dry weight as hexose, i.e. a ratio of chondroitin sulphate to keratansulphate of about 2:1. Higher concentrations are found by other workers using whole papain digests of the superficial cartilage rather than semi-purified extracts but chondroitin sulphate/keratansulphate ratios are similar. There appears to be some variation between joints, patellar and femoral condylar cartilage having lower glycosaminoglycan content than the femoral head (Venn & Maroudas, 1977). Abnormal loss of stainable ground substance ('regressive change'), found histochemically beneath smooth articular surfaces, may be a predisposing factor in cartilage fibrillation (Meachim, Ghadially & Collins, 1965).

Attempts have been made to analyse tissue lying closer to the articular surface. Because the collagen fibres are close packed near the articular surface, the space available for ground substance is limited, but this does not mean that glycosaminoglycans are absent. In the superficial 50 μm of femoral condylar cartilage (Stockwell, 1975*a*; Table 8) and of femoral head cartilage (Larsson & Lemperg, 1975) there are similar concentrations of hexosamine (though less than in the subjacent tissue) with a 2:1 ratio of chondroitin sulphate/keratansulphate. However, analysis of femoral condylar cartilage within 10 μm of the articular surface gives a rather different result. Using a combination of the cation tracer technique (Maroudas & Thomas, 1970) and X-ray microprobe analysis, Maroudas (1972) finds that there is a FCD of about 0.2 mequiv. g^{-1} dry weight, indicating appreciable amounts of glycosaminoglycan at the surface. Since the ratio of FCD to sulphur content (an index of the contribution made by sulphate groups to the total charge) is 1.97, this suggests that the superficial 10 μm of tissue contains pure chondroitin sulphate, with no hyaluronic acid and very little keratansulphate. Balazs, Bloom & Swann (1966), analysing the same lamina obtained by scraping bovine articular surfaces, also came to the conclusion that keratansulphate is too low to permit analysis. Nevertheless, since the FCD measured at 10-μm intervals remains practically constant in the adult for 100–200 μm from the surface (Maroudas & Thomas, 1970), the results imply that there must be a change in the chondroitin sulphate/keratansulphate ratio which in no way affects the charge density.

The precise nature of the articular surface itself at the ultrastructural level is not finally settled (Fig. 71). Meachim & Roy (1969) describe collagen fibres extending right up to the surface where an osmiophilic line 8 nm thick may be seen in transverse section. Meachim & Stockwell (1973) consider this to be an 'intact' surface. Others describe a narrow amorphous lamina 30–200 nm wide between the surface and the commencement of the fibrous matrix (Davies *et al.*, 1962; Millington & Clarke, 1973). It has been suggested that this might correspond to the so-called 'lamina splendens' (MacConaill, 1951),

TABLE 8

Hexosamine content in the superficial matrix of human femoral condylar articular cartilage (nine specimens aged 20–66 years), using the naphthyl isothiocyanate method of Scott (1962). See also Fig. 39

Approximate distance from articular surface (μm)	% dry weight			Ratio glucosamine/ galactosamine
	Total hexosamine	Glucosamine	Galactosamine	
0–50	0.75 ± 0.48	0.18 ± 0.12	0.57 ± 0.40	0.35 ± 0.14
50–100	0.80 ± 0.46	0.24 ± 0.19	0.51 ± 0.33	0.39 ± 0.25

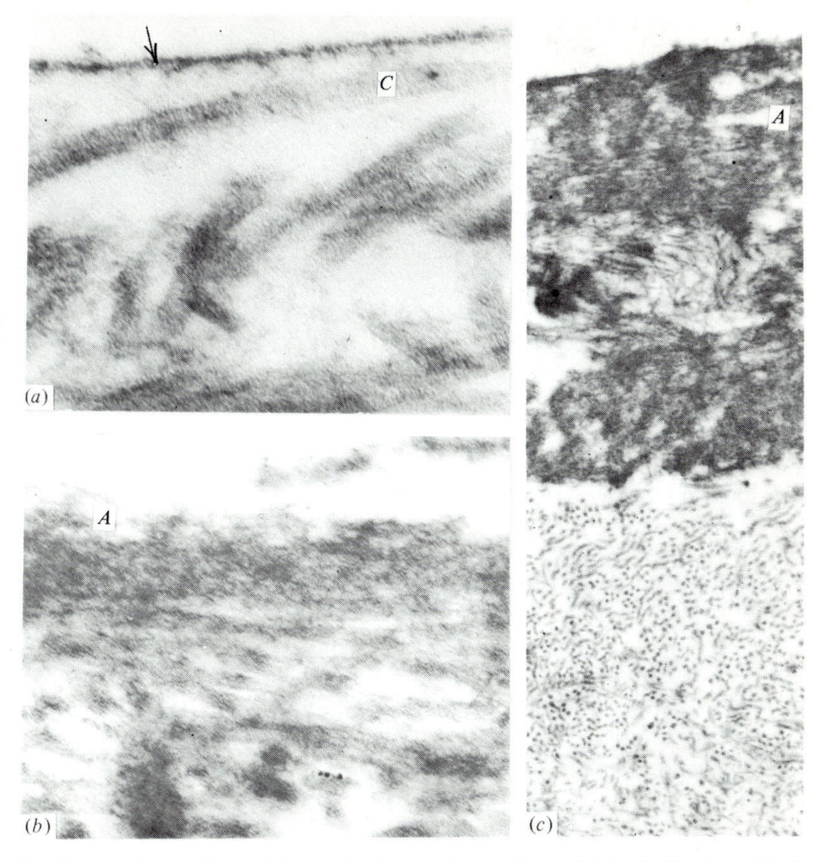

Fig. 71. Human femoral condylar articular cartilage, glutaraldehyde fixation. Various ultrastructural appearances of the articular surface. (*a*) A thin osmiophilic line (arrow) barely separated from subjacent collagen fibres (*C*). Aged 56 years, magnification ×113000. (*b*) A narrow amorphous layer (*A*) separating fibrous matrix from the joint space. Aged 46 years, magnification ×72000. (*c*) A thick, felt-like lamina of amorphous material (*A*) lying on the surface and infiltrating the cartilage. Aged 53 years, magnification ×18000.

a bright luminous layer immediately beneath the articular surface when viewed by phase contrast microscopy; more probably this phenomenon is associated with the highly orientated collagen fibres in this region (Meachim & Stockwell, 1973). A third structural variation at the surface consists of a felt-like lamina up to 4 μm or more thick containing fine filamentous and electron-dense amorphous material increasing in amount with age (Balazs *et al.*, 1966; Weiss *et al.*, 1968). This material is often considered to be a constituent of synovial fluid adsorbed or deposited on the articular surface; chemical analyses are consistent with it being hyaluronate (Balazs *et al.*, 1966). However, it is by no means certain whether the material is on the surface or

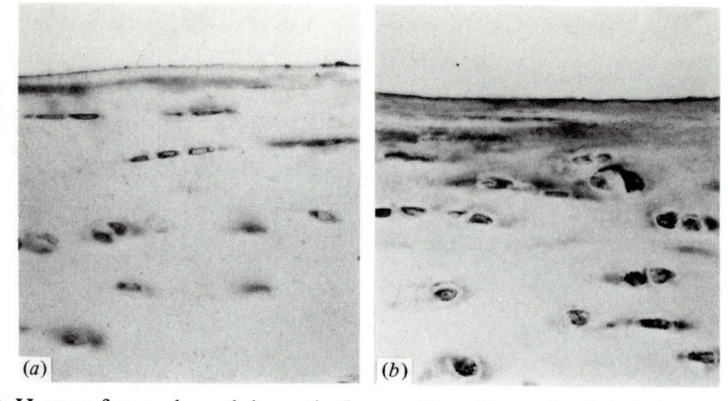

Fig. 72. Human femoral condylar articular cartilage. Formal-calcium fixation, frozen sections coloured with sudan black B, magnification ×250. Extracellular lipid in the superficial matrix in (*a*) young adult (20 years) and (*b*) ageing cartilage (65 years). Note the clear interval between the sudanophilia and the articular surface in (*a*). Deposits of lipid around the deeper cells are clearly demonstrated in (*b*).

within the cartilage itself. Maroudas (1972) was unable to find evidence of large amounts of hyaluronate in the matrix within 10 μm of the surface but other fibrillar materials, such as immune complex (Jasin *et al.*, 1973), can occur beneath the surface. The superficial matrix of fibrillated cartilage also becomes impregnated with electron-dense material (Meachim & Stockwell, 1973) and the fibre pattern becomes obscured (Redler, 1975).

The alcianophilic material present in the cartilage within 5–10 μm of the articular surface in some ageing adult human joints has been provisionally identified as keratansulphate (Stockwell, 1970, 1975). However, it is patchy in distribution, is not found in all specimens and seems to occur in conjunction with dense meshworks of 8-nm fine fibrils in regions of the matrix which give staining reactions indicative of amyloid. It is unlikely therefore, to be a normal part of the articular surface.

LIPID

In the superficial zone stainable extracellular lipid (Fig. 72) occurs both in a pericellular position and spread more diffusely in the matrix (Putschar, 1931; Schallock, 1942). The lipid in the matrix is often separated from the surface by a narrow unstained lamina. It is commonly present in adults (Ghadially *et al.*, 1965; Stockwell, 1965) though it varies in amount and depth from the surface. The lipid reacts histochemically for phospholipid and neutral lipid (Stockwell, 1965).

Lipid forms about 1% of the wet weight of whole human articular cartilage (Stockwell, 1967*c*) and is contained in the cells as well as in the matrix. Bonner, Jonsson, Malanos & Bryant (1975) have extracted the superficial 1 mm of the tissue separately and find that it is more abundant here than in the deeper

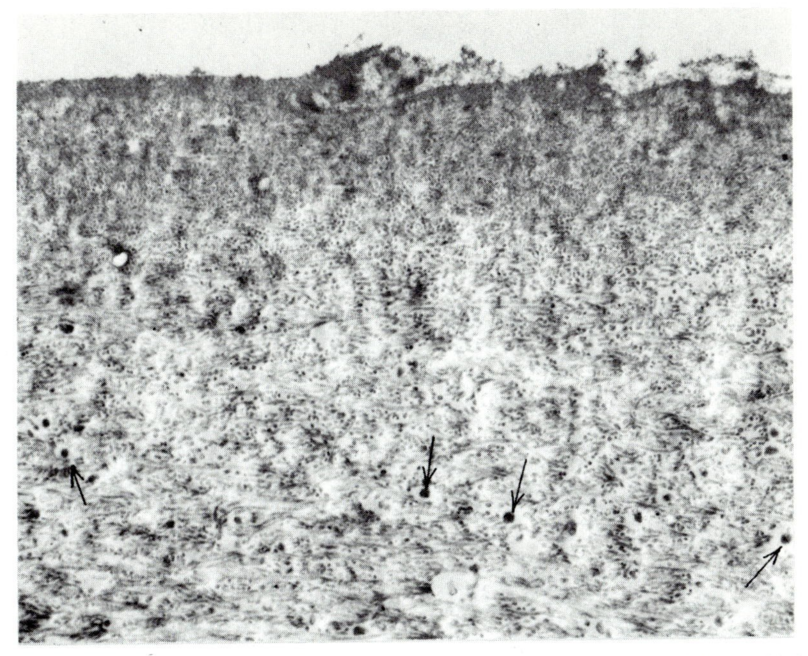

Fig. 73. Human femoral condylar articular cartilage (65 years). Glutaraldehyde fixation, magnification ×12000. Lipidic bodies (arrows) near the articular surface.

tissue; there is a modest age-related increase in total lipid during the period 15 to 66 years. Sphingomyelin and lecithin are the most abundant phospholipids; arachidonic acid (a precursor of the prostaglandins) is virtually restricted to the superficial zone and also increases with age.

Electron-dense membranous bodies, vesicles and granules (Fig. 73) are observed in ultrastructural studies of the region (Barnett *et al.*, 1963) and are thought to contain the stainable extracellular lipid. Meachim & Stockwell (1973) discuss the origin of the lipid and note that there are three main possibilities. First, the lipidic bodies could be the debris of cell necrosis (Barnett *et al.*, 1963), as nicely demonstrated at the margins of experimental surgical defects in cartilage (Fuller & Ghadially, 1972). Secondly, they might be formed in a more physiological fashion from intact cells, by continual detachment of the tips of cell processes (Ghadially *et al.*, 1965), as with the 'matrix vesicles' in the cell columns of the growth plate. Certainly pericellular lipid masses are more prominent at the poles of the superficial cells where cell processes tend to be longer. Thirdly, lipid might originate from the synovial fluid (Stockwell, 1965) which contains small amounts of lipid (Bole, 1962). Experimental production of lipoarthrosis shows that this mechanism is feasible, since superficial cells incorporate small amounts of the lipid (Fig. 16) injected into the synovial cavity (Ghadially *et al.*, 1970; Sprinz & Stockwell, 1976). Although in these experiments no concentration gradient

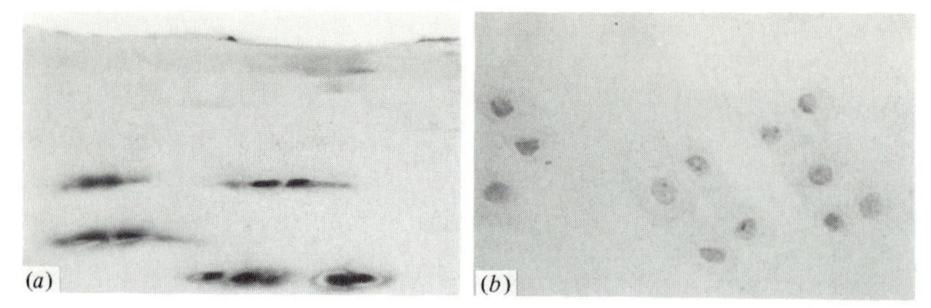

Fig. 74. Human articular cartilage (20 years). Formalin fixation, magnification ×410. Superficial cells in sections cut (*a*) normal and (*b*) tangential to the articular surface.

can be detected in the superficial matrix running from the surface inwards, over a long period in the natural state lipid might accumulate in the superficial matrix from this source, either directly or via the chondrocytes.

Whatever its origin, it is unlikely that the presence of extracellular lipid (like other extraneous materials) in the superficial matrix has a beneficial effect on diffusion through the matrix, although there is no evidence that fibrillation is more common in lipid-laden cartilage.

THE CELLS

The chondrocytes of the superficial zone are more numerous than elsewhere in the cartilage. They are discoidal cells, flattened in a plane parallel to the articular surface (Fig. 74), thus in sections cut normal to the articular surface they appear to be elongated and cigar-shaped. In tangential section they occur singly or in groups containing two to four cells. They are of a similar size in many joints and species, measuring about 14 μm in length and 3 μm in depth in human femoral condylar cartilage (Stockwell & Meachim, 1973). The number of layers of superficial cells varies in cartilages of different thicknesses; in normal tissue there is an acellular lamina of matrix 2–3 μm thick separating them from the articular surface.

Typically, the superficial chondrocytes (Fig. 75) present a relatively smooth cell surface on the side adjacent to the articular surface although on the deep aspect long processes may be found (Davies *et al.*, 1962; Weiss *et al.*, 1968). Numerous micropinocytotic vesicles line the cell surface; a small amount of granular ER and Golgi membranes are present and the cell possesses a number of small mitochondria. Although Collins & McElligott (1960) considered the superficial cells to be inactive in ^{35}S-sulphate incorporation, uptake varies from cell to cell; Maroudas & Evans (1974) find that sulphate incorporation is low near the surface, but that relative to the FCD of the matrix it is not very much smaller than in the deeper tissue. In the rabbit the superficial cells incorporate ^{3}H-cytidine, indicative of RNA synthesis where no DNA synthesis is in progress (Mankin & Baron, 1965); in addition, enzyme techniques for lactic and other dehydrogenases, and for glucosaminidase, show them to be active cells (Stockwell, 1966; Sprinz & Stockwell, 1976).

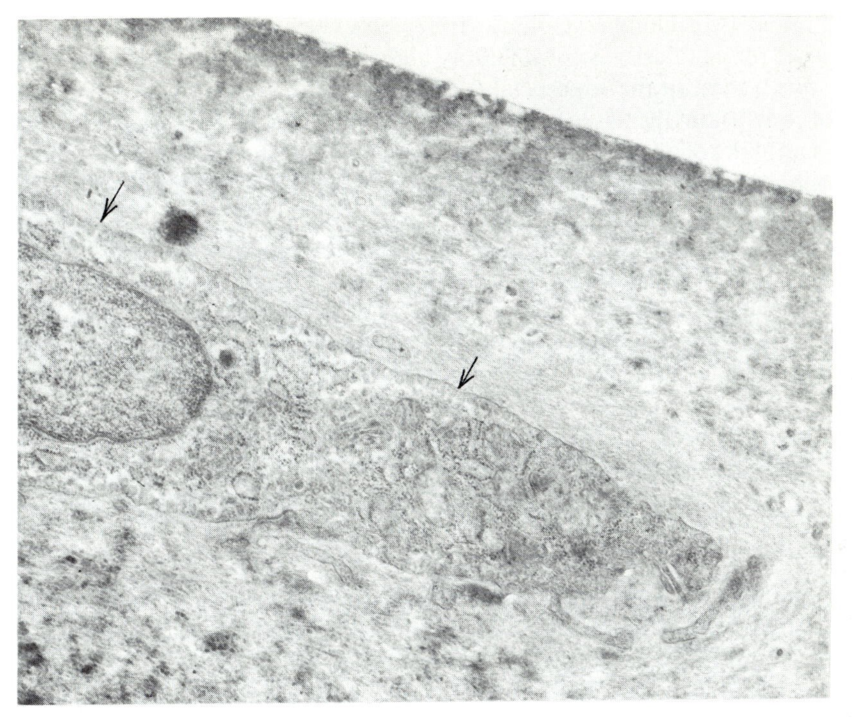

Fig. 75. Rabbit femoral condylar articular cartilage (nine months). Glutaraldehyde fixation, magnification × 10800. Part of a superficial cell. The cell margin next to the articular surface is relatively smooth. Note the abundance of micropinocytotic vesicles (arrows).

Despite the evidence of cell activity and their favourable position for nutrition from the synovial fluid the precise role of the superficial cells remains in doubt. In view of their low sulphate uptake and the dense fibre content of the superficial matrix, some have suggested that they might have a fibroblastic role (Weiss *et al.*, 1968); others that they might act as a metabolic 'sink' to ensure the correct metabolic environment for the deeper cells (Stockwell & Meachim, 1973). It is also possible that they might have a scavenging function to keep the superficial matrix free of extraneous matter. Whatever their function they are hardy cells, tolerating immunological insult more easily than do the deeper cells (Millroy & Poole, 1974) and surviving in a mechanically rigorous environment. There is no evidence that they are desquamated from the articular surface under normal conditions (Davies *et al.*, 1962; Weiss *et al.*, 1968). In some ageing joints, cell numbers in the superficial cartilage are reduced by death *in situ* (Barnett *et al.*, 1963); superficial cell loss is also a feature of regressive change (Meachim *et al.*, 1965).

In normal articular cartilage, cell density invariably increases towards the articular surface (Fig. 45). In the human knee and shoulder the cartilage within 0.25 mm of the surface is two to three times as cellular as the rest of the tissue (Meachim & Collins, 1962; Stockwell, 1967*a*). In the femoral condyle the cellularity is considerably higher in the superficial 0.1 mm (67000 per mm^3) than in the superficial 0.2 mm (45000 per mm^3).

In conclusion, the tissue adjacent to the articular surface shows adaptations, such as high collagen content and fibre orientation, designed to withstand the tensile stresses produced by joint motion and load bearing. It has some similarities to the perichondrium – the collagen content, the degree of hydration, the low glycosaminoglycan content – the last two characteristics enhancing its permeability. While the higher cellularity near the surface is compatible with the view that synovial fluid is a potent nutrient source, the high cell density might theoretically present an impediment to diffusion. Experimentally, however, the superficial zone is more permeable than the deeper tissue (see p. 166). Nevertheless, accumulation of lipid and abnormal materials is likely to have an adverse effect on diffusion, although this has not yet been investigated.

SURFACE CONTOUR

For many years this aspect of articular cartilage morphology was relatively uncontroversial. The articular surface at the microscopic level was considered to be smooth; any irregularities in its contour were assessed in the context of wear and tear and of fibrillation. Early studies with the transmission electron microscope (Davies *et al.*, 1962; Barnett *et al.*, 1963) agreed with this view, confirming that the surface was remarkably smooth although it was recognised that occasional irregularities existed of up to 0.2 μm in a length of 10 μm. Gardner (1972) attributes radical changes in this concept of the smooth articular surface to the advent of the scanning electron microscope.

Since scanning electron microscopic techniques necessarily involve dehydration of the tissue, studies of surface contour are likely to be susceptible to artefacts caused by shrinkage. Optimum techniques have now been developed, however, and these undoubtedly reveal a lack of smoothness in the surfaces of the specimens. Nevertheless the pattern of ridges observed by a number of early investigators (McCall, 1968; Walker *et al.*, 1969) appear to have been artefacts caused by tensions within the tissue near the cut and fractured edges of the block (Clarke, 1971*a*; Ghadially, Ghadially, Oryschak & Yong, 1976). Much finer ridges about 200 nm in depth which develop with advancing age cannot be attributed to this mechanism and may be due to the exposure of superficial fibre bundles as ground substance is gradually lost (Longmore & Gardner, 1975).

Articular surface morphology has been classified (Gardner, 1972; Longmore & Gardner, 1975) into four orders of anatomical irregularity:

(i) primary anatomical contours, where changes in curvature occur over several millimetres or centimetres;

(ii) secondary undulations 0.2–0.5 mm pitch;

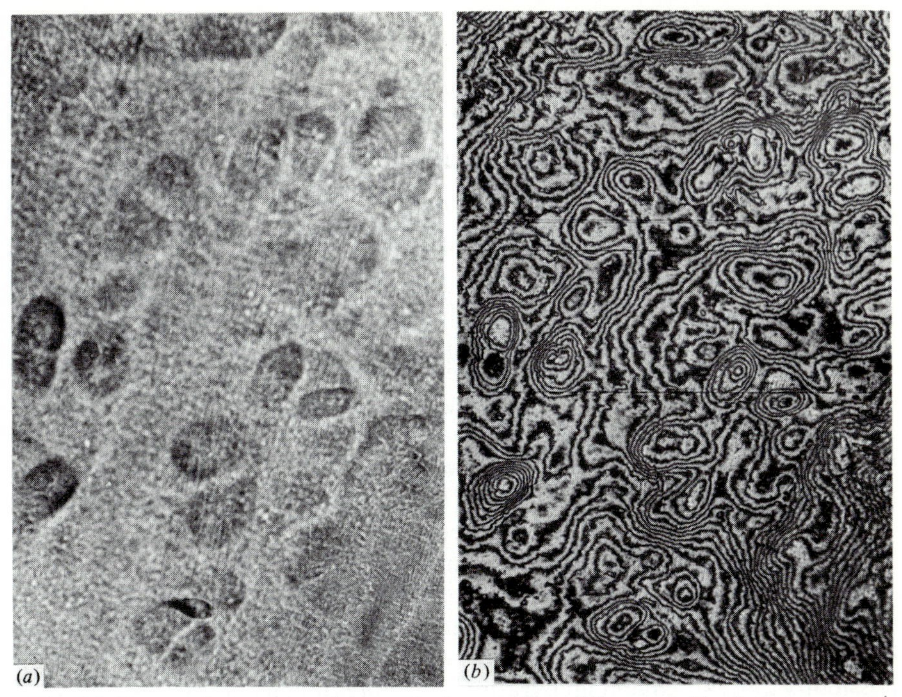

Fig. 76. Human femoral condylar articular cartilage (20 years). Fresh tissue, magnification ×200. The tertiary hollows of the articular surface examined *en face* (*a*) by incident light ($\lambda = 550$ nm) and (*b*) by reflected light interference microscopy. (Courtesy of Dr R. B. Longmore.)

(iii) tertiary hollows 20–30 μm in diameter and 0.5–2 μm deep;
(iv) quaternary ridges 1–4 μm in diameter and 130–275 nm deep.

The primary and secondary orders are amenable to inspection by the naked eye or hand lens; the quaternary ridges, thought to be the exposed fibre bundles, have already been mentioned. Most interest and some controversy has centred on the tertiary hollows.

The tertiary hollows have been observed and measured in a number of species and joints, including intra-articular fibrocartilages, and have been examined with incident light (in living and fixed tissue) as well as with the scanning electron microscope (Clarke, 1971*a*, *b*; Gardner & McGillivray, 1971). Reflected light interference microscopy (Longmore & Gardner, 1975) has been used to measure the dimensions of the tertiary hollows accurately (Fig. 76). In human femoral condylar cartilage, the depth of the hollows increases during maturation and ageing from 0.6 μm at 0–5 years to 1.4–1.7 μm at 30–50 years; their diameter increases from 20–25 μm to 35–45 μm over the same period. Clarke (1971*a*) finds that the pattern and diameters of the 'bowl-shaped' depressions correspond to those of the underlying superficial cell groups; furthermore, a comparison of frequency of superficial cell groups and of tertiary hollows (Clarke, 1971*b*) shows them to be similar. Clarke's conclusion that the surface depressions are produced by the underlying lacunae is given support by the changes that occur in their dimensions (Longmore & Gardner, 1975), and by the decline in their frequency (R. B. Longmore, personal communication) during maturation and ageing, which parallel similar changes in superficial cell frequency in femoral condylar cartilage (Stockwell, 1967*a*).

Since first reported, there has been some debate whether or not the tertiary hollows are present in life; in particular whether they are produced artefactually by lacunar collapse and/or cell volume changes during dehydration of the specimen. For example, transmission electron micrographs which indicate no distortion of the smooth surface immediately over chondrocytes lying within 3 μm of the surface (Davies *et al.*, 1962) have still to be reconciled with the surface irregularities observed using other techniques. On the other hand, the tertiary hollows have been seen in living articular surfaces (Gardner & McGillivray, 1971) albeit after a short delay between opening the joint and viewing the surface with the microscope. Ghadially *et al.* (1977) find no significant differences in articular surface appearance of fixed specimens following treatment of the cartilage with hypo- and hypertonic solutions. This suggests that alteration in chondrocyte volume is not involved in the production of the surface depressions, although it does not exclude the possibility that subsequent preparative procedures might override the osmotic effects or that matrix volume changes might occur *pari passu* with any cell changes.

Not all surfaces show hollows. Gardner and his colleagues note that it is sometimes difficult to be certain whether the features are in fact hollows or elevations. Clarke (1971*b*) finds that some areas contain quite shallow depressions with central protuberances and suggests that the latter correspond to cell nuclei. Ghadially, Moshurchak & Ghadially (1978) find that immature articular cartilage exhibits humps, while mature surfaces show numerous pits or humps arising from pits. Humps are attributed by the authors to the presence of more numerous, rounded superficial cells in immature cartilage. Whether or not these features are artefactual, the probability of obtaining a hump or a pit might also be related to the size of the lacuna or even to whether the cells occur singly or in groups, a wide lacuna allowing greater collapse with the formation of a hollow. Alternatively the cells might be either rounded or flattened: Gardner (1972) notes that the articular surface overlying the rounded superficial cells of rat tarsal joints is prominently raised. Joint exercise has also been found to cause swelling of superficial cells (Ekholm & Norbäck, 1951) which become flattened again after a short period of rest. Hence if indeed the hollows and humps are present during life, the predominance of one over the other might be related to recent joint activity. As Clarke (1971*b*) suggests, the precise contour of the articular surface may be continually changing over quite short periods of time.

It is relevant to consider what function the tertiary hollows might have in life. Lubrication theories based on the idea that thick pools of synovial fluid become caught between opposed loaded cartilage surfaces (Walker, Dowson, Longfield & Wright, 1968) have suggested that the tertiary hollows trap the lubricant; Gardner (1972) notes that the hollows are not obliterated under load. However, as Swanson (1973) comments, it is unlikely that pools of synovial fluid would be trapped continuously in the two sets of hollows on the opposed moving articular surfaces. At present it is difficult to see what role the pits or humps might have in joint lubrication. Are they then a structure without a function, occurring merely because a perfectly smooth articular surface is difficult to produce biologically and perhaps unnecessary? This may be the case, although it is not a satisfying conclusion. There remains one further possibility, that the hollows are useful in cartilage nutrition. The depth reached by solutes diffusing from the joint cavity into the cartilage is sensitive to the formation of stagnant fluid films (Maroudas, 1973) at the articular surface. By analogy with the golf ball surface, it may be that the tertiary hollows produce turbulence in the 'boundary layer' of synovial fluid at the articular surface during joint movement, so causing additional stirring and enhancing diffusion.

SYNOVIAL FLUID

The synovial fluid is derived from the synovial membrane and may be regarded as a modified intercellular fluid. It functions as a joint lubricant as well as a nutrient source for the articular cartilage. Its viscosity (non-Newtonian) and volume normally vary consistently from joint to joint in the same animal and in the same joint in different species. They are also altered in joint disease and trauma, and the volume is affected by joint exercise or immobility; only small quantities of fluid can be aspirated from normal human joints (Davies, 1967).

The synovial membrane covers those internal surfaces of the joint not formed by articular cartilage or intra-articular discs. It has an intimal lining of pleomorphic synovial cells resting on a subsynovial layer of less cellular connective tissue (Fig. 77). Electron microscopy demonstrates that there are appreciable intervals (0.05–1 μm) between the intimal cells, filled with moderately dense amorphous material and with few collagen fibres (Barland, Novikoff & Hamerman, 1962; Davies & Palfrey, 1966). The synovial membrane is highly vascular. A dense capillary network lies in contact with the basal (outer) surface of the intimal lining, approaching to within 10–20 μm of the joint cavity. Hence the blood plasma is separated from the joint cavity only by endothelium and a variable thickness of extracellular substance.

The results of chemical analysis are consistent with the view that the fluid is a dialysate of blood plasma with added hyaluronic acid (Bauer, Ropes & Waine, 1940). Hyaluronic acid is responsible for the viscosity of synovial fluid

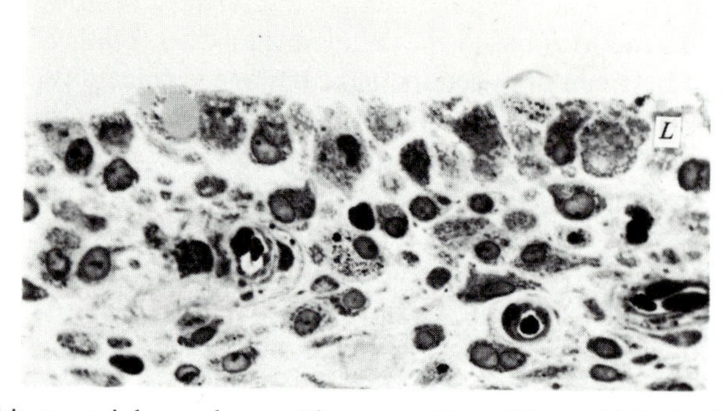

Fig. 77. Rabbit synovial membrane (three months). Glutaraldehyde fixation, 1-μm araldite section stained with toluidine blue and pyronin B, magnification ×410. The intimal lining cells (*L*) are close packed compared with the deeper vascular connective tissue.

though perhaps not for its lubricating properties under all conditions of joint function and loading (Wilkins, 1968; Swann, 1975). The hyaluronate content of synovial fluid is about 3.5 mg g^{-1} (Hamerman & Schuster, 1958) and the glycosaminoglycan is linked to about 2% of protein not derived from the plasma (Hamerman, Rojkind & Sandson, 1966). Synovial fluid is remarkable among connective tissue extracellular fluids in that it contains traces only of other glycosaminoglycans. Hyaluronate-protein is synthesised by cells of the synovial intima (Vaubel, 1933; Barland, Smith & Hamerman, 1968), which also have a 'scavenging' role in the phagocytosis of particles from the joint cavity.

The protein content (2.0 g 100 ml^{-1}) of synovial fluid is low and its albumin/globulin ratio (4:1) high compared with the plasma (7.0 g 100 ml^{-1}; 1.5:1). Proteins with a molecular weight over 160000, such as fibrinogen, are not found in synovial fluid owing to the permeability characteristics of the blood–synovial barrier and the excluded volume effect of the synovial mucin (Davies, 1967). The partition of electrolytes between plasma and synovial fluid is in accord with the Donnan equilibrium, although calcium is anomalously high; non-electrolytes occur in about the same proportions. However, the glucose content is lower (0.08 g 100 ml^{-1}) than in plasma (Ropes & Bauer, 1953), possibly due to its utilisation by the joint tissues, although it may also be a large enough molecule to be affected by exclusion, as in cartilage. A small quantity of lipid (0.02 g 100 ml^{-1}), principally lecithin, is present in the normal fluid (Bole, 1962). The gas tensions of fluids from normal and diseased human joints show wide variation. Thus oxygen has a partial pressure of between 1–10 kPa, although the more normal fluids show values of 5–8 kPa, about the same as in capillary blood. The pO_2 is lower in rheumatoid arthritis and in inflamed joints (Falchak *et al.*, 1970; Lund-Oleson, 1970; Treuhaft & McCarty, 1971) and is said to fall after exercise. Carbon dioxide content (pCO_2: 5–20 kPa) correlates well with lactate concentration and the pH level, which is normally slightly alkaline.

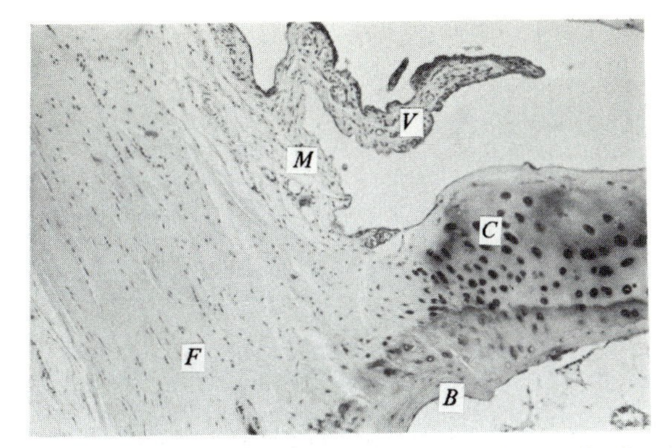

Fig. 78. Human distal inter-phalangeal joint (63 years). Formalin fixation, magnification ×50. The marginal transitional zone where the periphery of the articular cartilage (*C*) and subchondral bone (*B*) merge with the fibrous capsule (*F*) and the synovial membrane (*M*). A vascular fold (*V*) of synovial membrane projects into the joint space.

Thus synovial fluid is in equilibrium with the blood plasma. The whole of the articular surface is bathed by the fluid, which provides all the nutrient qualities of the plasma and eliminates the requirement for vascularisation of the superficial cartilage boundary, a mechanically impossible site for blood vessels. The fluid is, however, entirely dependent on joint movement for mixing and 'stirring'.

MARGINAL TRANSITIONAL ZONE

At the periphery of the articular area, the cartilage becomes continuous with the periosteum of the bone end, the deep fibrous layer of the synovial membrane and often the joint capsule, through a fibrocartilaginous region termed the marginal transitional zone (Fig. 78). This represents the circumferential boundary of articular cartilage which seems to have been studied little in recent years.

Traced from periphery towards the articular area, the cells in this zone show a gradual transition from fibroblasts with long processes to rounded cells in lacunae typical of cartilage (Key, 1932). The layer of tangential fibres in the superficial zone of articular cartilage becomes thicker (600 μm in the human femoral condyle) on passing toward the articular margin, and merges with the collagen in the periosteum (Benninghoff, 1925; Weiss *et al.*, 1968). The richly vascular synovial membrane extends over the superficial aspect of the marginal transitional zone as a wedge-shaped process, thinning out to a single layer of flattened cells and eventually becoming indistinguishable from cartilage. Its terminal part contains the *circulus articuli vasculosus* (Hunter, 1743), formed by wide bore anastomosing capillary loops which arch in the direction of the centre of the articular area (Fig. 79). Similar features are observed at the margin of intra-articular discs.

Fig. 79. Horse metacarpal cartilage following injection of articular vessels with carmine gelatin, magnification × 10. Capillary loops (circulus vasculosus articuli) of the periarticular synovial margin viewed *en face*.

The region is of some significance in pathology. Characteristically, age-related fibrillation is detectable initially at or near the periphery of the articular surface (Meachim & Emery, 1973). The marginal transitional zone reacts to the stresses resulting from the joint disturbances of osteoarthrosis by producing cartilage 'lipping' and osteophytes, which always originate in this zone unless they are of the small subarticular variety (Jeffrey, 1975). It is suggested that better nutritional conditions associated with the proximity of the circulus vasculosus (Fisher, 1923) account for, or permit, these local pathological growths. In view of its transitional histological character, the region may also contain fewer differentiated cells than elsewhere; certainly it is the remnant in the adult of the peripheral appositional growth zone of immature articular cartilage.

Although there is little evidence concerning the nutritional significance of this boundary except perhaps to the local peripheral tissue, it is doubtful if its blood vessels are of special importance to the nutrition of the bulk of the articular cartilage.

PASSAGE OF NUTRIENTS THROUGH CARTILAGE

The mechanism of solute transfer through the avascular matrix of cartilage presents an intriguing problem. The idea that nutrient fluid passed through pre-formed channels alongside the fibre bundles (Arnold, 1878) has long since been shown to be erroneous. In more recent years, it has become clear that solutes traverse the matrix by passive diffusion although mass transfer of fluid may be of some significance, for example in the intervertebral disc, and active transport by the cells has also been suggested. Some of the factors affecting diffusion were noted by earlier workers. Thus Harpuder (1926) found that cartilage had a low permeability and that diffusion rates varied with molecular size of the solute. Kantor & Schubert (1957) observed that blocks of nasal cartilage were readily permeable to cationic dyes but resisted the passage of anionic dyes, owing to the negative ionic charge of the glycosaminoglycans. Ogston & Phelps (1961) found that mucopolysaccharides in solution 'excluded' other large molecules from part of the solvent volume, suggesting that mucopolysaccharides can also alter permeability by this 'excluded volume' effect. However, more recently the most notable advances in knowledge and basic understanding of diffusion in cartilage have been made by Maroudas and her colleagues and reference should be made to her reviews on the subject (Maroudas, 1973, 1976).

The movement of solutes into and through cartilage from a well-stirred nutrient fluid depends on factors (Maroudas, 1973) whose magnitude is affected by the chemical nature of the tissue. These are:

(i) the diffusion coefficient of the solute in the cartilage;
(ii) the distribution coefficient of the solute between the cartilage and the nutrient fluid.

The flux of solute molecules into the tissue is dependent on the product of the two factors, and of course the surface area available for diffusion.

DIFFUSION

Although cartilage is a gel, i.e. a mass of water held in position and shape by its contained macromolecules, it has a maximum water content of about 80–85%. Since rates of diffusion depend on the water content of the tissue, the diffusion coefficients of solutes in cartilage must always be less than those in water, even though there is apparently very little bound water unavailable to solutes in cartilage. The effective rate of diffusion across a given thickness of cartilage is reduced even more because the solute molecules must follow a tortuous path through the matrix. Hence even with small molecules such as urea and glucose or ions such as sodium and sulphate whose interactions with the matrix are negligible, their diffusion coefficients in cartilage (Table 9) are only about 0.3–0.4 of those in water (Maroudas, 1970, 1973).

Larger molecules in any case have much lower diffusion coefficients in water than smaller molecules, but in cartilage are retarded even further as a result of friction between the solute and the matrix macromolecules. Thus high molecular weight dextrans have very low diffusion coefficients in cartilage, much less than a tenth of the aqueous values. However, molecular size and not molecular weight is the important factor: haemoglobin and serum albumin, which are more compact molecules, have diffusion coefficients as much as a quarter of those in water (Maroudas, 1973, 1976). The frictional retardation of the larger molecules is accentuated as a result of their 'excluded volume' which increases sharply with both solute size and concentration of matrix macromolecules. Thus the rate of diffusion of large molecules through cartilage is dependent on local variations of proteoglycan concentration. In

TABLE 9

Permeability of human articular cartilage slices with a fixed charge density (FCD) of 0.08 mequiv. g^{-1} (representative of tissue about 0.5 mm deep to the articular surface of femoral condylar cartilage) measured at 25 °C. $\bar{m}/m$ = ratio of solute concentration per unit weight of water in cartilage to that in Ringer's solution. Data from Maroudas (1970, 1973, 1976)

Solute	Molal distribution (partition) coefficient ($\bar{m}/m$)	Diffusion coefficient (cm^2 s^{-1}) in cartilage
Na^+	1.5	4.6×10^{-6}
Cl^-	0.75	7.25×10^{-6}
SO_4^{-2}	0.6	2.2×10^{-6}
Ca^{2+}	3.0	–
Urea	1.03	5.5×10^{-6}
Glucose	0.90	2.1×10^{-6}
Sucrose	0.57	1.3×10^{-6}
Dextran 10	0.23	(FCD: 0.09) 1.47×10^{-7}
Inulin	0.15	(FCD: 0.07) 2.5×10^{-7}
Dextran 40	0.07	(FCD: 0.09) 1.14×10^{-8}
Serum albumin	0.01	2.0×10^{-7}
Immunoglobulin G	0.01	–

general, they diffuse more freely where proteoglycan concentration is lower, for example near perichondrial and articular surfaces.

Since most nutrients and metabolites are of small molecular size, they should be freely diffusible in any zone of the cartilage. Variations in proteoglycan and collagen content have little effect unless they materially alter the free water content of the tissue or the tortuosity of the diffusion pathway. Thus in articular cartilage, the rather lower rate of diffusion in mature as compared with immature tissue (Makowsky, 1948; Stockwell & Barnett, 1964) is probably associated with differences in water content. Similarly the diffusion of dyes (McCutchen, 1962), of sulphate (Maroudas & Evans, 1974) and other solutes (Maroudas, 1968) is greater near the articular surface than in the deeper tissue, correlating with differences in water content. Local increases in dry mass, as for example in the 'border zone' around the chondrocytes in the central region of aged horse nasal cartilage (Galjaard, 1962), might also be expected to have some effect on diffusion. However, the most striking change occurs in calcified cartilage which may be more heavily mineralised than bone: here the free water and hence passage of nutrient must be reduced to a minimum, effectively forming a barrier to diffusion, as borne out by experiment (Maroudas *et al.*, 1968). It is noteworthy that diffusion of solutes may be enhanced where there is pathological degradation and loss of proteoglycans, as in early fibrillation of articular cartilage, especially if this is accompanied by abnormal overhydration of the tissue.

PARTITION OF SOLUTES BETWEEN CARTILAGE AND NUTRIENT FLUID

The ease of movement of solutes into cartilage is dependent on their interaction with the matrix macromolecules. Small non-electrolytes which have no interaction should partition equally between the cartilage water and the external solution and be unaffected by the proteoglycan concentration in the matrix. In practice, only urea shows these ideal characteristics (Maroudas, 1970). Larger solutes experience steric exclusion by the matrix proteoglycans, the partition coefficient being inversely proportional to the molecular size of the solute (Table 9); increase in proteoglycan content of the matrix causes a further diminution in the partition coefficient. It appears that substances of molecular weight *c.* 70000, for example haemoglobin and serum albumin, are about the largest which can enter and move through cartilage; however, very large molecules such as immunoglobulin G (molecular weight 160000) have a similar partition coefficient to serum albumin (Maroudas, 1976). While most nutrients are well below this molecular size, it is worth noting that glucose itself is large enough to suffer from the excluded volume effect since its partition coefficient (0.85) is less than unity and shows a small decrease with increasing proteoglycan concentration (Maroudas, 1970).

The distribution of ions is a consequence of the fixed negative charge on the matrix proteoglycans. If the external solution is equivalent to physiological saline, the partition coefficient of cations is three times greater than that of anions. The partition coefficient changes with increase in the fixed negative charge of the matrix, becoming higher for cations and lower for anions. It behaves similarly if the external solution is made very dilute, and penetration of anions into cartilage then becomes very small. As Maroudas (1973) points out, the use of dyes in very low salt solution probably accounts for Kantor & Schubert's (1957) finding that anionic dyes are unable to penetrate cartilage: anionic dyes used in Ringer's saline enter cartilage (Maroudas *et al.*, 1968). Thus essential anionic metabolites, such as free sulphate ions, though present in cartilage in very low concentration can have no difficulty in entering the tissue from physiological nutrient fluids.

OTHER MECHANISMS

It has been thought that cartilage nutrition might be enhanced by mass transfer of fluid through the matrix, resulting from mechanical compression under load, followed by re-expansion at rest (Ingelmark & Saaf, 1948; Ekholm, 1955). It is true that immobilisation of a joint leads to degenerative change (Evans, Eggers, Butler & Blumel, 1960; Sood, 1971) and there is also evidence suggesting that cartilage thickness increases owing to uptake of water following joint exercise (Ingelmark & Ekholm, 1948). However, it has been shown experimentally that intermittent compression of cartilage does not increase the passage of solutes into the tissue over and above that which occurs by passive diffusion (Maroudas *et al.*, 1968). This indicates that a 'pumping' mechanism is of little significance, at least in articular cartilage. It appears that the importance of joint movement in cartilage nutrition is in minimising the formation of a stagnant liquid film at the articular surface by 'stirring' the synovial fluid.

There seems to be no evidence that chondrocytes actively transport nutrients and by so doing build up concentration gradients to aid nutrition of tissue on their deep aspect (Maroudas *et al.*, 1968; Maroudas & Evans, 1974).

ARTICULAR CARTILAGE NUTRITION

The cartilage in the central part of the articular area can effectively obtain nutrients from only two sources:

(i) the synovial fluid via the articular surface;
(ii) the blood in the subchondral bone marrow.

The relative importance of the two sources has been obscured until recently by the changes which take place during maturation. It now seems clear that while there is a considerable nutritional supply to immature cartilage from the epiphysial marrow spaces, in normal adult joints articular cartilage derives its nourishment mainly via the articular surface. In large joints with thick cartilage, however, the deepest tissue may obtain its supply from the marrow spaces.

In view of the marked structural differences in the anatomical 'barriers' at the superficial and deep surfaces of articular cartilage, it is a little surprising that it was the marrow spaces which were first considered to be the nutrient source (Havers, 1691). This was a view held by Toynbee (1841) who observed the rich vascularity of the subchondral bone, although he recognised the barrier presented by the bone plate. Although Fisher (1923) was able to show that the bone plate was permeable, using carmine injections, other early experimental work using silver nitrate with human post-mortem material (Ishido, 1923) showed that diffusion from the marrow spaces was arrested at the calcified layer. On the other hand, silver nitrate injected into the joint cavity fully permeated the uncalcified articular cartilage as far as the calcified zone. This finding was in accord with the hypothesis proposed by Strangeways in 1920, at that time first being given serious consideration though suggested by Leidy in 1849, that the synovial fluid could nourish articular cartilage.

Strangeways showed that cartilaginous 'loose bodies' ('joint mice') lying free in the synovial cavity not only receive sufficient nourishment from the fluid to survive (Virchow, 1863) but also to increase in size. Other surgical observations also support the concept that synovial fluid has a nutrient role. Thus in avascular necrosis of the femoral head the cartilage survives although the subjacent bone dies (Muller, 1929; Gallie, 1956; D'Aubigne, 1964), a condition which can be reproduced experimentally (Nussbaum, 1923). Furthermore the survival of cartilage but not of bone in intra-articular osteochondral grafts (De Palma, Tsaltas & Mauler, 1963) is compatible with nutrition by the synovial route.

The anatomical demonstration of vascular perforations of the subchondral bone and calcified cartilage (Peterson, 1930, cited by Holmdahl & Ingelmark, 1950; Holmdahl & Ingelmark, 1950; Trueta & Harrison, 1953) stimulated renewed interest in the subchondral route of nutrition. Experiments carried out on rabbits, using starch grains (Ingelmark & Saaf, 1948) ^{198}Au-gold chloride (Ekholm, 1951), ^{35}P-phosphate (Ekholm, 1955) and fluorescent dyes (Brodin, 1955*b*), showed that solutes gain entry to the cartilage both from the superficial and deep aspects but that the subchondral route appeared to be dominant. In retrospect, it is clear that most of the animals used in these experiments were still immature, so accounting for the findings. Recent work has permitted the relative contributions of the subchondral and synovial routes to be evaluated.

In order to demonstrate diffusion from the marrow spaces unequivocally, all possibility of the availability of the synovial route must be excluded. Experiments so designed show little evidence of diffusion from the marrow spaces. Brower, Akahoshi & Orlic (1962), who used immature and adult rabbits, injected neutral red dye intra-epiphysially after the synovial membrane had been retracted and the articular surface covered with a saline sponge. Under these conditions, even in immature animals, only the chondrocytes in the calcified zone are stained; injections of dye into the intact synovial cavity quickly produces full thickness staining of the cartilage. In similar experiments in adult rabbits ^{35}S-sulphate readily enters the cartilage from the synovial fluid; however, where the synovial route has been minimised by continuous saline irrigation of the joint (Hodge & McKibbin, 1969) or eliminated by retraction of synovial membrane and saline sponge blockade (Honner & Thompson, 1971), sulphate uptake occurs only in the cells at the osseochondral junction. This demonstrates conclusively in the adult rabbit that ^{35}S-sulphate reaches the uncalcified cartilage only via the articular surface. However, in immature animals, the deeper part of the cartilage (which consists of unossified epiphysial cartilage) is nourished from the subchondral marrow spaces since sulphate uptake occurs following blockage of the synovial route. These results in the rabbit are in agreement with the sequelae of experimental subchondral infarction in immature (McKibbin & Holdsworth, 1966) and adult sheep knee joints (McKibbin & Holdsworth, 1968). In

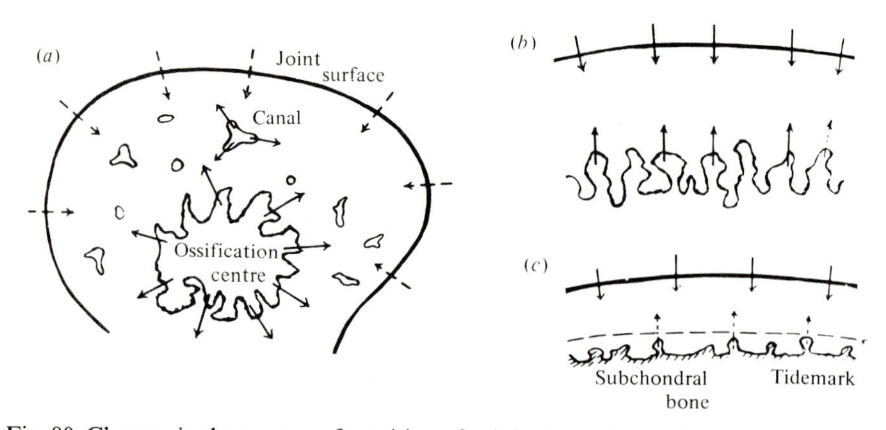

Fig. 80. Changes in the sources of nutrition of epiphysial and articular cartilage during development. (*a*) Prior to ossification nutrient supply comes from the synovial fluid, perichondrial vessels and cartilage canals. Subsequently, blood vessels of the ossification centre become a major source. (*b*) During late immaturity, the subchondral vessels in the marrow spaces are a major source. (*c*) In adult life, the synovial fluid is most important.

the lamb the subchondral bone and deeper cartilage degenerate and chondrocyte proliferation ceases, while in the sheep only the subchondral bone dies.

Such conclusions are fully compatible with changes in the degree of vascular perforation at the osseochondral junction during maturation (Fig. 80). While the epiphysial cartilage is undergoing rapid endochondral ossification, the eroding capillaries form numerous vascular contacts with the cartilage, a continuous calcified layer is lacking and there are only short chondrocyte columns in the cartilage. Later on, in the older immature rabbit femoral head, about 2.5% of the junctional area is still occupied by vascular contacts (Holmdahl & Ingelmark, 1950) some of which are said to penetrate the tidemark. In rabbits over a year old, canals 10–15 μm in diameter penetrate the calcified cartilage at regular intervals although they do not reach the tidemark (Lemperg, 1971*a*), while in 18-month-old rabbits no vascular channels are observed in the articular cartilage (Greenwald & Haynes, 1969).

Nourishment of adult articular cartilage from the synovial fluid appears to be consistent with measurements of the metabolic demand and the diffusion coefficient of the tissue. Bywaters (1937), deriving the rate of diffusion from observations of 'loose body' dimensions, calculated from his measurement of glycolytic rate that articular cartilage could be adequately supplied with glucose from the synovial cavity to a depth of 3 mm. Maroudas *et al.* (1968), measuring the diffusion coefficient of glucose in human femoral condylar cartilage *in vitro* and using Bywaters' glycolytic data, calculate a similar limiting depth, assuming that the synovial fluid source is well stirred. In accord with the *in vivo* experiments cited earlier, Maroudas *et al.* are unable

TABLE 10

Published values of extent of subchondral marrow contact with basal surface of articular cartilage. Values for human femoral head calculated from published data. References: Holmdahl & Ingelmark (1950); Enneking & Harrington (1969); Greenwald & Hayes (1969); Woods *et al.* (1970); Maroudas *et al.* (1975)

Specimen	Cartilage thickness (mm)	% osseochondral interface occupied by vascular contacts
Rabbit		
Femoral head		
10 months	0.6	2.4
18 months	0.4	Nil
Femoral condyle		
10 months	0.28	1.7
18 months	0.28*	Nil
Man		
Femoral head		
Superior surface	3.08	9
Inferior surface	1.94	2.5
Intervertebral joints (ninth thoracic)	0.27	Nil
Intervertebral disc (fourth lumbar)	5.2 (half thickness)	11–36

* Author's data.

to detect diffusion of glucose across the osseochondral junction in the adult although this is possible in immature specimens.

Although it might be reasonably inferred from these investigations that the subchondral route becomes redundant in the adult, observations in the human femoral head are not entirely in accord with this view. Using adult cadaver femoral heads, Greenwald & Haynes (1969) have shown that fluorescent dye can diffuse into the base of the cartilage from the marrow spaces. Although this observation may be criticised because of the long diffusion time employed and the possibility of tissue autolysis, it is consistent with the vascular contacts present in the adult human femoral head (Trueta & Harrison, 1953). A semi-quantitative analysis of the degree of vascular contact indicates that this is rather extensive: as much as 9% (an average of 14 contacts approximately 60 μm in diameter in a 9.5-mm length of the osseochondral junction) of the interface is occupied by patent vessels (Woods *et al.*, 1970). In addition there is evidence from pathology that some of the deep zone of the femoral head cartilage may be nourished via the subchondral route. Thus in avascular necrosis although the cartilage in general survives, chondrocytes in the deep zone degenerate (Axhausen, 1924; Catto, 1965). In

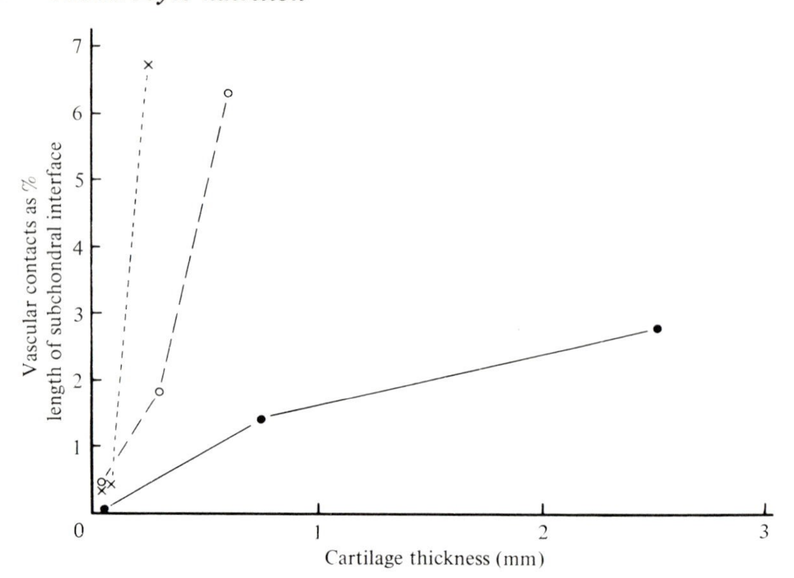

Fig. 81. The extent of contact between the subchondral marrow spaces and the deep surface of articular cartilage in the incudo-malleolar, inter-phalangeal and knee joints of three species; rat (×), dog (○) and man (●). (Courtesy of J. D. Connolly.)

normal femoral heads, cell density measurements indicate a rise in cellularity in the deep zone near the tidemark (Vignon *et al.*, 1976) from the low values found in the middle zone of the cartilage (a feature not observed in the femoral condyle). Both cellular phenomena are consistent with dependence on a nutritional source from the marrow spaces. It may also be relevant that the deeper cells of human articular cartilage (*c.* 3 mm thick) would have to exist in oxygen tensions approaching zero if diffusion from the marrow spaces were negligible, according to calculations published by Marcus (1973).

The observations in the human femoral head need not be regarded as contradictory to the results obtained in other joints. It is obvious that there is considerable variation in the degree of marrow contact between different cartilages, joints and species and even in different parts of the same joint (Table 10), quite apart from the changes which occur during maturation. Thus while vascular contacts are present in the human femoral head they are apparently absent in the rabbit femoral head (Greenwald & Haynes, 1969) and in some human vertebral articular processes (Enneking & Harrington, 1969). In the human femoral head, they are more numerous in contact areas (Woods *et al.*, 1970) where the cartilage is thicker. It is notable that the thickest avascular tissue in the body, the intervertebral disc, shows a large number of marrow contacts at the cartilage endplates, the fractional interfacial area available for diffusion being 11–36% (Maroudas *et al.*, 1975). From the published data, it is probable that the degree of vascular contact is related to the thickness of the cartilage (Woods *et al.*, 1970) and/or the cellularity

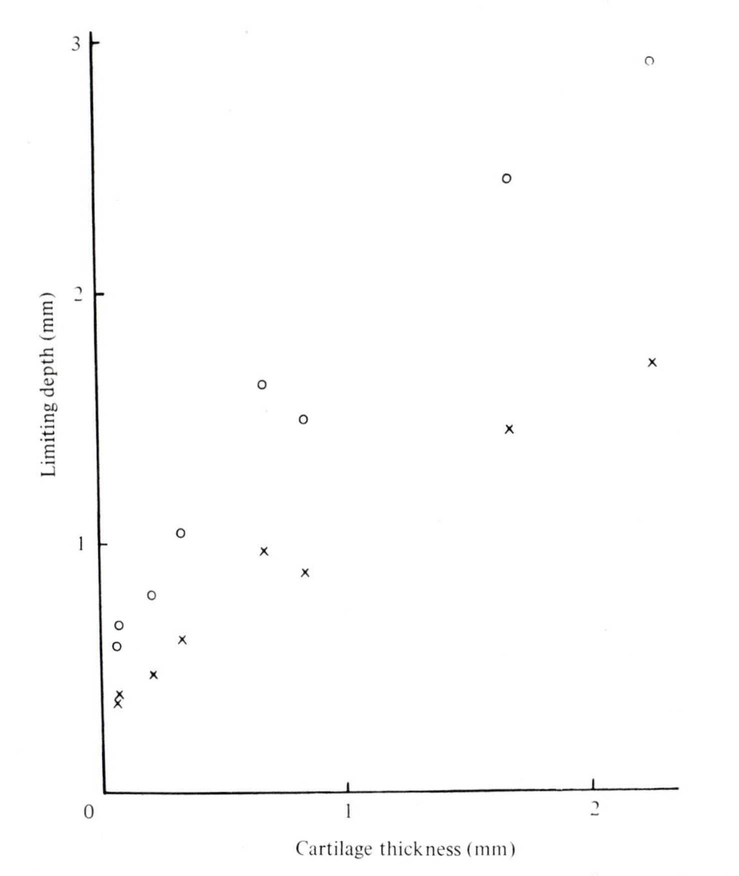

Fig. 82. Theoretical limiting depths of femoral condylar articular cartilage which can be nourished with glucose from the synovial fluid. Human experimental values (2.89 mm, stirred fluid (○); 1.7 mm, unstirred (×)) from Maroudas *et al.* (1968). Values for other species (cow, sheep, dog, rabbit, cat, rat and mouse) calculated from human values, allowing for differences in cell density but assuming no differences in glycolytic rate per cell or in diffusion coefficient, using the formula $d = \sqrt{(2PC_0/Q)}$, where d = limiting depth, P = diffusion coefficient, C_0 = concentration of glucose in synovial fluid, Q = glucose utilisation of the cartilage. Note that the unstirred limit falls below the actual thickness of the cartilage between 1.5 and 2.0 mm while the stirred limit is always in excess.

of the tissue. A survey of three joints of different size in each of three species (J. D. Connolly, personal communication) suggests that within each species the degree of contact, although small, increases with the thickness of the cartilage (Fig. 81). However, except where the cartilage is extremely thin, there are species differences in the interfacial contact area. If indeed the vascular contacts are concerned with nutrition, rather than, for example, basal remodelling of the cartilage (Lane, Villacin & Bullough, 1977), then the species differences might be related to differences in cellular oxygen utilisation.

Such small areas of vascular contact may be of little significance as regards the nutrition of the whole depth of the cartilage. Thus the total solute flux is proportional to the interfacial area between the cartilage and the contacting solution (Maroudas, 1973) as well as the diffusion characteristics of the tissue. Hence even in the human femoral head, flow through the deep surface will be limited to 9% of the flow through the articular surface. Nevertheless, the subchondral blood is normally always a 'stirred' source while the synovial fluid is dependent on joint motion for its agitation. It is possible that the actual thickness (and cell density) of cartilage found in animal joints is related more to the limiting depth determined by diffusion from 'unstirred' synovial fluid than that from 'well-stirred' fluid. Calculations suggest that the limiting depth for the supply of glucose from unstirred fluid becomes critical at a cartilage thickness of about 1.5–2 mm if cell density is taken into account (Fig. 82). This assumes that the diffusion coefficient and rate of glucose utilisation per cell of different species are similar. Thus it might be postulated that supply of glucose from the subchondral marrow spaces in large adult animals is required only in the bigger joints such as the human hip.

The limiting distance for nourishment from the subchondral source in the human femoral head is probably about 0.25 mm (i.e. less than 10% of the depth nourished via the articular surface) but this could become important when the joint is immobile and the limiting depth for supply via the articular surface thereby reduced (Maroudas *et al.*, 1968). If so, nutritional changes due to closure of the marrow contacts during ageing of the femoral head (Woods *et al.*, 1970) could be significant in relation to cell metabolism and, for example, the amount and type of matrix materials synthesised in the deep zone. Loss or changes (Stockwell, 1970) in these plasticising materials just above the tidemark might make this region more susceptible to the shearing stresses produced by joint loading and therefore to the horizontal splitting found pathologically (Fawns & Landells, 1953; Meachim, 1975; Sokoloff, 1973).

In most joints, however, supply comes effectively from the synovial fluid only. The interrelationship between articular cartilage thickness and cell density (Stockwell, 1971*b*) yields an approximately similar number of cells (25000) lying deep to 1 mm^2 of articular surface in all joints investigated. This suggests that the nutrient flux from the synovial fluid via 1 mm^2 of articular surface can support about 25000 cells in the adult. It is interesting that this is similar to the ratio of chondrocytes to perichondrial surface area in human adult costal cartilage (27000 cells per mm^2).

THE INTERVERTEBRAL DISC

Apart from its clinical significance, this avascular structure is of considerable interest in a nutritional context because of its large size. In man, for example, the intervertebral discs in the lumbar region are about 5 cm wide and 1.25 cm thick. Many authors have associated disc degeneration with inadequate nutrition (Ubermuth, 1929; Bohmig, 1930; Nachemson, 1969). Like other secondary cartilaginous joints, the intervertebral disc consists of a plate of fibrocartilage sandwiched between two thin sheets of hyaline cartilage, or endplates, which also act as growth cartilages during development. The peripheral part of the fibrocartilaginous plate, the anulus fibrosus, is more fibrous than cartilaginous. It consists of concentric collagenous lamellae which interlace obliquely as they spiral between their upper and lower attachments in the hyaline cartilage endplates, the external lamellae passing direct to bone or ligaments. The gelatinous nucleus pulposus is enclosed by the anulus peripherally; the endplates lie between the nucleus pulposus and the vertebral bodies.

BOUNDARIES

Cartilage canals are found in the anulus fibrosus during development (Bohmig, 1930) but in the adult the disc is avascular and entirely dependent on diffusion from blood vessels at its surfaces.

The fibrous surface at the margin of the anulus fibrosus is reinforced anteriorly and posteriorly by the longitudinal ligaments. Numerous blood vessels are present (Smith & Walmsley, 1951) equally distributed around the circumference, the smaller vessels tending to run horizontally around the disc in man (Maroudas *et al.*, 1975).

The cartilage endplates at the osseochondral interfaces are about 1 mm thick in man and are thinner at the periphery than at the centre. As in other similar boundaries, the cartilage immediately adjacent to the bone is calcified. The calcified layer is thinner (usually less than 50 μm) than in most articular cartilages (Nachemson, Lewin, Maroudas & Freeman, 1970; Maroudas *et al.*, 1975). The subchondral bone plate is perforated by numerous marrow spaces which often penetrate through the calcified cartilage in adult (Nachemson *et al.*, 1970) as well as immature specimens. Marrow contacts are both of the ampullary and canalicular variety with no difference between the two bone–cartilage interfaces of the disc (Maroudas *et al.*, 1975). However, consistent differences are found between the lateral and central portions of the endplate. The central portion in all normal specimens examined contains numerous marrow spaces while in only about one-third of specimens do the lateral portions show perforations (Nachemson *et al.*, 1970). In human discs about 10% of the bone–cartilage interface is occupied by contacts penetrating into the uncalcified cartilage; an additional 20–30% overlie marrow spaces separated from uncalcified cartilage by less than 25 μm of calcified tissue (Maroudas *et al.*, 1975). Permeability analysis shows that 0.36 of the interfacial area is available for diffusion.

The overall cellularity of the human disc is low, about 6000 cells per mm^3. The cell density is highest near the boundaries of the tissue, both at the anulus and the endplates, and rapidly falls to a low plateau within the substance of the disc (Maroudas *et al.*, 1975).

NUTRITION

Both the periphery of the anulus and the two bone–cartilage interfaces can act as nutrient sources. Thus in one-month-old rabbits, ^{35}S-sulphate injected intravenously is incorporated into the cells and the matrix, particularly in the peripheral parts of the nucleus pulposus (Souter & Taylor, 1970). Permeability experiments confirm that diffusion occurs through both types of surface.

In the immature rabbit, Brodin (1955*b*) finds that the cartilage endplate rapidly absorbs fluorescent dye injected intravenously, while the anulus fibrosus displays only a weak fluorescence. This suggests that the main nutritional route is through the endplate, a view expressed in several textbooks. However, in long bone epiphyses the subchondral marrow spaces are of much greater significance in immature than adult specimens and the same is probably true of the disc endplate. In adult human discs studied *in vitro*, Nachemson *et al.* (1970) confirm that the anulus fibrosus is permeable. However, the lateral parts of the endplate are less permeable than the central parts, showing a good correlation with histological evidence of perforations of the subchondral bone and calcified cartilage.

A quantitative study of human discs (Maroudas *et al.*, 1975) shows that the diffusion coefficient for glucose (2.5×10^{-6} cm^2 s^{-1}) in the anulus and the endplate cartilage is comparable to that in articular cartilage (2.7×10^{-6} cm^2 s^{-1}) although the bone–cartilage interface itself is much less permeable. The central part of the interface is about three times as permeable as the peripheral lateral portion, confirming the earlier studies. The proportion of the disc nourished from the endplate and from the anulus could be estimated tentatively by combining the permeability and cell density data with published values for chondrocyte glycolytic rate (Fig. 83). The calculations indicate that approximately equal volumes of the disc are nourished via the anulus and via the two bone–cartilage interfaces combined. Assuming low rates of glycolysis (Bywaters, 1937), the whole of the anulus is adequately nourished from the circumference of the disc. However, the calculations suggest that the centre of the nucleus pulposus is deprived of glucose: while this obviously cannot be true in life, these results do show that the nutritional conditions of the disc are precarious. It is possible that the diurnal fluctuation in disc thickness caused by expulsion and imbibition of fluid (Inman & Saunders, 1947) might aid nutrition. Nevertheless if the marrow contacts at the centre

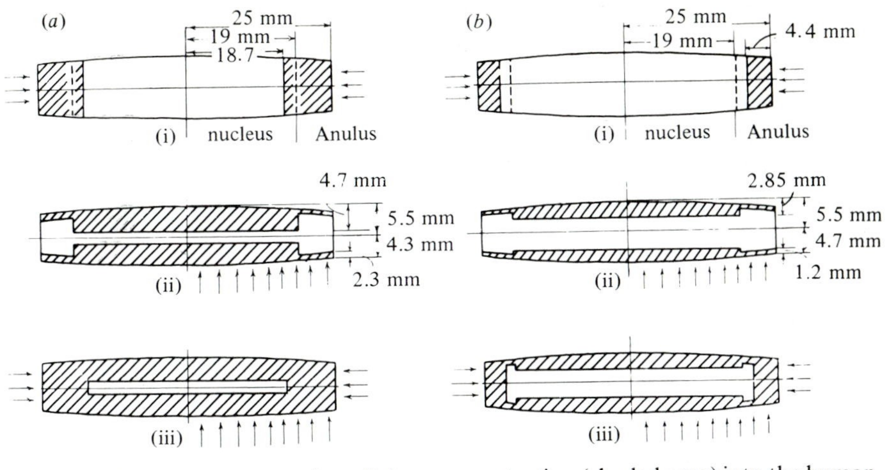

Fig. 83. Schematic representation of glucose penetration (shaded area) into the human lumbar intervertebral disc (L4–L5) in coronal section. (*a*) Where glycolytic rate = 1.9×10^{-11} mmol per cell per hour: (i) radial diffusion; (ii) diffusion through end plates; (iii) overall diffusion. (*b*) Where glycolytic rate = 5.2×10^{-11} mmol per cell per hour. With the lower glycolytic rate, nearly the whole of the disc can be adequately nourished. (From Maroudas *et al.*, 1975.)

of the endplate should become occluded an inadequate supply of nutrient might cause deterioration of disc tissue. However, although Nachemson *et al.* (1970) find a significant correlation between impermeability of the central portion of the endplate and disc degeneration, not all degenerate discs are impermeable.

Since the centre of the intervertebral disc apparently receives its nourishment only from the marrow spaces of the vertebra (Fig. 83), it follows that the number of cells (endplate and nucleus pulposus) nourished via 1 mm^2 of bone–cartilage interface can be estimated from the data supplied by Maroudas *et al.* (1975) for human discs.

$$\begin{aligned} \text{Cell number} &= \text{cells per mm}^3 \times 0.5 \text{ rostro-caudal disc thickness in mm} \\ &= 6000 \times 5.5 \\ &= 33000 \end{aligned}$$

However, the fractional interfacial area available for diffusion at the centre of the endplate = 0.36. Hence if 100% of the interfacial area were occupied by marrow spaces, about 90000 cells would presumably be accommodated.

Rather surprisingly, this figure is much *less* than the estimates obtained for the total cell number relating to 1 mm^2 of the surface area of the vascular cartilage canals in fetal and newborn epiphyses (about 0.3×10^6 cells). Furthermore, the values obtained for articular cartilages and for adult human costal cartilage (25000 and 27000 cells per mm^2 of boundary area) are also different from the above figure. Possibly the discrepancies can be accounted for by differences in the diffusion distance involved but there are many other variables which are difficult to quantitate. It would be satisfying to obtain a unified value for the different cartilages, immature and adult, with respect to the metabolic demand and the nutritional supply. Using a crude index – number of chondrocytes per unit area of nutritive cartilage boundary – it is possible only to give estimates for individual cases. More information is needed before this and other problems of chondrocyte nutrition can be resolved.

6

CHONDROGENESIS AND CHONDROCYTE DIFFERENTIATION

During early development many distinct cell types emerge to produce the tissues and organs of the body. The structural and functional characteristics of such fully differentiated cells involve gene regulation (Britten & Davidson, 1969), are inheritable and appear to be irreversible. However, under different environmental conditions certain differentiated cells, and connective tissue cells in particular, may undergo a process called modulation. This involves reversible changes both in structure and function. Thus modulated chondrocytes may resemble fibroblasts and synthesise Type I (skin type) collagen instead of the native cartilage Type II collagen (Mayne *et al.*, 1976*a*). Nevertheless it seems unlikely that chondrocytes ever proceed as far as true dedifferentiation, since it has yet to be demonstrated that cloned cartilage cells can be made to revert to a precursor clone which then changes to a clone of another differentiated cell type (Levitt & Dorfman, 1974). The relative contribution of intrinsic and extrinsic factors to chondrocyte differentiation during normal development remains unclear. Some progress has been made in recent years as to the nature of these factors and it is always to be hoped that further knowledge will lead to better understanding. However this may be, an essential preliminary requirement for the study of the differentiation of cartilage cells and of chondrogenesis is to define those characteristics which make the chondrocyte a unique cell type.

This is more difficult than would at first appear to be the case. Mature cartilage with its scattered cells lodged in their rounded lacunae surrounded by matrix is easily recognised. One might reasonably expect that the chondrocytes should show certain unique structural characteristics. True, several ultrastructural features may be described as typical, such as the scalloped cell margin and the prominent Golgi and rough ER membrane systems, but there is none that taken by themselves, or even as an ensemble, can be regarded as diagnostic. Inevitably, either consciously or subconsciously, histologists identify the chondrocyte by the nature of the surrounding materials; in practice the most reliable criterion is based on cell function, that is, evidence of the synthesis and secretion of matrix macromolecules. Indeed, modern concepts of differentiation appear to be based on the differing synthetic properties of cells especially with regard to proteins (Jacob & Monod, 1963). Holtzer, Weintraub, Mayne & Mochan (1972) regard differentiation as the emergence of daughter cells that synthesise molecules that their mother cells did or could not synthesise. In connective tissues, molecules such as collagen and proteoglycans are not confined to cartilage. Fortunately, cartilage matrix contains a collagen (Type II) which is unique to the tissue (Miller & Matukas, 1974) and a type of chondroitin sulphate proteoglycan which in aggregated form appears to be restricted to cartilage (McDevitt & Muir, 1976). Hence it may be possible to define a chondrocyte more precisely in biochemical than in morphological terms. However, in most studies of chondrogenesis the histologist has had to rely on the more empirical criterion of the presence of metachromatic matrix, although very recently immunofluorescent techniques have been developed for the localisation of Types I and II collagen (von der Mark *et al.*, 1976).

EARLY DEVELOPMENT OF THE APPENDICULAR AND POST-CRANIAL AXIAL SKELETON

Cartilage develops from mesenchyme which is usually mesodermal in origin although branchial cartilage can be derived from ectoderm via the neural crest (Horstadius, 1950). Regardless of its origin, pre-natal cartilage may have a variety of roles in development. Thus branchial cartilage may develop and persist into permanent cartilage such as the thyroid laminae (fourth and fifth arches), form ligaments as with the sphenomandibular (first arch) and stylohyoid (second arch) ligaments or form models for bones such as the auditory ossicles (first and second arches) which are the first bones to be fully ossified and to reach adult size (Anson, Bast & Cauldwell, 1948). Indeed apart from the permanent cartilage associated with the respiratory tract, the chief importance of cartilage is in providing temporary models for bone development and to participate in the formation of the various joints. Like the rest of the bony framework, the embryology of the appendicular and post-cranial axial skeleton has been studied intensively. It is helpful to consider the general pattern of cartilage development in these two regions since in recent years they have often served as models for experimental investigations of chondrogenesis and chondrocyte differentiation, particularly in the chick embryo. Larger works should be consulted concerning the comprehensive development of the skeleton.

DEVELOPMENT OF THE LIMB BONE BLASTEMA

The first morphological indication of the site of future cartilage is a condensation of mesenchyme. In the limb bud, the mesenchyme is produced by the rapidly proliferating tissue beneath the apical ectoderm (Fig. 84) and is ultimately derived from the lateral mesoderm of the body wall. The mesenchymal condensations for the bone primordia, or blastemata, are laid down in a proximo-distal sequence (Saunders, 1948) as the bud grows out from the body, during the third and fourth days of incubation in the chick embryo. Thus in the wing bud, mesenchymal condensations for the proximal

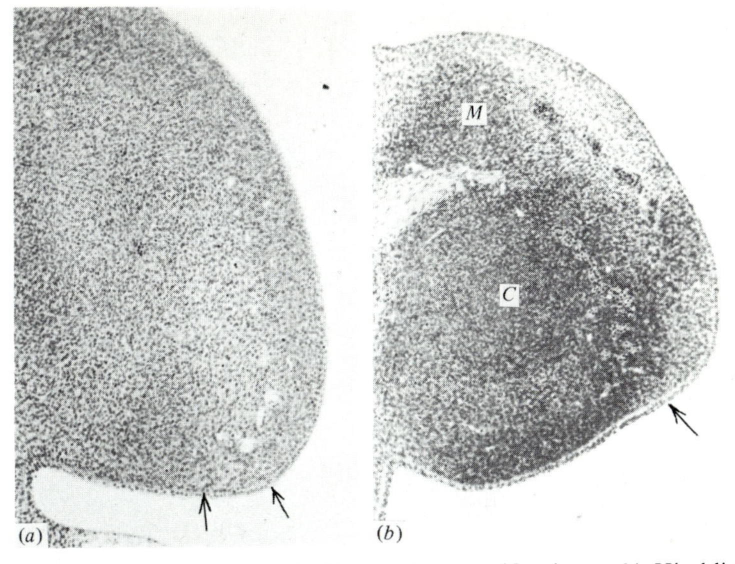

Fig. 84. (*a*) Human embryo (6 mm). Silver stain, magnification × 64. Hind limb bud prior to blastemal condensation, containing mesenchyme limited by the apical ectoderm (arrows). (*b*) Human embryo (10 mm). Iron haematoxylin, magnification × 64. Forelimb bud showing mesenchymal condensations forming the central chondrogenic blastema (*C*) and the peripheral myogenic tissue (*M*). At the tip of the limb bud is the apical ectoderm (arrow). (Material by courtesy of Professor G. J. Romanes.)

bone rudiments appear at Hamburger–Hamilton (Table 11) stages 21–22 (Gould *et al.*, 1972; Searls, 1973) and about one stage later in the leg bud (Fell & Canti, 1934). In human embryos, where development is slower, the blastemata are first recognisable histologically in the limb buds (humerus) at about the fifth week (Gardner, 1971) or Horizon XVI (Streeter, 1949) when the embryo is 7–10 mm long (Fig. 84).

Although the blastemata appear to be compact cell masses when viewed with the light microscope, ultrastructural studies of the chick wing have shown that the cells are not maximally close packed since up to half the volume may be intercellular space (Gould *et al.*, 1972; Searls, Hilfer & Mirow, 1972). However, in the chick hind limbs, although the cell density is no higher, Thorogood & Hinchliffe (1975) find a smaller extracellular space. The mesenchymal cells are round or spiky with many processes, and have a high nucleo-cytoplasmic ratio. As chondrification begins, at stages 24–25 (four to four and a half days) in the proximal part of the chick wing bud, metachromatic material appears in the intercellular space. As described by Fell (1925) and others, matrix first appears as a thin film over the cells which become separated as capsules form around them. Growth occurs by multiplication of cells, cell enlargement and secretion of matrix.

As a result the early cartilage mass presents a characteristic histological appearance (Fig. 85*a*). In longitudinal section, the central chondrocytes

TABLE 11

Development stages of chick embryo relevant to chondrogenesis: human equivalent stages are approximate only. Mesenchymal condensation and chondrification occur earlier in the wing than the leg bud. L = antero-posterior length of bud along body wall; W = distance from body wall to bud apex. References: Sensenig (1949); Streeter (1949); Hamburger & Hamilton (1951); Gardner (1971); Olson & Low (1971); Gould *et al.* (1972); Ruggeri (1972); Searls (1973)

Stage	Age	Number of somites	Wing	Leg	Axial skeleton	Human equivalent stage
7	23–26 hours	1	–	–	Notochordal sheath present at stages 8 and 9	21 days 2 mm
11	40–45 hours	13	–	–	Notochordal secretion of matrix material begins. Medial migration of sclerotome cells in anterior somites commences	Horizons XI–XII 2.5–3.5 mm
15	50–55 hours	24–27	Mesoderm present, limb area flat	–		
16	51–56 hours	26–28	Thickened ridge	Mesoderm present, limb area flat		
17	52–64 hours	29–32	Distinct swellings		Extracellular materials first seen near migrated sclerotome cells	
19	68–72 hours	37–40	L/W ratio: 4–6	L/W ratio: 4		
21–22	3.5 days	43–44	L/W: 2.3–2.7 Mesenchymal condensation begins in proximal rudiments	L/W: 2–2.5		Horizon XVI, fifth week, 7–10 mm
24	4 days	–	L/W: < 1 Digital plate not demarcated	L/W: < 1 Toe plate distinct		
25–26	4.5–5 days	–	Chondrocyte differentiation Chondrification of proximal rudiments		Chondrocyte differentiation Chondrification of vertebral body	Horizon XVII, sixth week, 10–12 mm

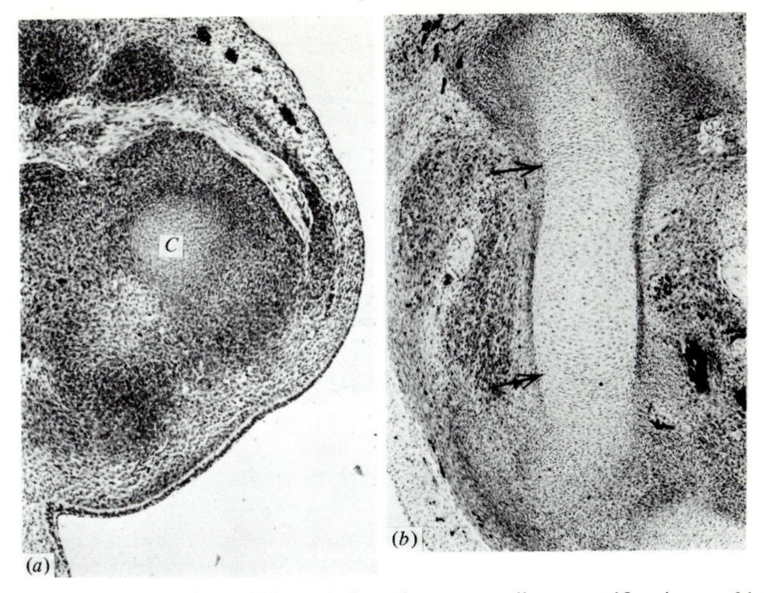

Fig. 85. (*a*) Human embryo (12 mm). Iron haematoxylin, magnification ×64. Early chondrification of the humerus (*c*). At the periphery of the blastema, a whorl of flattened cells is beginning to form around the central less cellular region. (*b*) Human embryo (15 mm). Iron haematoxylin, magnification ×51. Approximately longitudinal section of the cartilaginous rudiment of the femur. Note the elliptical orientation of the cells at the ends of the shaft (arrows) and the slight enlargement in the girth of the middle of the shaft.

appear to be elongated at right angles to the long axis of the blastema whereas at the margins of the cartilage the cells are condensed and appear to be elongated parallel to the long axis of the tissue, forming the rudiment of the perichondrium. A transverse section of the middle of the blastema shows a whorl-like formation of elongated cells around a less cellular central region containing large round chondrocyte profiles. Thus at first both the central and the peripheral cells have a discoidal shape, the central cells orientated transversely and the peripheral cells tangentially to the main body of the cartilage mass. Gould, Selwood, Day & Wolpert (1974), in agreement with Carey (1922) and Fell & Canti (1934), believe that the concentric orientation and apparent condensation of the peripheral cells is primarily due to the compressive effects of matrix secretion by the central cells, rather than aggregation due to a centripetal attraction of peripheral neighbouring cells as postulated by some authors. Gould *et al.* suggest that the transverse orientation of the central cells can be accounted for similarly in terms of the pressure effects of matrix secretions. This assumes that the resistance to expansion is greater along the long axis than the radial axis of the cartilage mass.

With further growth, the process of transverse orientation of the central

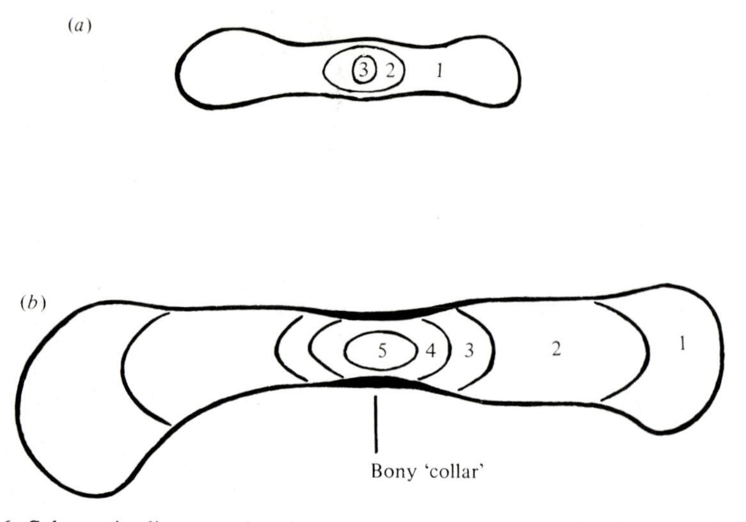

Fig. 86. Schematic diagram showing two stages in the growth of the cartilaginous rudiment of a long bone (*a*) before and (*b*) after the formation of the bony collar. The lozenge-shaped zones contain chondrocytes in different phases (1–5) of development.

cells spreads toward the ends of the blastema (Carey, 1922; Fell, 1925) where the long axes of the cells lie along ellipses convex to the future expanded ends of the cartilage model (Fig. 85*b*). Thereafter the perichondrium becomes bilaminar with an outer fibroblastic layer containing blood vessels and an inner layer which is chondroblastic, enabling the cartilage mass to increase its girth by appositional growth. In the centre of the blastema the cells enlarge, become polyhedral and cease to divide; some show signs of degeneration (Fell, 1925). Most of the chondrocytes enlarge to about three times their initial size before degenerating.

Similarly, in the human embryo, Streeter (1949) describes five phases in the life of the central cells of the blastema during chondrification of the humerus, which commences at Horizon XVII (about 10–12 mm). In phase 1, the cells acquire an environment of basophilic intercellular substance and in the next phase become orientated transversely to the long axis of the blastema. In phases 3 and 4 (Horizon XXI, 24 mm), the amount of matrix increases, the cells hypertrophy to three or four times their former size and their cytoplasm becomes vacuolated. In phase 5 (Horizon XXIII, 32 mm) the cells disintegrate and liquefy. There is no evidence of cell migration; mitoses occur in the first three phases but the cell density falls progressively. These changes in the chondrocytes extend to neighbouring regions of the cartilaginous diaphysis, forming 'concentric' lozenge-shaped areas containing cells of the various phases (Fig. 86). Thus it is possible to see all phases at a single chronological stage of development of the rudiment.

As the central cells hypertrophy(phase 3) a collar of bone is laid down deep to the perichondrium (periosteum) around the middle of the shaft (Fig. 87)

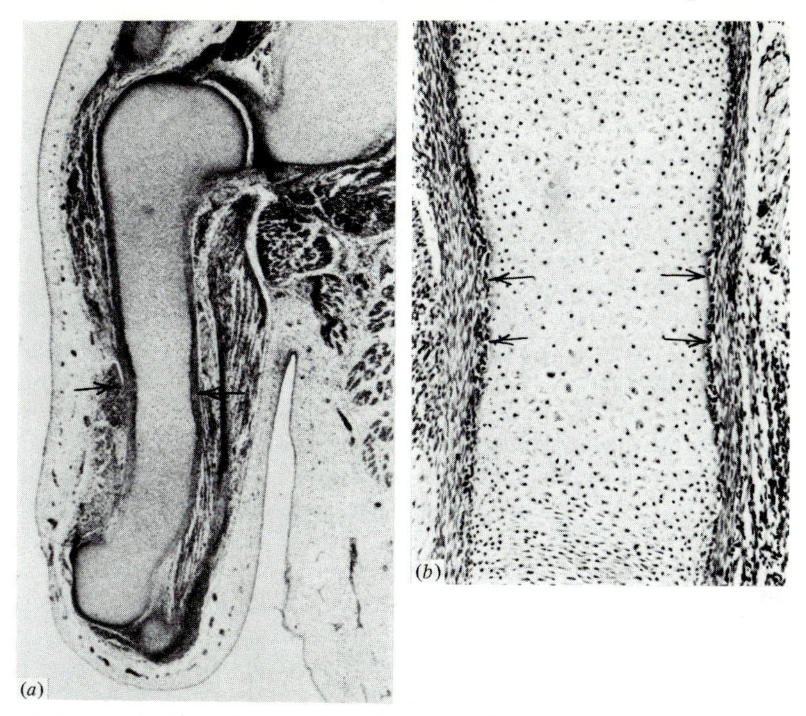

Fig. 87. Human embryo (25 mm). Iron haematoxylin. (*a*) Longitudinal section of the arm, showing rudiment of humerus at beginning of subperiosteal ossification. The middle of the shaft shows a constriction (arrows). Magnification × 20. (*b*) Middle of shaft at higher magnification. A narrow lamina of osteoid (arrows) has been deposited on the surface of the cartilage, which contains hypertrophied cells. Magnification × 80.

and in most long bones (Bloom & Bloom, 1940; Gardner, 1971) the cartilage matrix calcifies subsequently. With the entry of vascular mesenchyme from the periphery into the calcified cartilage, endochondral ossification begins.

DEVELOPMENT OF THE SCLEROTOME

The mesoderm of the somites is derived from the basal aspect of the ectoderm of the primitive streak and from the sides of Hensen's node (Hamilton, Boyd & Mossman, 1972). The newly formed primary (Hay, 1968) mesenchymal cells pass laterally and forwards between the columnar epithelium of the ectoderm or epiblast and the squamous epithelium of the endoderm or hypoblast to form the paraxial mesoderm on either side of the notochordal process. The mesoderm becomes segmented into compact masses, the somites, commencing at a level just behind the cephalic end of the notochord and progressing caudally. The anterior (cephalic) somites form at stages 7–8 in the chick embryo (25 hours of incubation) and by the twenty-first day in man (about 2 mm). The epithelioid cells of each somite enclose a cavity (the myelocoele) which becomes obliterated by cell proliferation. The ventromedial cells of the somite are known as the sclerotome (Goodsir, 1857) and give rise to connective tissue and cartilage. These cells proliferate rapidly and break away from the rest of the somite (containing myogenic and dermatogenic tissue), forming the secondary mesenchyme (Hay, 1968).

Separation occurs in the anterior somites at about stage 11 (45 hours) in the chick embryo. During stages 12–17 (2–2.5 days) in the chick and Horizons XI–XIII (2.5–3.5 mm) in human embryos (Sensenig, 1949) the sclerotome cells migrate medially toward the notochord (Fig. 88*a*). Cartilaginous matrix is first observed between sclerotomic cells aggregated around the notochord (Ruggeri, 1972; Minor, 1973; O'Hare, 1973), from stages 17–21 (2.5–3.5 days) onwards in the chick. Until this occurs, the skeletal axis of the body is represented solely by the notochord. For a time the 'specialised embryonic cartilage' of the perinotochordal region is said to form a continuous cartilaginous column (Peacock, 1951). In the chick, chondrification extends laterally to form the primordia of the vertebral bodies (Strudel, 1973).

At about the fourth to fifth week (6 mm) in man, each segment of sclerotome tissue has a caudal condensed part believed to migrate cranially (Prader, 1947*a*), which differentiates into the perichordal disc. The 'specialised embryonic cartilage' of this region contributes to the inner part of the anulus fibrosus (Prader, 1947*b*; Peacock, 1951). Lamellae of the outer part of the anulus fibrosus develop as early as the 15-mm stage (Walmsley, 1953). The nucleus pulposus is derived initially (commencing at 9 mm) from an enlargement of the notochord and later (from about 35 mm) by liquefaction of the inner anulus (Luschka, 1858; Sensenig, 1949; Peacock, 1951; Walmsley, 1953; Taylor, 1973).

The remaining caudal tissue of each segment fuses with the less condensed cephalic portion of the succeeding axial sclerotome to form the vertebral body (Remak, 1855, cited by Walmsley, 1953). Hence at this stage the axial skeleton consists of alternating blocks (Fig. 88*b*) of dense (the perichordal disc) and loose (the vertebral body) tissue, at first of equal thickness. The vertebral body is cartilaginous from the sixth week (11–12 mm) of development (Wyburn, 1944; Sensenig, 1949). Chondrification begins in centres on either side of the notochord but without the usual prior blastemal condensation (Sensenig, 1949). However, the formation of the secondary mesenchyme from the epithelioid somite may be regarded as equivalent to the condensation stage in the limb bud blastema (Searls, 1973). The notochord atrophies at the level of each vertebral body, and becomes the mucoid streak which if persistent appears to inhibit the growth of cartilage (Walmsley, 1953). The neural arches chondrify slightly later, in the more dorsal cells of the sclerotome alongside the neural tube; the transverse and costal processes form laterally. In the vertebral body, calcification of the cartilage matrix precedes the deposition of periosteal bone (Walmsley, 1953; Gardner, 1971) initially observed posteriorly. Centres of ossification for the body and each side of the neural arch appear at the eighth week.

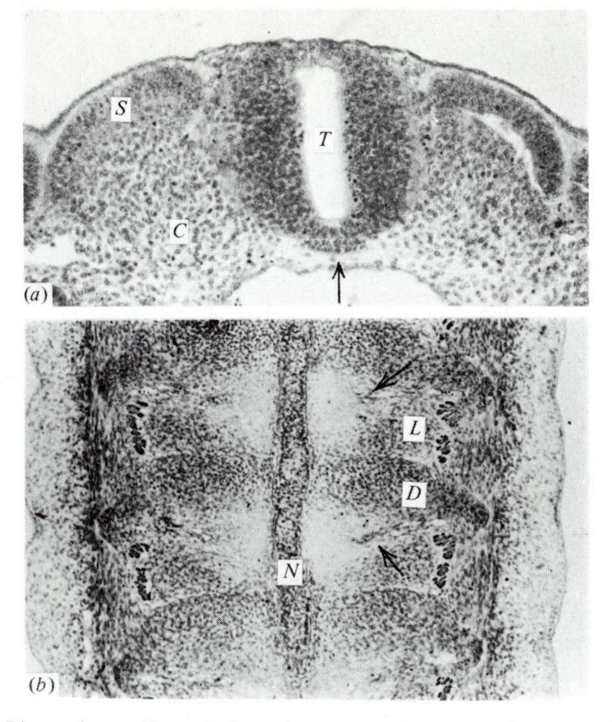

Fig. 88. (*a*) Pig embryo (5 mm). Iron haematoxylin, magnification ×100. Caudal region (transverse section) illustrating neural tube (*T*) with notochord (arrow) in close proximity. Sclerotome cells (*C*) have moved medially toward the notochord from the somites (*S*). (*b*) Human embryo (6 mm). Silver stain, magnification ×64. Caudal region showing the notochord (*N*) and its sheath in longitudinal section. Laterally the intersegmental arteries (arrow) lie between the sclerotomes which show alternating caudal condensed (*D*) and cephalic less condensed (*L*) portions. (Material by courtesy of Professor G. J. Romanes.)

CHANGES PRIOR TO CHONDRIFICATION

These accounts of the well-known developmental changes in the somites and in the limb bud blastema place the event of chondrification in perspective but tell us little of the transformations in the mesenchyme which result in the differentiation of chondrocytes. However, they pinpoint the stage at which cartilaginous matrix first appears and indicate structures such as the notochord which may have an influence on cartilage differentiation. They also show that there are similarities and differences in the events in the two regions, one obvious difference being the migration of the sclerotome cells and an obvious similarity the close association of mesenchymal cells at some stage prior to cartilage formation. It is now of interest to consider the changes which take place at the cellular level before chondrification and where possible to compare and contrast the two regions.

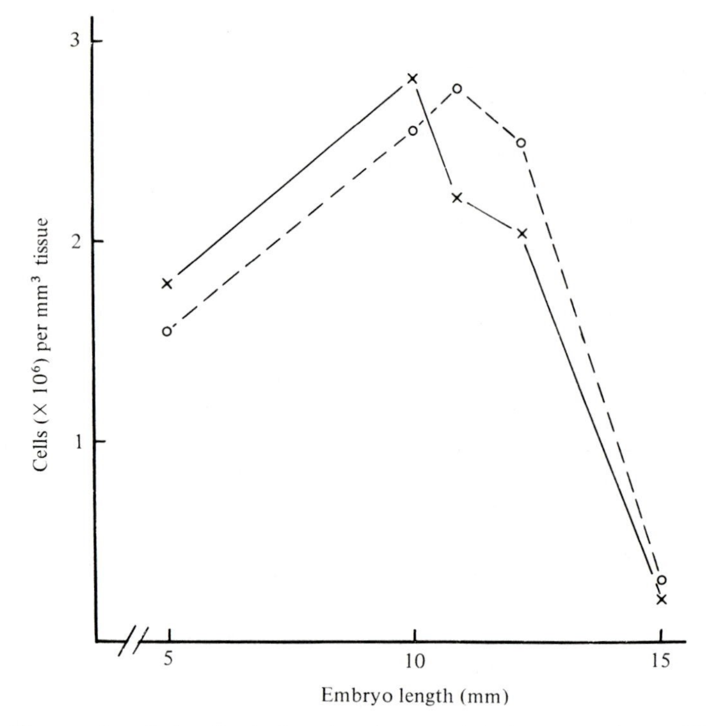

Fig. 89. Changes in cell density during chondrogenesis in human embryos. The values at 15 mm are for the mid-shaft of the rudiment. × = humerus; ○ = femur.

CELLULARITY

At mesenchymal condensation, cell density rises to a peak of about 2.7×10^6 cells per mm^3 as seen at the fifth week in human embryo limbs (Fig. 89). This value corresponds to those found by Gould *et al.* (1972) – 15 cells per 1000 μm^2 in an ultra-thin section – and by Searls (1973) for stages 22–24 (3.5–4 days) in the chick wing bud – 180–225 cells per 0.01 mm^2 in a 5 μm section – which both approximate to $2.0–2.5 \times 10^6$ per mm^3 if calculated per unit volume. A similar figure for cell density is found in the chick hind limb bud (Thorogood & Hinchliffe, 1975), about 70 cells in an area 0.069×0.05 mm in a 6-μm section, which is equivalent to about 2.0×10^6 cells per mm^3. In the earliest stage of chondrification *in vitro* in re-aggregating dissociated stage 26 chick limb buds, Ede & Flint (1972) find a rather lower cell density – 225 cells per 10^4 μm^2 in an 8-μm section – which approximates to about 1.5×10^6 cells per mm^3. These values are for fixed sectioned tissue and can take no account of possible shrinkage. However, it is interesting that this order of cell density is close to that which would be obtained at maximum packing if the cells were rigid congruent spheres. The relative volume (d) at maximum packing $= \pi/\sqrt{18} \simeq 0.74$ (Saaty & Alexander, 1975): assuming spheres 8 μm in diameter (the approximate diameter of the cells at condensation), the theoretical maximum 'cell' density is 2.76×10^6 per mm^3. In reality the cells are not spherical but polygonal and stellate and are also deformable, so it would be possible to have both a high cell density and a larger extracellular space than allowed for sphere packing. Despite this high cell density, which may be regarded as a maximum for cartilage or pre-cartilage, the extracellular space is considerable, at least in chick wing condensations, possibly owing to a smaller cell volume. There appear to be no comparable figures for sclerotome cellularity.

The increase in cell density during condensation is relatively small, rising by about 30–50% (Gould *et al.*, 1972; Searls, 1973) in the chick wing and about 60–70% in the hind limb (Thorogood & Hinchliffe, 1975); changes in the human embryo are on a similar scale (Fig. 89). In agreement with Searls (1967), Gould *et al.* believe that the modest increase in cell density found in mesenchymal condensation is not due to cell aggregation (Thorogood & Hinchliffe, 1975) by centripetal migration but rather arises from mitosis and the failure of the cells to move apart. Over the period of mesenchymal condensation Searls (1973) finds that there are considerable changes in rates of cell division. Until stage 20, there is a high mitotic rate in the 'central proximal' (chondrogenic) region of the wing bud: using tritiated thymidine the labelling index is about 40%. During stages 21–23 (at condensation) the labelling index falls to less than 20%, a much greater reduction than in the myogenic region. Searls calculates that 75% of the cells in the chondrogenic region withdraw from the cell cycle, the first non-dividing cells appearing at stage 21. Summerbell & Wolpert (1972) have attributed the fall in mitotic rate to the increase in cellularity ('density dependence').

Thus at condensation there is only a moderate increase in cell density, associated with a falling mitotic rate, but nevertheless reaching values close to the theoretical maximum.

ULTRASTRUCTURAL CHANGES

A number of studies have been made of the fine structure of chondrogenesis both in the limb bud (Godman & Porter, 1960; Goel, 1970; Gould *et al.*, 1972; Searls *et al.*, 1972; Thorogood & Hinchliffe, 1975) and in the axial mesenchyme (Hay, 1968; Olson & Low, 1971; Ruggeri, 1972; Minor, 1973). These provide considerable information about the cells and cell contacts and about the incidence of structures in the extracellular space.

In the chick wing bud prior to condensation, stages 18–19, mesenchymal cells (Fig. 90*a*) are stellate or polymorphic and randomly orientated, with many filopodia 0.1–0.4 μm in diameter (Gould *et al.*, 1972; Searls *et al.*, 1972). They have large round nuclei with a well-developed nucleolus. Golgi lamellar membranes, short channels of rough ER and many free ribosomes and glycogen granules are present in the cytoplasm. Mitochondria are quite

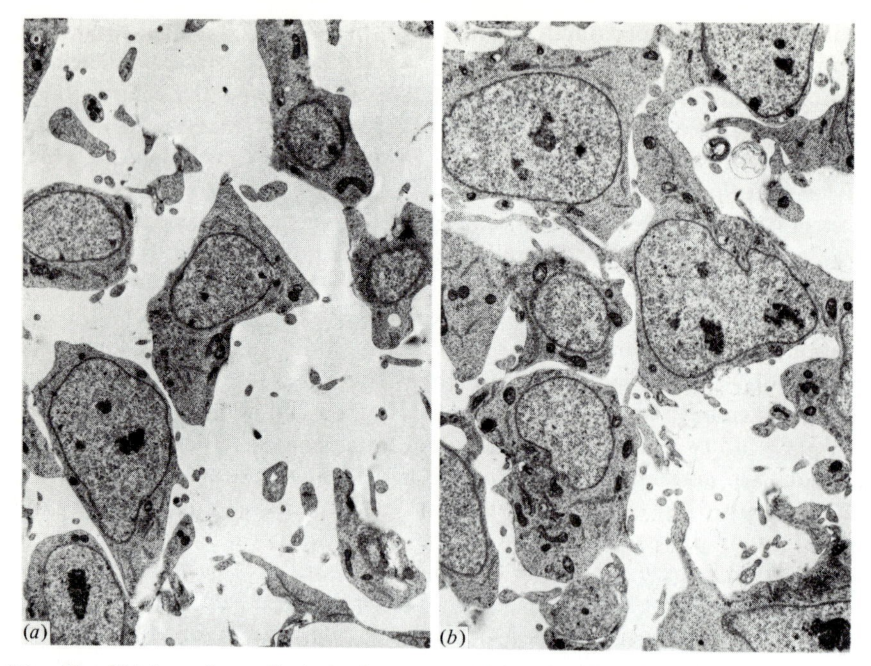

Fig. 90. Chick embryo limb buds (stage 24). Formalin–glutaraldehyde fixation, magnification × 3200. (*a*) Pre-condensation mesenchyme from distal part of limb bud. (*b*) Chondrogenic mesenchymal condensation from proximal part of limb bud. (From Gould *et al.*, 1972. Courtesy of Dr R. P. Gould.)

numerous, have a pale intramitochondrial matrix and well-formed cristae. Microtubules are present and microfilaments 5–7 nm in diameter are seen at the periphery of the cell in close relation to the plasmalemma. Although no specialised gap or tight junctions are described, the cells exhibit two types of contact – short lengths via the filopodia and broad contacts where two adjacent cells may be in apposition over 4 μm or more of their surfaces, with a space of 8–20 nm between the membranes. Similar features are seen in hind limb mesenchymal cells prior to condensation (Thorogood & Hinchliffe, 1975), but tight and gap junctions are also observed.

When condensation occurs in the wing bud (stages 21–22), the cells appear to have less cytoplasm (Searls *et al.*, 1972) but increase the number of filopodia, remaining rather spiky in form (Fig. 90*b*). The nuclei may be slightly smaller and more oval but no changes take place in the cytoplasmic organelles. Although the number of cells per unit area increases, there may be an increase also in the extracellular space, presumably associated with a decrease in cell volume (Searls *et al.*, 1972). In the distal part of the chick hind limb at stage 26, however, Thorogood & Hinchliffe (1975) find that the intercellular space is reduced, the cells becoming rounder with fewer filopodia. Such differences from the wing bud studies may be the result of improved fixation, although since the various investigators are in agreement about other aspects and other phases before and after condensation, it is odd that only the condensation phase should be affected. Possibly the selective nature of electron microscopy is involved. All are in accord as regards cell contact, however. Searls *et al.* (1972) note that the number of broad contacts is reduced although the short filopodial contacts increase; Gould *et al.* (1972), who find occasional gap junctions, comment that only occasional long contacts are seen; Thorogood & Hinchliffe (1975) find no wide areas of direct cell contact, although the cells lie close together.

At stage 25 in the wing bud and later in the distal hind limb, when metachromatic material is first detectable with the light microscope in the intercellular space, the cells become more elongated and have oval nuclei. A few large vesicles are now associated with the Golgi complex but other cytoplasmic organelles show little change. The number of filopodia increases still further, cell contacts are still observed but are rarely longer than 0.5 μm. Subsequently, the number of large Golgi vesicles increases, the rough ER becomes prominent with dilated cisternae, the cell contour becomes scalloped resembling that of the typical mature chondrocyte and cell contacts are reduced or absent.

As regards extracellular material, Searls *et al.* describe several types of structure. These include amorphous substance found especially on the cell surface, matrix granules (proteoglycan) 25 nm in diameter, beaded strands 3.5–6.5 nm in diameter and fine non-banded (< 30 nm in diameter) and thicker fibrils (35–70 nm in diameter) with cross-banding typical of collagen. Throughout the period studied, stages 18–27, there appears to be little qualitative change in the types of structure present but rather a gradual increase in their frequency, accelerating at stage 25 in the wing bud.

Similar changes occur in the sclerotome although here in addition the cells migrate toward the notochord. At the epithelioid stage of the somite, a basal lamina develops and the cells are joined by zonulae occludentes and other types of close junction (Hay, 1968). As the secondary mesenchyme of the sclerotome breaks away from the somite at stages 11–12 and migration commences, the cells retain vestiges of basal lamina material and broad tight and gap junctions are still seen. Until stages 17–18 (60–70 hours), prior to cytodifferentiation, the sclerotome cells are oval or stellate with prominent nucleoli, many free ribosomes and mitochondria but with sparse Golgi and rough ER membranes (Olson & Low, 1971; Minor, 1973). Apart from the remains of basal lamina material (Hay, 1968), no formed elements in the extracellular space are observed. By the third and the fourth days (stages 20–24) some sclerotome cells have aggregated around the notochord, becoming flattened as they orientate concentrically to it. The first extracellular structures are found between these cells, consisting of amorphous material, matrix granules and microfibrils, initially 5 nm, increasing to 15 nm in diameter but with no periodicity (Olson & Low, 1971; Minor, 1973). After the fourth day, cytoplasmic changes occur in the sclerotome cells. The free ribosomes dissipate and plentiful rough ER and Golgi membranes with secretory vacuoles appear; the cells begin to take on the appearance of chondrocytes (Olson & Low, 1971; Minor, 1973; Ruggeri, 1972). Although short cell contacts are still apparent at stage 24, their number has become considerably reduced.

Thus the ultrastructural features and changes in the sclerotome prior to chondrification closely resemble those of the limb mesenchyme, apart from the anatomical relationship to the notochord. These studies show that there are few cellular changes until metachromatic material appears in the matrix, when the organelles relevant to its synthesis and secretion become prominent. At mesenchymal condensation in the wing bud (and probably at formation of the secondary mesenchyme of the sclerotome), the cells become more stellate and very likely reduced in volume. The most consistent change is a progressive reduction in cell contact, apparently continuing through the period of condensation. It is noteworthy that despite the high cell density found at condensation in the limb bud, cell contacts are relatively sparse, including those of the gap junction type which are believed to facilitate electrical coupling and other forms of communication between cells. Similarly, cell contacts reduce in number as the sclerotome cells detach from the somite.

INCIDENCE OF MATRIX MACROMOLECULES

The distribution and characterisation of matrix macromolecules in mesenchymal tissues have been more intensively studied in the limb bud than in the somites. Although stainable metachromatic materials do not appear in the limb blastema until stage 25 (Fell, 1925), ultrastructural studies show that a few matrix granules and thin fibrils are present in the extracellular space as early as stage 18. Autoradiographic and biochemical studies confirm that polysaccharides and collagen occur before mesenchymal condensation.

Deposition of ^{35}S-sulphate in embryonic tissues can be detected as early as stage 7 when the first somite forms (Johnston & Comer, 1957), while sulphate uptake in the limb bud can be demonstrated during stages 15–19 (Searls, 1965*a*). Uptake in the limb bud is uniform until stage 22 when the level in the central (pre-cartilage) region begins to increase and that in the peripheral (soft tissue) area decreases. The sulphate is incorporated into chondroitin sulphates A and C (Searls, 1965*b*), which are detectable in limb bud tissues from stage 15 (Franco-Browder, de Rydt & Dorfman, 1963; Medoff, 1967) and into a third minor component identified as chondroitin sulphate B. Similarly, there is biochemical evidence that chondroitin sulphates A and C are synthesised in the somite at or before stage 17 (Franco-Browder *et al.*, 1963; Kvist & Finnegan, 1970). These substances have been detected even earlier (stage 14) histochemically, using the alcian blue–CEC method (O'Hare, 1973). Medoff (1967) finds appreciable activities of enzymes involved in glycosaminoglycan synthesis in cell suspensions of stage 19 limb buds; Lash

(1968) has found evidence of enzyme activity in stage 17 somites also. Thus substances related to those found in cartilage matrix and relevant synthetic enzymes are present long before chondrification begins.

Enzyme activity is considerably greater (two to four times) at stage 27 than at stage 19 (Medoff, 1967). Searls (1973) calculates that if the enzyme activity found at stage 27 is substantially restricted to the cartilage region of the limb, the enzyme level in the chondroblasts must be some 20–40-fold greater than in the mesenchyme of stage 19, on a per cell basis. Nonetheless, Searls suggests that both enzyme activity and the amount of sulphated polyanion increase steadily over stages 22–26: metachromasia occurs abruptly at stage 25 because the glycosaminoglycan (synthesised at an enhanced rate) is no longer being 'diluted' by cell divisions and so accumulates around each cell.

However, although cartilage-related polyanions are present at an early stage, considerable changes occur at chondrification. Thus in the somites there is an increase in the ratio of chondroitin sulphate to hyaluronate at stage 27 as compared with stage 17 (Kvist & Finnegan, 1970; Toole, 1972). Similar changes occur in the limb bud (Toole, 1972). Goetinck, Pennypacker & Royal (1974), using dissociated limb bud cells in culture, find that sulphate is incorporated into two types of proteoglycan. From stages 18–23, 90% of the proteoglycan has a peak elution volume of 40 ml (Peak II) on agarose columns, the remaining 10% eluting at 20 ml (Peak I). At about stages 23–24, the proportions of Peaks I and II change and, if stage 24 cells are cultured for nine days until cartilage nodules appear, undergo reversal (ratio of I/II, 89:11). Peak I is dissociated by 4-M guanidinium chloride and Goetinck *et al.* propose that it represents the cartilage-specific proteo–chondroitin sulphate aggregate, while Peak II is the ubiquitous form of the molecule. If this is so, the small amount of Peak I present in early limb bud tissue is consistent with the scanty matrix granules seen in electron micrographs of the mesenchyme (Searls *et al.*, 1972). Similar changes in proteoglycan quality have been noted in somite tissue before and after differentiation (Okayama, Pacifici & Holtzer, 1976).

These results suggest strongly that there is a change in the quality of the macromolecules secreted as cartilage differentiates: even though chondroitin sulphate is synthesised prior to chondrogenesis it may not be part of the correct macromolecule. It is possible that the change in quality is associated with the core protein part of the proteoglycan molecule. Levitt & Dorfman (1974) have tested the reaction of cultured limb bud cells to xylose. The presence of xylose in the medium is thought to eliminate the necessity for core protein, which under normal conditions is synthesised before the polysaccharide side chains of the proteoglycan. They find that pre-cartilaginous cells of stage 24 fail to respond to xylose, while differentiated chondroblasts react by enhanced synthesis of chondroitin sulphate. Levitt & Dorfman suggest that this is because the mesenchymal cells lack the organization of the ER required to accommodate the necessary multi-enzyme complexes.

In limb bud tissue, collagen is not detectable with routine histological techniques nor hydroxyproline with biochemical analytical techniques until metachromatic material has been deposited at stage 27 (Mottet, 1967). However, as noted in ultrastructural studies (Searls *et al.*, 1972), collagen is undoubtedly present before chondrogenesis although only occasional fibrils with the collagen periodicity and rather more numerous unbanded fine fibrils are observed at stages 18–19 in the mesenchyme. Chemical analyses of cultures of limb bud cells have yielded interesting information about the collagen content. Linsenmayer, Toole & Trelstad (1973*a*) find that Type I $((\alpha_1 I)_2\alpha_2)$ is predominant at stages 23–24 but at stages 25–26 (when metachromatic material appears), cells from the core of the limb begin to synthesise an $(\alpha_1)_3$ type (probably α_1II chains). The outer part of the limb (the myogenic portion) continues to form $(\alpha_1 I)_2\alpha_2$ collagen. Similar evidence has been obtained by Levitt & Dorfman (1974). The production of $(\alpha_1)_3$ continues to increase until ossification begins when Type I collagen reappears. This sequence of change has been confirmed using newly developed immunofluorescent histological techniques for Type I and Type II collagen (von der Mark *et al.*, 1976). In somite tissue also, a change in collagen type occurs at differentiation (Okayama *et al.*, 1976).

In conclusion, it is apparent that although sulphated glycosaminoglycans and collagen are present prior to cartilage formation, little if any is in the tissue-specific form, i.e. Type II collagen and proteoglycan aggregate, until just before metachromatic material becomes stainable.

INFLUENCE OF THE NOTOCHORD

Several investigators have established beyond doubt the inductive relationship between the notochord (and neural tube) and the somites *in vivo* (Holtzer & Detwiler, 1953; Grobstein & Parker, 1954; Watterson, Fowler & Fowler, 1954; Avery, Chow & Holtzer, 1956; Lash, Holtzer & Holtzer, 1957). If the notochord/neural tube are removed the vertebral chondrocytes fail to differentiate, while if these organs are grafted into parts of the somite which are normally pre-cartilaginous, the somite cells proliferate and cartilage is formed. Similar observations have been made *in vitro* (Grobstein & Holtzer, 1955; Avery *et al.*, 1956; Hay & Meier, 1974; Kosher & Lash, 1975). The somites are the only tissue to respond to the notochord by producing cartilage (Holtzer & Matheson, 1970), although there is evidence that limb bud ectoderm can also promote somite differentiation and that spinal cord can stimulate chondrogenesis in lateral plate mesoderm (O'Hare, 1972*b*, *c*).

ULTRASTRUCTURAL CHANGES IN THE NOTOCHORDAL REGION

In the chick notochord (Fig. 91) at stages 8–13 (Ruggeri, 1972; Bancroft & Bellairs, 1976) the cells contain abundant rough ER with some dilated cisternae and a prominent Golgi complex; a basal lamina 25 nm thick

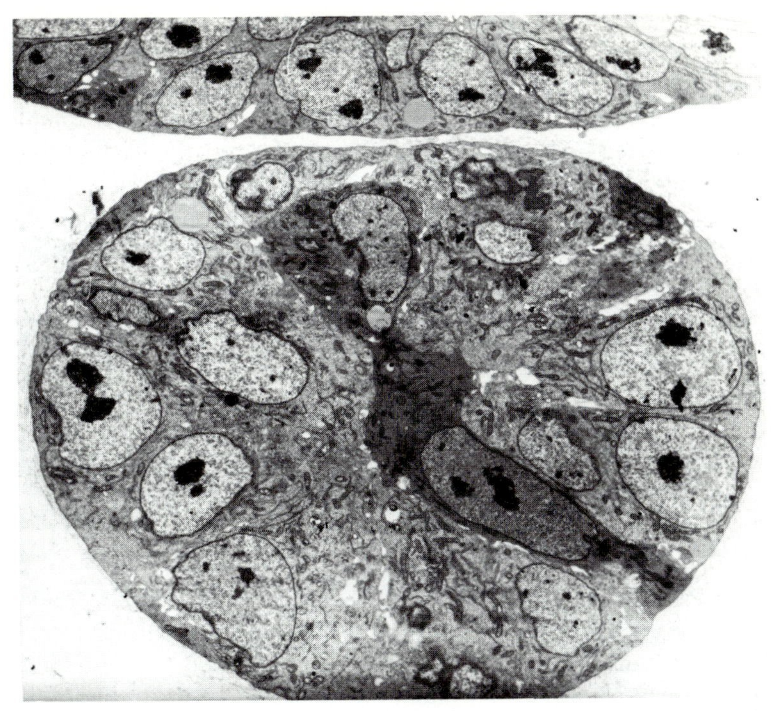

Fig. 91. Chick embryo (stage 15). Magnification ×1300. Transverse section of notochord close to the basal part of the neural tube (top of picture). The notochordal sheath is just visible. (Courtesy of Dr Ruth Bellairs.)

surrounds the notochord. External to the basal lamina, a fibrillar (Duncan, 1957) sheath forms at about stages 8–9, containing amorphous material, microfibrils 5 nm in diameter and banded (20-nm periodicity) fibrils 15 nm in diameter. As late as stage 17, extracellular structures such as matrix granules and fibrils are still separated from the nearest migrating sclerotome cell by 4–10 μm (Minor, 1973). During stages 10–17, histochemical stains for polyanions, using ruthenium red and alcian blue, also indicate that chondroitin sulphate is present in the basal lamina and in the notochordal cells but not in and around the cells of the neighbouring sclerotome (Ruggeri, 1972).

After stages 14–17, the notochordal cells enlarge and become highly vacuolated, with distended rough ER and many secretory Golgi vacuoles (Ruggeri, 1972; Bancroft & Bellairs, 1976). The basal lamina becomes thicker (40 nm) and externally the notochordal sheath becomes enriched with fibrils and granules which spread out toward and among the sclerotome cells. Histochemically, chondroitin sulphate is seen initially as a halo around the notochord, expanding thereafter around the nearest sclerotome cells, while a hyaluronidase-resistant polyanion with a high CEC (probably heparitin sulphate) is confined to the basal lamina. Autoradiographic studies are in agreement; ^{35}S-sulphate is localised to the notochordal cells and the expanding notochordal sheath (Olson & Low, 1971; Ruggeri, 1972; Minor, 1973). Similar observations have been made in rabbit embryos (Leeson & Leeson, 1958). Biochemical analyses have confirmed that the polyanions consist of equal amounts of chondroitin sulphate A and C and a smaller quantity of heparitin sulphate (Kosher & Lash, 1975). As regards the fibrillar material, there seems no doubt that it is collagenous, though the fibrils lack the typical periodicity. Deuchar (1963) obtained autoradiographs showing proline uptake around the notochord at the third day (stage 18) and amino acid analyses (Minor, 1973) confirm the presence of collagen. Furthermore, analyses both of neural tube (Trelstad, Kang, Cohen & Hay, 1973) and of notochord (Linsenmayer, Trelstad & Gross, 1973*b*) provide evidence that it is Type II rather than Type I collagen.

Thus the secretion by the notochord of matrix materials ultrastructurally and biochemically of the cartilage type precedes their appearance in the vicinity of the sclerotome cells.

EFFECT OF NOTOCHORDAL MACROMOLECULES

A number of experiments indicate that the perinotochordal polyanions and collagen have a cartilage-promoting role. Thus using stage 17 somites as the assay tissue, it can be shown that living notochordal cells need not be present (Kosher, Lash & Minor, 1973). Similar conclusions apply to the stimulation of chondrogenesis in the somites of stages 9–12 by lethally irradiated spinal cord (O'Hare, 1972*c*), and to scleral cartilage formation in neural crest cell cultures by killed retinal pigmented epithelium (Newsome, 1976). The evidence suggests that extracellular macromolecules are responsible, contained in or near the basal lamina (O'Hare, 1972*b*, 1973). First, proteoglycans extracted with 4-M guanidinium chloride from 10–15-day chick vertebral cartilage promote chondrogenesis *in vitro* in stage 17 somites (Kosher *et al.*, 1973). Such proteoglycans had a high content of chondroitin sulphate although they also contained another glycosaminoglycan and one or more glycoproteins. Conversely, if notochords are treated with enzymes which degrade and extract proteoglycan, the stimulatory effect on stage 17 somites is markedly impaired (Kosher & Lash, 1975). Secondly, again using stage 17 somites, chondrogenesis is stimulated by both Type I procollagen (skin fibroblast) and collagen (calf skin) and by Type II collagen (lathyritic chick sternal cartilage) (Kosher & Church, 1975), agreeing with observations on failure of stimulation by collagenase-extracted notochords. Since notochordal collagen appears to be Type II (Linsenmayer *et al.*, 1973*b*) it is significant that in these experiments Type II had a greater effect than Type I on the somites; procollagen was less effective than either.

Thus chondroitin sulphate-containing proteoglycans and collagen secreted by the notochord appear to be an important factor in somite chondrogenesis. It should be remembered, however, that stage 17 somites have been closely related to the notochord for many hours *in vivo* prior to the experiments *in vitro*.

POSSIBLE MODE OF ACTION OF MACROMOLECULES

It appears probable that the proteoglycans and collagen exert their effects on somite cells by an interaction with the cell surface (Kosher *et al.*, 1973), as suggested by Nevo & Dorfman (1972) for the stimulation of proteoglycan synthesis in mature chondrocytes by proteoglycan. In an analogous system, Hay & Meier (1976) have used trans-filter induction of corneal epithelium by lens capsule containing ^{3}H-proline-labelled collagen. They find that the stimulatory effect is related to the surface area of epithelial cell processes in contact with the capsule; collagen does not enter the cell. They suggest that a specific receptor glycoprotein in the cell membrane could be involved as part of a 'second messenger'. Kosher (1976) believes that cyclic AMP is the 'second messenger'. He finds that notochord and collagen-induced somite chondrogenesis is inhibited by derivatives of cyclic AMP. He postulates that (i) somite chondrogenesis is antagonised by high intracellular levels of cyclic AMP, and (ii) interaction of collagen or proteoglycan with the cell surface results in a lower level of intracellular cyclic AMP, possibly by affecting adenylate cyclase. It is known that somatomedin ('sulphation factor'), which stimulates the synthesis of proteoglycans by chondrocytes (Salmon & Daughaday, 1957), inhibits adenylate cyclase in membrane preparations of chick embryonic chondrocytes and other tissue cells (Tell *et al.*, 1973).

However, not all observations are compatible with this concept. Thus Toole (1973) finds that cyclic AMP prevents the inhibition of cartilage expression by hyaluronate in dissociated stage 26 somite chondrocytes. Nevertheless it should be mentioned that several other nucleotides, including cyclic GMP, were equally effective; furthermore, although thyroxin, growth hormone and calcitonin prevented the hyaluronate inhibition, a number of hormones known to stimulate production of cyclic AMP failed to do so. Kelley & Palmer (1976) suggest that differentiation is associated with increased levels of cyclic AMP, since treatment of limb bud mesenchymal cells with hyaluronidase increases the generation time, stimulates synthesis of RNA and raises the level of adenylate cyclase activity. Further work is needed on this aspect of the problem.

STATUS OF MACROMOLECULES IN CHONDROGENESIS

A number of polyanions, including the proteoglycans though not hyaluronic acid, stimulates differentiated cartilage cells to synthesise and secrete cartilage proteoglycan by a positive feedback mechanism, probably involving interaction with the cell surface (Nevo & Dorfman, 1972; Wiebkin *et al.*, 1975). Proteoglycans and collagen elaborated by the notochord and neural tube enhance chondrogenesis in stage 17 somites; much of the collagen appears to be of the cartilage type and cartilage chondromucoprotein from external sources will also stimulate chondrogenesis. It is of interest, therefore to consider the significance of collagen and proteoglycan in chondrogenesis.

The simplest view would be that, just as with the fully differentiated cells, collagen and proteoglycan cause the synthesis of matrix by stimulating and amplifying enzyme pathways which already exist. This could merely reflect the presence of an increasing population of 'active' cells, present in only small numbers prior to stage 25. Heterogeneity of this kind, however, is unlikely. Thus there are no intensely active cells in either the mesenchyme as a whole or in the central (pre-cartilaginous) region of the limb bud as observed in autoradiographs following ^{35}S-sulphate uptake (Searls, 1965*a*). In stage 17 somites also, there seem to be no chondroblasts present (Abbott, Mayne & Holtzer, 1972) since inhibition of chondrogenic expression by 5-bromo-2-deoxyuridine is permanent, unlike the reversible effects on functional chondrocytes.

Alternatively, prepared enzymatic pathways might be present in each and every cell. That mesenchymal cells prior to cartilage formation can synthesise and secrete molecules akin to those found in cartilage itself is suggested by many of the early results. The possibility that pre-chondrogenic cells already possess the requisite enzyme pathways is consistent with the ideas of Kosher & Lash (1975) who suggest that chondrogenesis is a two-step process. First the somite cell, for example, must acquire a 'bias' toward differentiation into a chondrocyte and secondly the 'biased' somite cell must interact with extracellular macromolecules of the cartilage type. A 'bias' is suggested because the somites are the only tissue to respond to the notochord by forming cartilage. Similarly, only somites will respond to irradiated and killed neural tube, although living neural tube can stimulate lateral mesoderm to form cartilage as well (O'Hare, 1972*c*). Other workers also consider that the interaction with proteoglycan and collagen has a 'stabilising' rather than an initiating role (Levitt & Dorfman, 1974).

It is probable of course that cartilage macromolecules confer stability in a passive sense. Thus Searls & Janners (1969) showed that when the central regions of limb mesoderm are transplanted into host limb buds (stages 22–27) the donor tissue does not usually form cartilage unless it is taken from stage 25 or older embryos, leading to the conclusion that the pre-cartilage blocks of tissue 'stabilise' as regards cartilage formation at stage 24–25. Searls & Janners suggest that stabilisation, which occurs at the time that the cells begin to be surrounded by metachromatic material, is indeed due to the accumulation of the proteoglycans. Stabilisation in these circumstances may be no more than the result of the seclusion, from soft tissue influences, afforded to the chondroblasts by the matrix. In a general sense it may be analogous to the protection given by the matrix in adult cartilage against the effects of antisera (Millroy & Poole, 1974).

However, it has been shown quite clearly that the type of proteoglycan and collagen change at the time of differentiation, i.e. when extracellular metachromatic matrix appears, both in the limb bud and in the sclerotome. Hence initially at least, the 'stabilising' effect of the macromolecules would have to be mediated in a more active sense, probably through an interaction with the cell membrane, resulting in a qualitative change in the pattern of synthesis. Some degree of membrane reorganisation may take place in the cell, possibly to facilitate multi-enzyme complexes. Certainly a quantitative change takes place with respect to the amount of rough ER, but whether such change constitutes 'stabilisation' or whether more fundamental changes occur is not clear.

Not all the evidence, however, suggests that prior interaction of the undifferentiated cell with proteoglycans or collagen is necessary for new cartilage formation. Thus limb buds explanted as early as stage 17 and cultured without addition of such molecules eventually produce cartilage (Dienstein, Biehl, Holtzer & Holtzer, 1974). Even in the case of the axial mesenchyme, stage 17 (Abbott *et al.*, 1972) or earlier somites (Ellison, Ambrose & Easty, 1969; O'Hare, 1972*a*) form cartilage when cultured alone. It is obvious of course that such cultured somites have already interacted or at least have had the opportunity to interact with the notochord (or neural tube) and possibly also with its proteoglycans and collagen. However, Okayama *et al.* (1976), in contra-distinction to Kosher's findings (Kosher *et al.*, 1973; Kosher & Church, 1975; Kosher & Lash, 1975), observe that addition of proteoglycan and collagen to cultures of somites or of limb buds does not induce the secretion of cartilage-type proteoglycans and collagen. Okayama *et al.* suggest that the effects of these molecules may merely reflect a 'non-specific enrichment' of the environment and that the real purpose of the interaction with the notochord is to permit preprogrammed presumptive chondroblasts to undergo mitoses and differentiate into definitive chondroblasts.

While the idea of 'stabilisation' by proteoglycans and collagen may be plausible in the case of the sclerotome cells, it appears to be rather improbable in the limb bud. Here, there is no bountiful source of 'stabiliser' comparable to that offered by the notochord. It is true that the transition at stages 24–25 may be in the *proportions* of the cartilaginous and non-cartilaginous types (which are already both present) of proteoglycan and collagen, rather than an absolute change from an all non-cartilaginous to an all cartilaginous variety. Therefore a small amount of the cartilaginous variety of these molecules could be present earlier in the blastema and act as the 'stabiliser'. However, Strudel (1973) has shown that limb buds explanted at 2.5–3 days (stage 17) and cultured in the presence of collagenase and hyaluronidase will still produce cartilage, although somites taken at the same stage of development and cultured under the same conditions fail to do so. Thus in the limb bud at least, it would seem that the presence of macromolecules of the cartilage type is unnecessary for chondrocyte differentiation.

Hence the role of the proteoglycans and of collagen in chondrogenesis creates something of a dilemma. If they have a significant inductive and 'stabilising' effect in the case of the sclerotome, why does this not happen in the limb bud? What is the equivalent mechanism in the limb bud? On the other hand, if they are of little significance as regards 'stabilisation' in the axial mesoderm, merely providing 'enrichment', why does the notochord produce them in such abundance? It is possible of course that the notochord and its secretion products have a chemotactic role directing the medial migration of the sclerotome cells, as suggested by Bancroft & Bellairs (1976), so that the cells or their progeny will form cartilage in the right place. The mode of growth of the limb bud should not require a similar guiding mechanism and hence no production of cartilage-type macromolecules prior to differentiation. However, the problem of whether or not the macromolecules act as 'stabilisers' in the axial tissues, mechanisms for the initiation of differentiation, or 'biasing' (Kosher & Lash, 1975), still requires elucidation in the sclerotome and also particularly in the limb bud.

CONTROL OF CHONDROGENESIS IN THE LIMB BUD

Wolpert (1976) states that in the limb bud there appear to be two general possibilities.

(i) The mesenchymal cells may all be similar (all 'limb bud cells', a view taken by Levitt & Dorfman, 1974) and labile, capable of differentiating into cartilage or other tissues up until stage 24. At stage 25 the cells become either chondroblasts or soft tissue cells, due to controlling factors which are probably environmental.

(ii) The mesenchymal cells are heterogeneous, having entered two separate cell lineages (cartilage and muscle) at an early stage. Although environmental factors may be involved during the evolution of the lineage, the 'steps' in differentiation are essentially autonomous.

These two alternatives differ as to the significance attributed to environmental control and as to the stage at which a cell lineage becomes committed to cartilage expression.

ENVIRONMENTAL FACTORS

Searls (1973) discusses a number of possible sources of control, including predetermination of cartilage according to the mosaic theory (Murray & Huxley, 1925), control by proximal influences, control by nerves and blood vessels, cell density, nutritional control by metabolites and oxygen tension and control by distal influences. The last two factors – i.e. anoxia and the influence of the apical ectodermal ridge – are likely to be the most significant in cartilage formation. As to the other possible types of control, he concludes that the limb mesenchyme appears to be completely regulative and is not in any sense committed to a cartilage phenotype prior to stage 22; in any case it is difficult to adhere to a rigorous mosaic concept in the face of the proximo-distal sequential formation of the blastemata (Zwilling, 1961). Although the axial mesenchyme is involved in the initiation of limb outgrowth, Searls finds no evidence that it is involved in cartilage differentiation. Nor are nerves and blood vessels required since early limb bud (stages 15–18) grafts *in vivo* and *in vitro* can produce cartilage and these are not invaded by blood vessels. Although some experiments indicate that a high cell density is favourable (Bassett & Herrmann, 1961; Ede & Agerbak, 1968), Searls points out that the cell density by itself is not critical since at mesenchymal condensation it is not greatly elevated, and, moreover, that the myogenic blastema is just as cellular.

Nutritional factors and anoxia

The avascular state of cartilage has naturally prompted interest in the role of oxygen in chondrogenesis. Thus Bassett & Herrmann (1961) note that cultured connective tissue cells produce cartilage at low oxygen tension (5%) but bone or fibrous tissue at high tension. Using dissociated chick embryonic sternal chondrocytes, Pawalek (1969) found that cartilage formation and sulphate uptake are stimulated by low oxygen tension (5%) provided that thyroxin is present. Similar effects of low oxygen tension in stimulating proteoglycan synthesis have been reported by others (Nevo, Horwitz & Dorfman, 1972). Although studies on adult articular chondrocytes (Marcus, 1973; Lane *et al.*, 1976) suggest that low oxygen tension is *tolerated* more successfully by chondrocytes than by other cells, it is nevertheless possible that anoxia plays a key role in cartilage differentiation (Hall, 1970).

Extracellular material at stages 25–27 is most abundant at the middle of the blastema perhaps because of low oxygen levels there. According to Caplan & Koutroupas (1973), as the blood vessels form in the chick limb buds during stages 21–22 the prospective cartilage-forming areas are less vascularised than the muscle areas; at stage 24 the core of the limb is avascular while the muscle blastema is well vascularised. The anatomical difference in vascular arrangement slightly precedes the differential uptake of ^{35}S-sulphate at stage 25. Since the cartilage blastema is only about 0.1 mm in diameter, it might be thought to be too thin for oxygen not to penetrate. However, it is worth recalling, for a moment, certain observations on articular cartilage. In the adult the number of cells accommodated in a full depth column of articular cartilage of 1 mm^2 cross-sectional area is remarkably constant in all joints and species, which suggests that there may be limits set on that number by nutritional requirements and diffusion conditions (see Chapters 3 and 5). The thickness of articular cartilage of the adult mouse femoral condyle is of the same order (0.05 mm) as the radius of the chick limb bud blastema (Fig. 92), yet the cell density of the blastema is about seven times that of the mouse tissue and the oxygen utilisation of embryonic chondrocytes is much greater

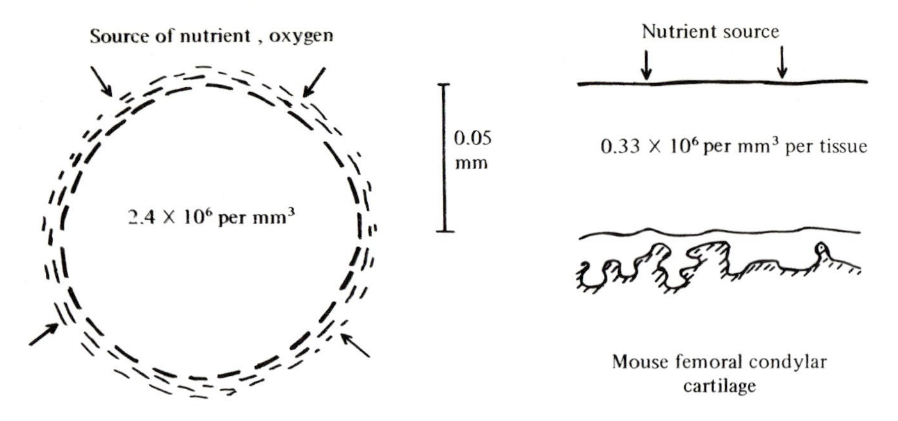

Fig. 92. In adult articular cartilage the number of cells beneath a unit area of articular surface is remarkably constant and may be limited by nutritional factors. The dimensions of the blastema and of adult mouse femoral condylar cartilage are comparable but the blastema has a much higher cell density; embryonic cells also have a much higher respiratory rate than adult cells. Hence if there is a limitation on the dimensions or cellularity of adult cartilage due to exhaustion of oxygen or nutrients, this is also possible in the case of the blastema. Therefore a low oxygen tension could occur at the centre of the blastema and enhance cartilage expression.

than that of adult cells (Boyd & Neuman, 1954). It is therefore conceivable that there is a low oxygen tension at the centre of the blastema. Again, the enhancement of spontaneous cartilage formation in cultured somites covered with light mineral oil (Ellison *et al.*, 1969) may be due to the restriction of gaseous exchange.

Changes occur in the NAD (nicotinamide adenine dinucleotide, coenzyme I) level of limb bud cells prior to chondrification (stages 23–24). It is proposed by Caplan (Caplan, 1970; Rosenberg & Caplan, 1974) that initiation of chondrification is associated with low levels (50% reduction) of NAD and myogenesis with high levels of NAD. As outlined above for oxygen tensions, it is postulated that the levels of coenzyme (or of a precursor, nicotinamide) result from the anatomical relationships of the blastemata to the blood vessels of the limb bud (Caplan & Koutroupas, 1973). Support for this hypothesis comes from experiments using 3-acetylpyridine (3-AP – a teratogenic pyridine analogue) which enhances cartilage formation in cultured stage 24 limb bud cells (Caplan, 1970). Searls (1973) considers that this effect of 3-AP may be the result of an induced anoxia. Thus 3-AP is incorporated into 3-AP-NAD which neither acts as a hydrogen acceptor so readily as does NAD with most dehydrogenases (Kaplan, Ciotti & Stolzenbach, 1956) nor transfers electrons to the cytochromes so efficiently as NAD (Weber & Kaplan, 1957). Searls' view is compatible with the low NAD levels normally found in chondrogenic areas (Rosenberg & Caplan, 1974). Nevertheless, at first sight this interpretation appears to be at variance with the results of analysis of limb bud tissue following incubation in 3-AP, which suggest that 3-AP causes a decrease in NAD levels (Rosenberg & Caplan, 1975). However, this apparent diminution in NAD may again be explicable in terms of the properties of 3-AP-NAD, this time in relation to the catalytic cycling technique (Matchinsky, 1971) used to measure NAD levels. This essentially measures the amount of glutamate formed by reduction of α-ketoglutarate (NADH → NAD) in the cycling stage. Since the oxidation reaction (glutamate → α-ketoglutarate) is stoichoimetric using 3-AP-NAD, the equilibrium constant being several orders higher than with NAD (Kaplan *et al.*, 1956), little glutamate would be formed. Thus after incorporation of 3-AP into NAD by the tissue it is possible that the analytical technique used would indicate false low levels of 3-AP-NAD.

However, the precise significance of diminished levels of NAD in normal stage 23–24 limb buds (Rosenberg & Caplan, 1974) in initiating chondrification remains to be established. Rather than acting directly by enhancing glycolytic pathway and glycosaminoglycan synthesis, Rosenberg & Caplan (1975) suggest that low cellular levels of NAD are sensed by the genes coding for chondrogenic properties.

Oxygen tension and perhaps other associated factors, such as NAD levels, dependent on diffusion may be important in determining the central site of cartilage matrix deposition in the blastema once the vascular pattern has been established in the limb. Of course, since cartilage contains a factor which inhibits the growth of blood vessels (Eisenstein *et al.*, 1975; Langer *et al.*, 1976), it may be the differentiation of the cartilage mass which determines the vascular arrangement. In any event it is unlikely that formation of the chondrogenic region at stage 22 is due to anoxia, since the many investigations involving chondrogenesis in culture are difficult to relate to oxygen tension (Searls, 1973).

The apical ectodermal ridge

There is little doubt that ectoderm and endoderm have a fundamental role in the determination of cartilage. Aspects of the relationship of the notochord (and neural tube) to somitic cartilage have already been discussed. Branchial cartilage is itself mesectodermal (Horstadius, 1950); these and other cartilages, such as the ear capsule, develop in close relation to epithelia (Jacobson & Fell, 1941; Holtfreter, 1968). In culture, unmigrated neural crest cells form cartilage in the presence of pigmented retinal epithelium, which normally induces the avian scleral cartilage *in vivo* (Newsome, 1976). Thus, it is not surprising that ectoderm is implicated in limb bud chondrogenesis.

Numerous experiments have demonstrated the importance of the apical ectodermal ridge (AER) in the regulation of the limb mesenchyme (Saunders, 1948; Zwilling, 1961; Saunders, 1972). Experimental removal of the AER results in terminal limb deficiencies: the older the embryo at operation the

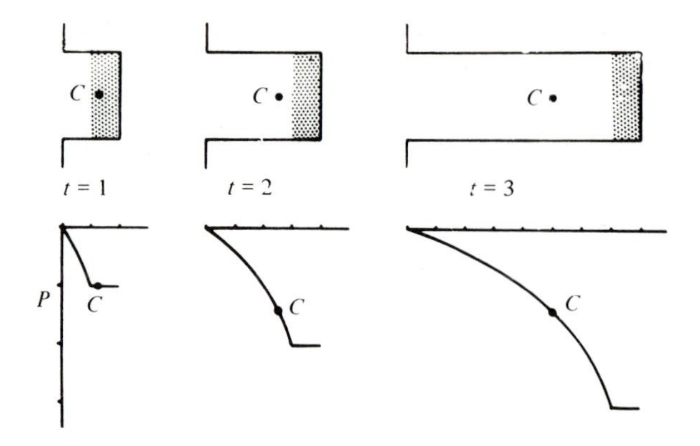

Fig. 93. Schematic diagram of three successive stages in the growth of a limb bud, showing the progress zone stippled. At time $t = 1$, the cell (*C*) is resident in the progress zone, having the 'positional value' (*P*) indicated in the graph beneath. At $t = 2$ and $t = 3$, the cell has left the progress zone and, as shown in the graphs, has a fixed positional value equivalent to that of the progress zone at the time of exit. The value in the progress zone has continued to decrease, however. (From Wolpert *et al.* (1975).)

more distal the level of deletion. During limb growth, mesoderm proliferates in the region deep to the AER and the chondrogenic (and myogenic) blastemata are derived from the proliferating tissue of this 'progress zone' (Wolpert *et al.*, 1975) in a proximo-distal sequence. At the time of exit from the progress zone each element along the proximo-distal axis is about 300 μm long and the amount of growth thereafter varies according to whether the humerus or the carpus, for example, is to be formed. Wolpert *et al.* (1975) propose that the organisation of the proximo-distal axis of the limb is mediated by 'positional information' obtained while the cells are resident in the progress zone (Fig. 93). The 'positional value' of cells in the progress zone decreases autonomously with time but becomes fixed (at whatever level reached at that time) when the cells leave the progress zone.

In addition to the influence of the AER on the proximo-distal axis, a second zone prescribes the orientation of-the antero-posterior axis of the limb. This is situated in a region of mesenchyme at the posterior margin of the limb bud, initially near its junction with the body wall (Saunders & Gasseling, 1968); it was originally called the 'posterior necrotic zone' (Saunders, 1966) but is now referred to as the 'zone of polarising activity' (ZPA). Only cells in the progress zone can respond to the influence of the ZPA (Summerbell, 1974) and there is experimental evidence (MacCabe & Parker, 1976) of a diffusion gradient with its maximum in the ZPA, the labile diffusible morphogen probably passing from cell to cell via gap junctions (Tickle, Summerbell & Wolpert, 1975), according to theoretical models proposed by Crick (1970). It appears that the ZPA also brings about change in the extent of the AER since the ectoderm pre-axial to a ZPA graft becomes thicker (Saunders & Gasseling, 1968) thus affecting the outgrowth of the limb.

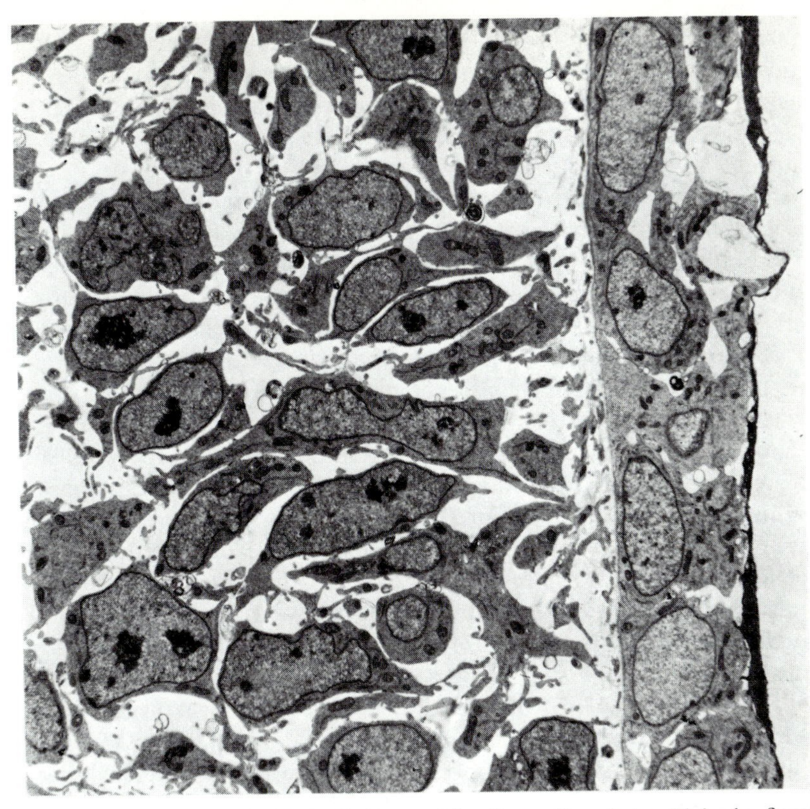

Fig. 94. Chick embryo limb bud (stage 24). Formalin–glutaraldehyde fixation, magnification ×2000. Although the mesenchymal cells tend to become orientated radially near the ectoderm (right of picture), they do not appear to come into contact with it. (From Gould *et al.*, 1972.)

The nature of 'positional information' is not specified precisely but it is postulated that the main signal for orientation of the proximo-distal axis is between the AER and the progress zone of sub-ridge mesenchyme (Fig. 94). However, although the mesenchymal cells form a palisade-like arrangement near the ectoderm, ultrastructural studies reveal no evidence of cell contact between the ectoderm and the mesenchyme (Jurand, 1965; Gould *et al.*, 1972; Kelley & Fallon, 1976). There are numerous cell contacts, including tight and gap junctions, between cells of the apical ectoderm itself, though not in the adjacent epithelium (Kelley & Fallon, 1976). Jurand (1965) notes that the Golgi complex and cytoplasmic RN A are most highly developed at stage 24 in chick ectoderm, which coincides with the maximum development of the AER. Contacts, including gap junctions, are also described between the mesenchymal cells. These cells are characterised by a relative scarcity of cytoplasmic organelles, apart from a profusion of free ribosomes and many short profiles of rough ER observed in continuity with the outer nuclear membrane, a feature said to be associated with cytodifferentiation. Although Jurand notes discontinuities in the basal lamina beneath blade-like extensions of the ectodermal cells in stage 22–26 chick tissue, the lack of direct cell contact suggests that ectodermal–mesenchymal signals must pass via the basal lamina.

It is relevant to consider the comparable situation in the early development of the axial tissues, where the relationship of epithelium and mesenchyme may be different. At stage 8, tight junctions (postulated to control cell movement) occur between the primary mesenchyme and both the overlying ectoderm and underlying endoderm (Hay, 1968) and also between cells of the notochord itself (Bancroft & Bellairs, 1976). There appears to be no direct evidence, either at the somite or the secondary mesenchyme stage, of cell contact between the mesenchyme and the notochord. However, during stages 8–9, filopodia of notochordal cells extend through the basal lamina to abut on the basal lamina of the neural tube and, up to stage 12, short cell processes are sometimes seen extending from the notochord into the surrounding tissue space (Hay, 1968; Bancroft & Bellairs, 1976). Similarly, cell processes from somite cells at the epithelioid stage pass through the somite basal lamina toward the notochord and also abut on the basal lamina of the neural tube (Hay, 1968). Thus although interaction between sclerotome and ectoderm/endoderm (notochord and neural tube) may be limited to indirect contact via the basal lamina, as with the apex of the limb bud, there appears to be a structural probability that direct cell contacts could occur.

Whether or not cell contacts between ectoderm and mesenchyme are involved, or whatever the mode of control by ectodermal structures, it is possible that contacts between cells of the early mesenchyme facilitate their interaction and the coordination of their stage of differentiation. The apparent progressive diminution in the degree of cell contact from an early stage, even through the mesenchymal stage in the limb bud, suggests that local coordination may also diminish progressively, possibly enabling the emergence of different cell lines.

In the limb bud, the AER and the ZPA appear to confer morphogenetic patterns on the developing limb, presumably including chondrogenetic information. This is essentially an environmental influence which, according to Wolpert *et al.* (1975), acts on cell lineages, interpretation through cytodifferentiation commencing when the cells leave the progress zone. This hypothesis therefore suggests an early separation of cell lines.

CELL LINEAGES

The main protagonists for the concept that the limb bud (like the somite) contains separate cell lineages from an early stage have been Holtzer and his colleagues. Holtzer *et al.* (1972) consider that differentiation involves many successive events occurring over many generations in a cell lineage, although the last transitional stage always appears to be abrupt because it is not until then that the phenotype is recognisable by its products. Such ideas appear to be compatible with theories of gene regulation as proposed by Britten & Davidson (1969) and Paul (1972). Holtzer suggests that in fact there are no 'undifferentiated' cells, if considered in molecular terms, since even early precursor cells may be functionally specialised and possess unique messenger RNA molecules. A subsidiary hypothesis is that at cell division either (i) the two daughter cells may be identical to the mother cell, i.e. the progeny of a 'proliferative' cell cycle or (ii) one or both daughter cells may exhibit synthetic pathways very different from those of the mother cell, i.e. the progeny of a 'quantal' cell cycle. Thus in a 'quantal' cell cycle, modification of the genome control occurs, permitting a 'step' in differentiation.

Although Cooke (1973) has produced evidence that differentiation can progress under conditions adverse to mitosis, it is still true that differentiation in a cell lineage is essentially time-dependent, even if not necessarily measured in terms of quantal cell cycles. As described above, both somite and limb bud mesenchyme at early stages of their development are closely related to ectoderm (if not in direct cell contact) when the cell lineage might be 'programmed'. It follows that in the somite or the limb mesenchyme, there are precursors of chondroblasts whose descendants may be capable of differentiating into the phenotype without any further inductive stimulus. There is indeed much evidence that either early somite or limb bud tissue will eventually form cartilage when explanted *in vitro* alone. Furthermore, with stage 17–25 limb buds it is found that the older the tissue at explant, the higher the proportion of cells (i.e. precursors) whose progeny are capable of terminal differentiation (Dienstein *et al.*, 1974) and similar evidence has been obtained in the sclerotome (Abbott *et al.*, 1972).

Cell multiplication is a prominent early feature both in the somite (Holtzer & Detwiler, 1953) and in the limb mesenchyme (Searls, 1973), especially in the progress zone (Summerbell & Wolpert, 1972). It is suggested that the change in 'positional value' may be linked to cell division (Wolpert *et al.*, 1975); such a mechanism may control morphogenesis in insect epidermis (Lawrence, Crick & Munro, 1972). This is consistent with the idea that steps in differentiation are linked to quantal mitoses, provided that differentiation can be expressed in terms of change in 'positional value'. While the mesenchymal cell is resident in the progress zone the influence of the AER may be restrictive, imposing stringent controls on the cell to prevent the premature emergence of mechanisms which would lead to the synthesis of 'luxury' molecules of the cartilaginous or muscular variety.

Having left the progress zone, cytodifferentiation in the cell lineage can proceed (Wolpert *et al.*, 1975). In the context that 'no cell is undifferentiated' (Holtzer *et al.*, 1972), it may be that the 'luxury' task (or one of the tasks) of the cells at early stages of differentiation is to create tissue bulk appropriate to that particular portion of the limb. Prior to chondrification, this is achieved largely by cell multiplication (and some remodelling by cell death). Hence it is advantageous that extracellular macromolecules should facilitate diffusion and perhaps lubricate the cells. Low molecular weight proteoglycans of the type found, including hyaluronate-protein, could fulfil these needs.

If indeed there are separate cell lineages in the limb bud, problems then arise due to heterogeneity of the mesenchyme population: at some stage (e.g. 22–25) the cells must segregate topographically into muscle and cartilage (Wolpert, 1976). Segregation by active migration does not seem to occur (Searls, 1967). Therefore, the emergence of the definitive cartilage and muscle blastemata would have to be mediated by mitosis and/or cell death, presumably occurring over several stages prior to chondrogenesis. Environmental factors may or may not be involved in the evolution of the lineage, local conditions perhaps favouring one of the cell lines at the expense of the other. As emphasised by Dienstein *et al.* (1974) the evidence obtained from studies of nicotinamide and NAD levels (Rosenberg & Caplan, 1975) can be interpreted as selective against different cell lines in a mixed population, as well as positively inducing cartilage or muscle; this also applies to considerations of anoxia. Thus, some insight may be gained by examining factors which mitigate against the differentiated state of the cell.

MODIFICATION OF THE DIFFERENTIATED STATE

A number of agents and procedures can affect the differentiated state of the chondrocyte, particularly in dissociated cells cultured *in vitro*. Probably all culture conditions are adverse to a certain degree and it is obviously necessary that the medium should supply the basic requirements of the cell, including factors such as ascorbic acid (Levenson, 1969) particularly in cells which synthesise collagen. Much depends on the mode of culture of the cells. Thus cartilage expression, assessed by the types of macromolecules secreted, is hindered when the chondrocytes are grown as monolayers instead of in suspension ('spinner') cultures (Srivistava *et al.*, 1974; Sokoloff, 1976), although the effect is reversible. The density of the cell inoculum is also important. Lavietes (1970) found a level of 10^6 cells per 60-mm dish more favourable than lower densities, although cartilage is also formed in clonal culture (10^2 cells per dish). Thus even in monolayers, high-density culture produces Type II cartilage collagen (Handley, Bateman, Oakes & Lowther, 1975) where low-density culture yields Type I skin-like collagen (Layman, Sokoloff & Miller, 1972).

Cultures tend to become fibroblast-like when the differentiated state of the chondrocytes is modified. This occurs even in clonal cultures (Holtzer *et al.*, 1970) where it is difficult to attribute the phenomenon to overgrowth by fibroblasts. It has been emphasised by Srivistava *et al.* (1974) that 'dedifferentiation' does not take place using 2–3-month-old rabbit articular chondrocytes, yet certain phenotypic properties appear to be lost in cultures of embryonic chondrocytes. Mayne *et al.* (1976*a*) have confirmed that cloned chondrocytes are unstable when grown *in vitro*, subcultures eventually losing the capacity to divide (after about 30 generations) and becoming fibroblast-like. Similar fibroblast-like changes occur when embryo extract is included in the medium, although the capacity for cell division is not lost. A high molecular weight heat-labile fraction is responsible (Coon & Cahn, 1966) and the effect is reversible; a cell surface interaction is suspected (Mayne, Vail & Miller, 1976*b*).

The same effect can conveniently be produced by a thymidine analogue, 5-bromo-2-deoxyuridine (BrdU). This substance is known to be incorporated into DNA in competition with thymidine (Dunn, Smith, Zamenhoff & Griboff, 1954; Bardos, Levin, Herr & Gordon, 1955) during 'S' phase (Stellwagen & Tomkins, 1971). The chromosomes of dividing mammalian cells in culture become heavily labelled with BrdU when the analogue is included in the medium (Schneider, Chaillet & Tice, 1976). Since BrdU is thought to act at the gene level (Stellwagen & Tomkins, 1971) it is helpful to consider briefly concepts of gene regulation before discussing the effects of BrdU on the cell.

GENE CONTROL AND THE DIFFERENTIATED STATE

The genome is the same in all the cells of a multicellular organism yet it is expressed in different ways in various cell types. Hence differentiation of cells must involve regulation of some of the genes, some repressed and others potentiated. The genome is much larger in multi- than in unicellular organisms although nearly all the biosynthetic pathways are present in unicellular organisms. Hence it is suggested that much of the large amount of DNA in multicellular organisms (eukaryote DNA) compared with that in unicellular organisms must be concerned with gene regulation (Britten & Davidson, 1969; Paul, 1972). Eukaryote DNA contains large fractions of moderately (10^2–10^4 copies per genome) and highly repetitive nucleotide sequences, intimately interspersed with non-repetitive unique sequences (MacLean & Hilder, 1977). The unique sequences are the structural genes which control the synthesis of proteins, while the moderately repetitive sequences are involved in gene regulation. Thus it is suggested that the repetitive sequences provide mechanisms (Fig. 95) for the recognition (address site), binding (promoter site) and control (regulator site) of RNA polymerase molecules which then transcribe the unique sequence to form messenger RNA (Paul, 1972).

There is much evidence that basic histone proteins cover up recognition and regulator sites and so repress gene expression (Stedman & Stedman, 1943; Allfrey, Littau & Mirsky, 1963). Acidic non-histone proteins made in the cytoplasm are believed to attach to such specific sites and, becoming phosphorylated, compete with the anionic groups of DNA for the basic

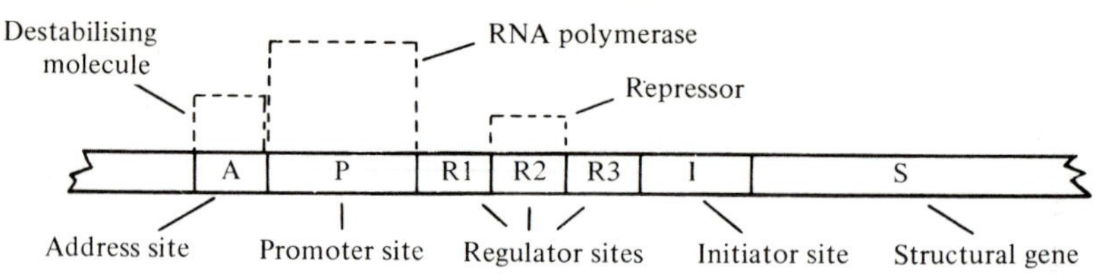

Fig. 95. Schematic diagram of the organisation of the repetitive sequence parts of the DNA molecule involved in gene regulation. (Redrawn and modified from Paul, 1972.)

histone. This displaces the histone from that portion of the DNA, leading to de-repression of the gene (Stein, Spelsberg & Kleinsmith, 1974; Kleinsmith, 1975). Thus brain chromatin (which does not normally transcribe the globin gene) reconstituted in the presence of the non-histone protein fraction of liver chromatin (in which DNA globin sequences are transcribed) acquires the specificity for transcription of the globin gene sequence (Gilmour, Windass, Affara & Paul, 1975). Hence specificity for expression of a differentiated cell type (in synthesising the appropriate 'luxury' molecules) may reside in the interaction of appropriate acidic non-histone proteins with chromatin.

EFFECTS OF 5-BROMO-2-DEOXYURIDINE

The suppressive effect of 5-bromo-2-deoxyuridine (BrdU) is not confined to cartilage since many differentiated tissues and cells are affected in an analogous fashion (Abbott *et al.*, 1972). Embryonic chondrocytes have been extensively investigated, however. Examining cultured dissociated 10-day chick vertebral chondrocytes with the electron microscope, Anderson, Chacko, Abbott & Holtzer (1970) find that the cells become flattened, fibroblast-like and amoeboid. The rough ER membranes and Golgi complex are much less prominent, while free ribosomes, microtubules and microfilaments become more numerous. The cells are not surrounded by metachromatic matrix and matrix granules are few or absent. Studying high-density cultures of dissociated stage 23–24 chick limb bud cells, slightly different observations have been made (Levitt & Dorfman, 1974; Levitt, Ho & Dorfman, 1975). Thus in normal media, typical chondrocytes are found by the fourth day of culture with prominent rough ER, Golgi membranes and secretory vacuoles. When incubated in BrdU, the cells appear to be very similar to the controls, except that the mitochondrial cristae are said to appear less deranged than in the control. The lack of change in cell ultrastructure in limb bud cells when cultured in BrdU may be associated with the fact that high-density cultures were used (Levitt & Dorfman, 1974). However, although fibres with periodic banding were abundant, no matrix granules were present in the extracellular space, in agreement with Anderson *et al.*'s findings. This correlates with a profound inhibition of sulphate uptake.

These ultrastructural observations suggest that the principal effect of BrdU is on the synthesis and secretion of cartilage proteoglycan. However, although collagen appears to be unaffected, Levitt & Dorfman (1974) find on analysis that there is a much higher proportion of α_2 chains in BrdU than in normal limb bud cultures. Using sternal chondrocytes, Mayne *et al.* (1975) confirm that Type I rather than the cartilage-specific Type II is formed in the presence of BrdU; a Type I trimer $((\alpha_1 I)_3)$ is also present. It is interesting that a similar Type I trimer, as well as Type I collagen, is also produced by chondrocytes after protracted subculturing and after growth in embryo extract (Mayne *et al.*, 1976*a*, *b*). Since the trimer is not produced by fibroblasts, this suggests that although the chondrocytes modulate they are not transformed into true fibroblasts. Thus although after BrdU treatment the cells survive and continue to synthesise many proteins, the 'luxury' molecules (Holtzer *et al.*, 1972) which express the state of differentiation of the cell (i.e. Type II collagen and cartilage proteoglycan) are not synthesised.

The cell population must be actively dividing if BrdU is to be effective; thus, Mayne, Abbott & Holtzer (1973*a*) find that dissociated cells but not organ cultures are suppressed. The state of cell differentiation is also important. BrdU suppression of fully differentiated cells is reversible in their progeny after three to five cell divisions both in somite cells (Holtzer *et al.*, 1970) and in limb bud tissue (Levitt & Dorfman, 1973). However, if undifferentiated dissociated stage 17–18 somite cells (Mayne, Schiltz & Holtzer, 1973*b*) or stage 24 limb bud cells (Levitt & Dorfman, 1973) are exposed to BrdU, the effect is irreversible and they fail to differentiate. Nevertheless, Levitt & Dorfman (1974) suggest that BrdU may permit some degree of differentiation, as shown by the response to xylose stimulation; xylose is thought to stimulate chondroitin sulphate synthesis by eliminating the need for core protein. They treated stage 24 limb bud cells with BrdU and then cultured them for a time period normally sufficient for differentiation to occur. Addition of xylose to the culture medium resulted in a considerable elevation in chondroitin sulphate synthesis, while untreated mesenchymal cells prior to differentiation showed no response to xylose. This suggests that some of the multi-enzyme systems required for proteoglycan synthesis had differentiated in the BrdU cells although perhaps not those required for the core protein. Stage 24 limb buds contain cells which according to either concept (early or late separation) of cell lineages are in the penultimate stage of differentiation. Investigation of xylose stimulation of BrdU-suppressed limb bud cells at stages earlier than stage 24 might help to elucidate the cell lineage problem.

Using synchronised stage 22 limb bud mesenchyme, Flickinger (1975) finds that BrdU is taken up into DNA in early 'S' phase and that it causes more inhibition of messenger RNA than of transfer RNA or ribosomal RNA synthesis. Since DNA containing BrdU has a higher melting point than normal DNA, he suggests that there will be a lesser degree of strand separation at the initiating point for transcription (and replication). Thus, as proposed by Stellwagen & Tomkins (1971), he concludes that BrdU will diminish the binding of RNA polymerase, resulting in a lower transcription rate. There is evidence that BrdU may modify the repressor–DNA interaction. Thus the *lac*-repressor is bound much more tightly to the BrdU-substituted *lac*-operator than to the normal operator (Lin & Riggs, 1972); histones similarly show increased binding to BrdU-substituted DNA (Lin, Lin & Riggs, 1976).

One of several questions concerns the apparent selectivity of BrdU for systems which control the synthesis of 'luxury' molecules in the differentiated cell. Possibly, sites in DNA controlling the synthesis of 'luxury' molecules are unusually rich in thymidine (Levitt & Dorfman, 1974), but more needs to be learned about such regions. Recently, Strom & Dorfman (1976) have obtained results with stage 24 limb bud cells which may be relevant to the problem. They find that while ^{3}H-thymidine is taken up into repetitive, moderately repetitive and unique sequences of DNA, BrdU is preferentially incorporated into the late, moderately repetitive region. They postulate that in pre-differentiated cells (Stage 24), these sequences are being amplified and that this may be facilitated by a separate polymerase that has a higher affinity for BrdU than for thymidine. Following amplification (with BrdU) these sequences are broken down and the cells irreversibly prevented from differentiating.

More information is required concerning the mode of action of BrdU, which is of course an 'unphysiological' molecule. We know even less about normal extracellular factors which presumably act via the intermediary 'second messenger' systems involving cyclic AMP, although other possibilities exist. Thus Rosenberg & Caplan (1975) have investigated an NAD-related enzyme system, synthesising poly-(ADP-ribose), which is closely associated with nuclear chromatin and histone. They suggest that cellular NAD levels (which are affected by environmental relationships) are 'sensed' by this system, potentiation of cartilage expression correlating with a high rate of synthesis of poly-(ADP-ribose) and a low level of NAD. Much remains to be learned about such extracellular influences or how intracellular and intranuclear mechanisms (perhaps associated with 'preprogramming') affect chromatin to modify genome control during differentiation. Perhaps at the level of the DNA molecule the distinction between extrinsic and intrinsic factors in chondrogenesis becomes a mere semantic problem.

7

CHONDROCYTE PROLIFERATION, CARTILAGE GROWTH AND REPAIR

Growth of cartilage occurs by cell reduplication, cell hypertrophy and matrix secretion. Although in adult permanent cartilage, the bulk of the tissue is due to matrix rather than cell volume, cell proliferation is important in the early stages of growth of all cartilage and is part of the reaction to injury. In growing cartilage many factors are relevant to cell multiplication. These include species differences, nutrition, hormonal influences, the type of cartilage and the position of the cell in the tissue. Chondrocyte proliferation need not cease at full maturation in permanent cartilage and there are circumstances in which mitoses are observed in chondrocytes of adult tissue.

Chondrocytes divide by mitosis both in immature and adult tissue. This is worth emphasising since it was for many years believed that adult chondrocytes divided amitotically. Particularly in articular cartilage, the postulate of amitosis in the adult (Nowikoff, 1908) gained credence for three reasons. First, the apparent necessity for cell replacement – it was believed that cells multiplied in the middle zone of the tissue and then migrated toward the articular surface where they became dislodged in the course of joint movement, a mechanism analogous to that of the epidermis (Ogston, 1875). Secondly, it was difficult to detect mitoses in the normal adult tissue. Thirdly, the presence of dumb-bell-shaped nuclei and other unusual features which appeared to be compatible with the view (Elliott, 1936) that amitosis occurred. However, although the cells of ageing cartilage exhibit irregular and folded nuclei, there is now substantial evidence (Barnett *et al.*, 1963) that cells are not normally 'desquamated' at the articular surface; hence a steady rate of cell replacement is not required and this explains the absence of mitoses in the normal tissue. Even where cell loss occurs, associated with early stages of joint pathology, chondrocyte death occurs *in situ* and initially at least there is little evidence of cell replacement, the cell density of the superficial zone falling. In addition, the concept of amitosis is difficult to relate to modern genetic theory and to the necessity to replicate DNA (Mankin, 1963*a*).

Although cell multiplication is important in the growth of cartilage, relatively few sites have been studied quantitatively and knowledge is rather scanty about the permanent cartilages apart from articular cartilage. In most sites, evaluation of the contribution of cell proliferation to growth is made difficult by the large volume of matrix subsequently secreted by the new cells.

INITIAL GROWTH OF EMBRYONIC LONG BONE RUDIMENTS

As the mesenchymal cells of the pre-cartilage blastema differentiate into chondrocytes, the rate of cell proliferation falls. Studying the chick embryo, Searls (1973) finds a labelling index of 20% at chondrification, using ^{3}H-thymidine. Thereafter the cells pass through Streeter's phases 1–5, cell division terminating at the end of phase 3 when the cells become hypertrophied (Streeter, 1949). Cells destined to become part of articular and other permanent cartilages remain smaller and do not normally progress to the degree of hypertrophy (e.g. phase 4) seen in temporary cartilage, although considerable enlargement is seen at the centre of adult permanent cartilage. As the process of chondrification extends to either end of the blastema (Fell, 1925), the cells successively pass through phases 1–5 in regions of the tissue sequentially more distant from the centre of the blastema (Fig. 86). Cell division seems to occur randomly within the tissue (interstitial growth) in each region, the daughter cells moving apart as they secrete matrix, although two or more cells often appear to remain in the same lacuna. Mitoses are also found at the margin of the blastema in what becomes the perichondrium, facilitating appositional growth (Ham, 1932). In each concentric region of the blastema, mitoses cease at the end of phase 3.

This is followed by degeneration of the cells and vascular invasion of the calcified tissue at the centre of the blastema by the periosteal bud. The calcified cartilage is partly eroded and bone deposited on the remaining tissue by cells carried in by the vascular mesenchyme. This process of endochondral ossification then spreads to either end of the shaft. The expanded ends of the cartilaginous mass, whether or not an epiphysis forms subsequently, do not undergo these changes at first, the cells remaining in the proliferative and pre-hypertrophied state of phases 1 and 2. Hence there is at either end of the long bone rudiment a zone of transition from phase 2 through to phase 5 chondrocytes adjacent to the ossification front of erosion and bone deposition on the diaphysial side. This arrangement of the cartilage resembles closely that of the definitive growth plate established after the epiphysis has ossified.

THE EPIPHYSIAL CARTILAGINOUS GROWTH PLATE

Long bones grow at their ends. In the early eighteenth century the Reverend Stephen Hales, minister of Teddington, demonstrated this now well-known fact as a corollary to his exhaustive studies of growth in vine shoots (Hales, 1727). He bored two holes half an inch apart in the shaft of the tibia of a half-grown chicken. Two months later, the interval between the holes was unchanged although the bone had grown an inch in length. He rightly attributed this phenomenon to the 'wonderful provision...at the glutinous serrated joining of the heads to the shanks of the bones'. As recounted by Keith (1919), similar investigations were carried out by Duhamel during the years 1739–1743 and a series of later workers including John Hunter, 1760–1770 (Palmer, 1835) and Flourens and Ollier in the nineteenth century, who used metal marker techniques and madder feeding (earlier shown by Belchier to stain growing bone), all substantiating the conclusion that growth in length occurs at the ends of the shaft.

It has long been known that the growth cartilage of the epiphysis persists until cessation of growth and bony union of the epiphysis and diaphysis. The presence of cartilage at the ends of the diaphysis is a consequence of three factors: (*a*) the appositional nature of bone deposition; (*b*) the capability of cartilage for rapid growth; and (*c*) the necessity to retain a resilient cartilaginous lining for the synovial joints at the ends of long bones during the growth period and thereafter. Most of the bone of the shaft is formed initially by the periosteum: if the periosteum covered the cartilaginous end so that appositional bone deposition might add to the length of the bone, resilience would be lost and longitudinal growth would be slow. These defects are circumvented by retaining a cartilaginous end (the 'chondro-epiphysis') to the bone and later, when the bony epiphysis forms, an articular cartilage persisting throughout life, and a growth cartilage until maturity.

A comprehensive early account of the growth plate is given by Todd & Bowman (1845). It has been understood since Muller (1858) gave the first full description of the mechanism of growth that cell proliferation in the cell columns occurs on the epiphysial side of the plate and that erosion of the hypertrophied and degenerate cells followed by bone deposition takes place simultaneously on the diaphysial (or metaphysial) side (Fig. 64). Thus, so long as chondrocyte proliferation on the one side and erosion and bone deposition on the other side are in equilibrium, diaphysial elongation will continue. However, it is comparatively only recently that it was pointed out by Sissons (1956), following the analogy of bean root growth, that the *rate* of elongation of the bony diaphysis is the product of the cell proliferation rate and the length of the hypertrophied cells next to the zone of erosion. Since the diameter of the hypertrophied cell varies within fairly narrow limits, the rate of bone growth is a crude index of the rate of proliferation of chondrocytes in the growth plate.

ELONGATION OF THE DIAPHYSIS

Longitudinal bone growth rate is more easily assessed than chondrocyte proliferation. There are many studies of bone growth in man and other animals. Growth rate varies widely (Table 12). There are obvious species differences: for example, the lower end of the femur grows 0.31 mm per day in the young rabbit (Sissons, 1953) which is about 10 times faster than that in the young child (Harris, 1933; Gill & Abbott, 1942; Kember & Sissons,

TABLE 12
Species differences in metaphysial growth rate. Data from: Dubreuil (1913); Payton (1932); Harris (1933); Sissons (1953); Walker & Kember (1972*b*); Kember & Sissons (1976)

Species	Approximate age	Growth of distal femur (mm per day)	Growth of proximal tibia (mm per day)
Rat	5–10 weeks	0.18	0.24
Rabbit	9 weeks	0.31	0.36
Pig	12 weeks	0.4	0.48
	6 months	0.18	0.13
Man	3.5 years	0.044	0.025

1976). Species differences must be considered in the context of the duration of the growth period as well as of the adult size of the species. Thus the growth rates at the lower end of the femur are very similar (about 0.18 mm per day) in the six-week-old rat (Sissons, 1953) and the six-month-old pig (Payton, 1932). Differences also occur within the same animal, the most important of which is the deceleration of growth rate during maturation. In most species, this occurs in a gradual exponential fashion (Fig. 96) although the primates exhibit a slight acceleration – the 'growth spurt' – at adolescence (Tanner, 1962). During development, there are also differences between the growth patterns of the upper and lower limbs, compensating for the initial cranio-caudal gradient, and also between the appendicular and axial parts of the skeleton.

In a single long bone at any given age, differences in growth rate are found between the two ends of the shaft. Thus in the upper limb the proximal end of the humerus and the distal ends of the radius and ulna grow most rapidly, while in the lower limb the bone ends near the knee joint predominate (Table 13). This phenomenon was noted by Stephen Hales in the chick tibia experiment and has since been observed repeatedly using madder feeding and metallic marker techniques with and without the use of X-rays. In human bones, ingenious use has also been made of the direction of the nutrient canal and the position of its medullary end (Digby, 1916), said to indicate the original site of ingrowth of the vascular mesenchyme when the primary ossification centre is established. Although the validity of this method as applied to lower animals has been questioned by Payton (1934), the human data appear to be correct (Bisgard & Bisgard, 1935; Gill & Abbott, 1942). Differential growth rate in man can also be deduced from the correlation of the clinical history with the intervals between 'growth arrest' lines at the two bone ends (Harris, 1933).

As might be expected, differences in growth rate between the two ends are less pronounced in the earlier stages of development. As Serafini-Fracassini

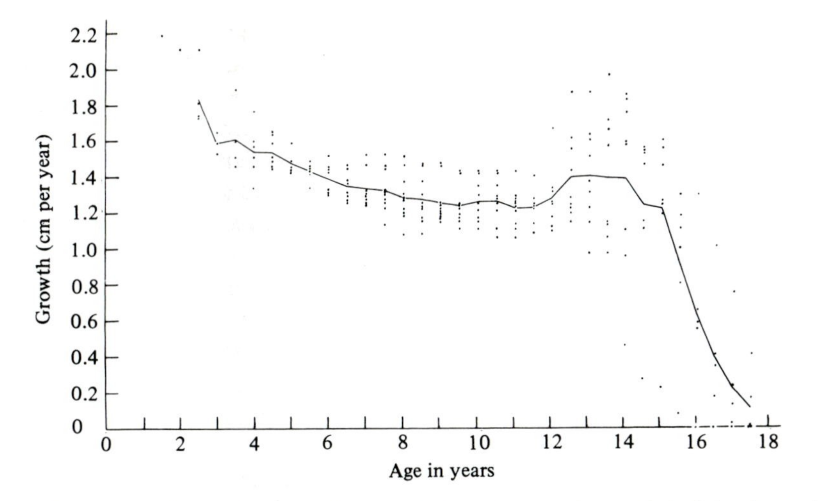

Fig. 96. Incremental growth of the distal end of the human femoral shaft in the male. (From Kember & Sissons, 1976.)

& Smith (1974) point out, the elongation at the upper end of the shaft of the human femur, based on the evidence of growth arrest lines (Harris, 1933), is nearly equal to that of the lower end in the first year of life although thereafter about 70% of elongation occurs at the lower end. Of course, as development nears completion, growth may take place at one end only following closure of the epiphysis at the opposite end. Although in the lower limb maximum growth occurs near the knee joint, in man the differences between bone ends are larger in the femur than in the tibia (Table 13). This is particularly noticeable in growth arrest line intervals (Harris, 1933, Fig. 3). However, in other species, growth at the upper end of the tibia is markedly greater than that at the lower end. This is certainly true in the rat where the proximal tibia has been widely used for direct studies of cell proliferation.

If growth at one end of a long bone is arrested experimentally by destruction of the growth plate or by stapling, compensatory growth occurs at the opposite end. The amount of compensatory growth is proportional to the reduction in rate at the other end and changes in growth rate do not occur in any other bone of that limb (Hall-Craggs, 1969; Hall-Craggs & Lawrence, 1969). The mechanism of compensatory growth is not understood although changes in metabolites or inhibitory agents have been suggested. However, in view of the postulated mechanical interaction of growth plate and periosteum (Lutfi, 1974*b*) it seems probable that changes in tension in the periosteum may mediate compensatory growth.

TABLE 13

Difference in growth rate between the ends of long bones, expressed as proximal/distal ratios. References: Digby (1916); Payton (1932); Bisgard & Bisgard (1935); Gill & Abbott (1942); Dawson & Kember (1974)

Species	Femur	Tibia	Humerus	Radius	Ulna
Rat	–	87:13	–	–	–
Goat	36:64	55:45	82:18	26:74	19:81
Pig	37:63	72:28	79:21	21:79	31:69
Man	31:69,	57:43,	81:19	25:75	19:81
	30:70	55:45	–	–	–

CELL KINETICS OF LONGITUDINAL GROWTH

Cell proliferation occurs in the flattened cell portion of the cell columns, where the stacks of cells resemble piles of coins (Fig. 64): hence growth is interstitial. Cell division is seen best in transverse sections of the growth plate, the mitotic spindle lying transverse to the long axis of the cell column (Dodds, 1930). The daughter cells become wedge-shaped, permitting overlap, further transverse elongation and thickening of the cells producing a discoidal shape. Results of investigations using ^{3}H-thymidine (Kember, 1960; Messier & Leblond, 1960; Tonna, 1961; Belanger & Migicovsky, 1963; Lutfi, 1970*b*) confirm the earlier studies of mitosis, showing that initial uptake of the nucleotide takes place in the classical zone of proliferation. After three to four days the label sequentially appears in the hypertrophic zone and eventually disappears from the cell columns. Rigal (1962) noted that all the cells in a column might be labelled as early as one hour after administration of the isotope, both *in vivo* and *in vitro*; an unusual result which cannot be entirely attributed to the use of *in vitro* techniques.

The first quantitative though indirect estimate of cell proliferation in the growth plate preceded the use of thymidine. To assess the rate of formation of new cells per day (n) in a column, Sissons (1956) established the rate of elongation of the bony shaft (nD), the number of cells in a column (nt) and the longitudinal diameter of the hypertrophied cell (D). In the upper tibial growth plate of four-week-old rats, Sissons calculated that 7.5 cells are produced per day per column (nD/D) and that five days elapse between formation and replacement of individual cells (nt/t). Cell enlargement takes about four days and degenerate cells persist for about one day prior to erosion. In older, more slowly growing rats, the results indicate a decline in the rate of cell formation but little change in the lifespan of the chondrocyte.

Kember (1960) obtained direct confirmation of these values observing ^{3}H-thymidine incorporation in the proximal tibia of six–eight-week-old rats. In a 'standard' column of length 25 cells, he observes a proliferation rate of

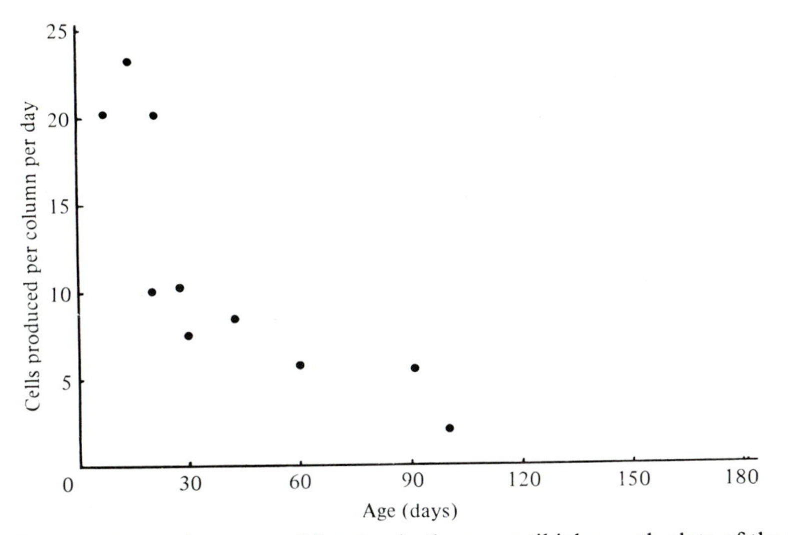

Fig. 97. Rates of chondrocyte proliferation in the upper tibial growth plate of the rat. (Data from Sissons, 1956, Walker & Kember, 1972*b* and Thorngren & Hansson, 1973.)

5 cells per day: this is consistent with the rate of shaft elongation found using X-rays and, allowing for the use of older rats, agrees with the values found by Sissons (1956), in four-week-old rats. More recently, Kember & Sissons (1976) have had the opportunity to study the distal femoral growth plate in man (Fig. 96). The growth rate of 0.038 mm per day between five and eight years of age corresponds to a cell proliferation rate of about 1.2 cells per column per day. This is much slower but of course lasts for much longer than in the rat.

In the rat proximal tibia, Kember (Kember, 1971; Walker & Kember, 1972*a*, *b*; Dawson & Kember, 1976) gives values for the cell cycle time (about 55 hours), 'S' phase (6.5 hours) and 'G' phase (3.2 hours). The generation time is longer in cells of the resting or reserve zone, since they are not labelled so frequently. In man, Kember & Sissons (1976) estimate that the cell cycle time in the proliferation zone of the distal femoral growth plate is about 20 days.

During maturation the rate of chondrocyte proliferation declines (Fig. 97), as is predictable from the changes in shaft elongation. In the proximal tibia of the rat during the period 20–100 days, the hypertrophic cell size declines from about 35 to 21 μm; the length of the proliferation zone is reduced in about the same proportion (Walker & Kember, 1972*b*; Thorngren & Hansson, 1973). As in the mouse lower femur (Tonna, 1961), the labelling index with thymidine in the rat is also halved or reduced even more (Walker & Kember, 1972*b*). Hence the major factor in growth deceleration, as found indirectly by Thorngren & Hanson (1973), is the decline in cell proliferation. Similar conclusions apply to the results of human studies (Kember & Sissons, 1976).

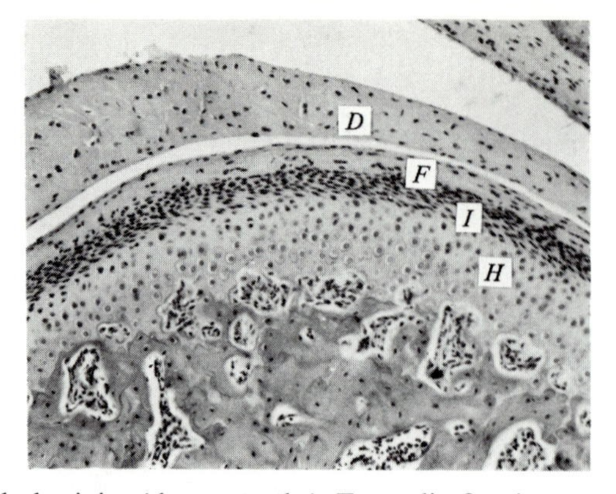

Fig. 98. Rat mandibular joint (three months). Formalin fixation, magnification ×80. In the mandibular condylar cartilage, chondrocytes of the hypertrophic zone (*H*) are produced in the intermediate zone (*I*). Cell multiplication in the superficial fibrocartilaginous zone (*F*) appears to be independent of that in the intermediate zone. Note the intra-articular disc (*D*) of fibrocartilage. (Material by courtesy of Dr R. Sprinz.)

The slight increase in growth rate at adolescence is again accounted for by a rise in cell proliferation. Regarding the diminution in the rate of cell proliferation, Walker & Kember (1972*b*) state that it is not possible to distinguish between an increase in the cycle time of a proliferating cell and a change in the fraction of non-cycling cells; there is no change in the duration of 'S' phase over the period 4 to 13 weeks in the rat proximal tibia. The mean cycle time in neonatal rat vertebral cartilage is only 22 hours (Dixon, 1971), as compared with 55 hours in the proximal tibia at six weeks, although the duration of 'S' and 'G_2' phase is about the same. However, this difference may be related to regional variation as well as to changes during maturation.

In six-week-old rats, Kember (1972) finds that differences in the labelling index and in the number of cells in the proliferating zone correlate well with differences in growth rate of various bones of the lower limb. In the mandibular condyle (Fig. 98), where there is no definitive growth plate, the cells of the intermediate layer form a proliferation zone for the hypertrophic cartilage (Blackwood, 1966; Frommer, Monroe, Morehead & Belt, 1968). Growth and cell proliferation rate are only about one-third of that in the proximal tibia.

As in other situations, there is a diurnal variation in cell proliferation in the growth plate. Simmons (1964) finds that the mitotic index shows a peak at 1800 hours in the rat distal femur. However, Walker & Kember (1972*a*) obtain different results, finding that the mitotic index is highest at 0800 and lowest at 2000 hours in the rat proximal tibia: labelling index maxima and minima were consistent with these data. Differences in diurnal exposure to light in the two sets of experiments were not sufficient to account for the conflicting findings.

It has long been known that there is a correlation between the rate of longitudinal bone growth and the thickness of the growth plate. The thickness is related to the number of cells in a column and the longitudinal diameter of the hypertrophic cell. Thus there are more cells per column, especially in the proliferation zone, at rapidly than at slowly growing ends of bones (Harris, 1933; Kember, 1972). During maturation the thickness of the plate, the number of cells in a column and the hypertrophic cell diameter diminish (Sissons, 1956; Johnston, 1972; Walker & Kember, 1972*b*; Kember, 1973; Thorngren & Hansson, 1973).

Although rate of bone growth and plate thickness are normally correlated, the width of the plate is not a good indicator of growth rate in abnormal circumstances. After hypophysectomy (Kember, 1971; Thorngren & Hansson, 1973) the labelling index is reduced (80%) much more than either the width of the proliferative zone (25%) or the diameter of the hypertrophic cell (33%). Thus the reduction in diaphysial growth rate, to about 10% of the normal, is much greater than the loss of plate width (about 50%). Low doses of radiation (400 rad) have little or no effect on the plate thickness, the number of cells per column or the maximum size of the hypertrophic cell, yet the growth rate (Sissons, 1956) and the labelling index (Kember, 1971) are reduced. As is well known, rachitic plates increase in thickness; this is due to an increase in the width of the hypertrophic zone, there being little change in the width and labelling index of the proliferative zone (Kember, 1971).

HORMONAL EFFECTS ON CELL PROLIFERATION

Culture of explants *in vivo* and *in vitro* shows that the growth plate has considerable intrinsic growth capacity (Lacroix, 1951). Normal growth requires an adequate diet with the requisite vitamins, although deficiencies of vitamins C and D do not primarily affect cell proliferation (Sissons, 1971; Mankin & Lippiello, 1969*a*). Several hormones exercise a control over the rate of cell proliferation: the subject is reviewed comprehensively by Silberberg & Silberberg (1971).

As already described, hypophysectomy causes a marked reduction in the labelling index, due to the loss of stimulation by growth hormone. In the body, growth hormone acts via the somatomedins produced in the liver (Salmon & Daughaday, 1957; Sledge, 1973), types A and C stimulating activity in cartilage (Hall *et al.*, 1975). Thus growth hormone administered to hypophysectomised rats reverses the decline in proliferation rate after a delay of 12 hours (Kember, 1971). Somatomedin *in vitro* causes a marked increase in ^{3}H-thymidine uptake in the proliferative but not in the hypertrophic zone of rabbit growth plates, suggesting that there is a reduction in response to the hormone as the chondrocyte passes into the hypertrophic phase (Ash & Francis, 1975).

Other hormones which act synergistically with growth hormone in the intact animal tend to have a direct action on the growth plate, causing reversal of stimulation of cell multiplication. Thus in hypophysectomised rats, oestradiol not only reverses the stimulation by growth hormone but has a direct depressant effect on chondrocyte mitosis (Silberberg & Silberberg, 1971; Strickland & Sprinz, 1973; Gustafsson *et al.*, 1975). Similarly, thyroxine, which is required for adequate growth, inhibits the stimulation of growth plate cell proliferation *in vitro* by somatomedin (Ash & Francis, 1975). The growth-promoting action of testosterone is thought to be independent of the hypophysis (Silberberg & Silberberg, 1971). It enhances endochondral ossification but cell proliferation is less affected than cell hypertrophy and calcification. In short-term experiments, Fahmy *et al.* (1971) find that cell multiplication is stimulated but they consider that in the long term testosterone would inhibit cell proliferation. Cortisone and related preparations inhibit skeletal growth (Sissons, 1971), including mitotic activity in the growth plate (Young & Crane, 1964; Simmons & Kunin, 1967).

Together with other factors, the levels and interplay of hormones must have an important role in the complex mechanisms involved in the cessation of growth. In the rat proximal tibia where the growth plate persists throughout life, Kember (1973) notes that the width of the proliferative zone (and of the whole plate) becomes reduced and the labelling index declines by 20 weeks of age; the stem cell (reserve) zone also becomes reduced or absent. Such changes are interrelated and it is difficult to establish which, if any, of them is the primary event in termination of growth. Discussing the reduction in stem cells, Kember remarks that it takes 40 cell divisions of stem cells to grow a rat ribia and that this is similar to the number said by Hayflick (1965) to be the limit for a cell; however, he notes that reduction in vascularity of the epiphysial side of the plate, changes of permeability due to calcification and other factors must also be taken into account.

Studying epiphysial closure in man and dog, Haines (1975) finds nodules of cartilage forming between the epiphysial bone marrow and the cartilage of the plate. He deduces that cartilage can be recruited from marrow cells and therefore that growth cannot cease because of lack of reserve cells in the plate. Kember & Sissons (1976) find no progressive diminution in thickness of the reserve zone of the human distal femur during the growth period. This also suggests that loss of stem cells cannot be material to cessation of growth, but the argument is complicated by the authors' belief that this reserve zone is in any case inert, making no contribution to longitudinal growth.

Toward the end of growth, multiple 'tidemarks' of calcification are found both on the epiphysial and the diaphysial sides of the growth plate and fusion of calcified tissue occurs across the plate over limited areas (Haines, 1975). Bony union may commence either as single or multiple perforations of vascular tissue, the single perforation expanding and eroding the basal plate of epiphysial bone which is spared in multiple perforation.

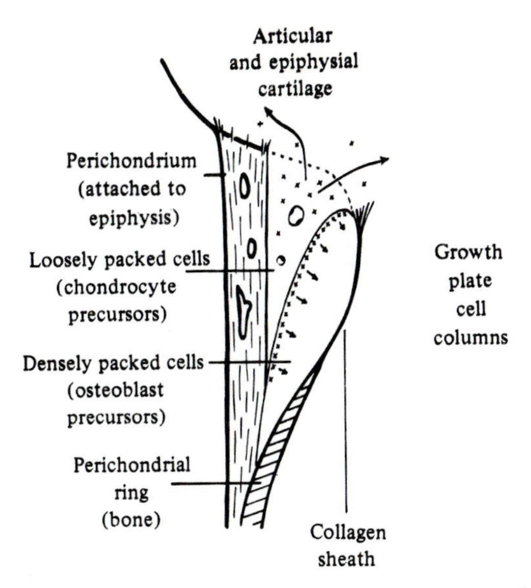

Fig. 99. Structure of the ossification groove around the growth plate. The arrows indicate the directions of cell growth and mitoses are indicated by the distribution of small crosses. Note the vascularity of the perichondrium and of the loosely packed cells. Compare with Fig. 65.

TRANSVERSE EXPANSION OF THE GROWTH PLATE

During longitudinal growth, the correct proportions of the bones are preserved by enlargement of the diameter of the shaft and the epiphysis. Remodelling facilitated by reabsorption of shaft wall at the metaphysis is also required (Keith, 1920; Brash, 1934). At the level of the growth plate, the transverse diameter is increased, involving a multiplication of the number of longitudinal cell columns (Dodds, 1930).

It is generally believed that extension of the diameter of the growth plate occurs at the margin of the columnar portion. Although growth in the more central parts has not been excluded, Lacroix (1951) suggests that this is unlikely because the longitudinal bars of the growth cartilage are fixed in position by calcification and remain parallel. It was once assumed that the ossification groove prevented transverse growth but the reverse appears to be true. It is here necessary to distinguish between the fibrocellular tissue of the groove and the bony perichondrial ring (Figs. 65, 99).

The fibrocellular tissue is associated with growth. Evidence of cell proliferation, both from mitotic figures and thymidine labelling, is found in this region, confirming Lacroix's (1951) views on the appositional nature of transverse expansion of the plate. Thus Tonna (1961) observes that labelled cells migrate from the perichondrial region into the periphery of the growth plate (Fig. 99), the new cells becoming longitudinally aligned and elongating in the metaphysial direction, so contributing to growth in length as well as diameter (Shimomura, Wezeman & Ray, 1973). However, Rigal (1962) claims that the growth is interstitial since the labelling is found in the proliferative parts of the cell columns lying at the junction of the growth cartilage with the perichondrial ring and the articular cartilage, producing new columns there. In practice, there seems to be little difference between the results, in either case the zone of proliferation essentially lying beyond the epiphysial border of the bony perichondrial ring (Fig. 99). Recent results indicate that the fibrocellular tissue contains separate groups of progenitor cells (Shapiro *et al.*, 1977). The densely packed cells adjacent to the bony ring and to the collagen lamina are alkaline phosphatase positive and are the source of the osteoblasts. The zone of loosely packed cells is chondrogenic, providing for the transverse expansion of the growth plate and the growth of articular cartilage (Fig. 99).

Subsequent hypertrophy of the newly formed cells is presumably the cause of the extension of the bony ring although the cells immediately adjacent to the inner aspect of the perichondrial ring are narrower than most hypertrophied chondrocytes. Although Lacroix (1951) finds that grafts of the growth plate will continue columnar growth for a time in the absence of a perichondrial ring before a new one is formed, it is suggested that the bony ring acts as a 'supporting corset' (Rigal, 1962) for the growth plate, perhaps directing growth into the longitudinal channel.

There are obviously difficulties in locating precisely the site and source of cell proliferation relating to transverse growth. These may be partly attributed to the slow rate of plate expansion relative to that of diaphysial elongation. The increase in radius of the ground plate appears to be between one-tenth to one-fifth of the diaphysial elongation rate. Kember (1972) mentions a value of 0.025 mm per day based on radiographic studies of the rat tibia. This is similar to estimates obtained by measurements of published micrographs of the rat tibia (Becks & Evans, 1953) and the rabbit second metatarsal (Lacroix, 1951) during development. Thus the periphery of the growth plate receives an annular increment of width equivalent to one hypertrophic cell diameter or less per day, compared with 5–10 cells per column per day in the longitudinal axis.

GROWTH AT THE CARTILAGINOUS ENDS OF THE BONE

There is less information on cell proliferation in the chondro-epiphysis and articular cartilage than in the growth plate. This may be partly owing to the considerable importance of the growth plate in bone development and its clinical concern; possibly the greater difficulty of three-dimensional analysis in the epiphysis is due to less rapid cell proliferation and greater separation of the cells by matrix than in the growth plate. The chondro-epiphysis enlarges both by interstitial and appositional growth and although there are few quantitative data on their relative contribution, the evidence suggests that the pattern of cell proliferation does not remain constant throughout the whole period of development. It is convenient, therefore, to consider the growth of the epiphysis and of articular cartilage in three stages. The stages are not rigidly demarcated and anatomical and developmental differences may modify the sequence in various epiphyses.

STAGE 1: THE EARLY CHONDRO-EPIPHYSIS

This period lasts until the related synovial joint undergoes cavitation. It is helpful to recall briefly the early development of synovial joints (Fig. 100). An interzone of avascular condensed tissue marks the position of the synovial joint between the related chondro-epiphyses of successive long bones (Fig. 101*a*). This tissue is continuous peripherally with the vascular perichondrium covering the sides of the chondro-epiphysis. The joint capsule, of the knee joint for example, is first apparent in man at 16 mm crown-rump length as a distinct curving layer of condensed tissue joining the bone ends and lying outside the perichondrium (Haines, 1947, 1953). Although not all authorities agree (Andersen, 1962), it is generally believed that vascular mesenchyme lying external to the interzone comes within the confines of the joint owing to the increase in size of the bone blastemata and by capsular enclosure. The capsule also encloses part of the vascular perichondrium. Subsequently, at about 26 mm in the human knee joint, the interzone becomes three-layered; the intermediate layer is continuous peripherally with the synovial mesenchyme and the outer layers with vascular perichondrium (Fig. 101*b*).

Appositional growth of the chondro-epiphysis takes place both at the interzone and the perichondrium (Fig. 100*a*, *b*). After lamination, only the outer layers of the interzone are chondrogenous and make the first contributions to the appositional growth of the articular parts of the chondro-epiphyses, while the vascular perichondrium continues to add to the main cartilaginous mass. Bruch (1852) appears to have been the first to note the transition in cell shape on passing from the perichondrium into the chondro-epiphysis. He concluded that the rounded chondrocytes are derived from the fibroblast-like cells of the perichondrium, or more correctly the inner layer of subperichondrial tissue (Stump, 1925) where abundant mitoses may be seen (Andersen, 1962).

While many authors do not specify any other mode of growth at this stage, there seems no doubt that mitoses occur within the substance of the cartilage and at all stages thereafter.

STAGE 2: POST-CAVITATION PRE-OSSIFICATION

Development of the joint and of the epiphysis is considered to be under genetic control until the beginning of this stage (Drachman & Sokoloff, 1966),

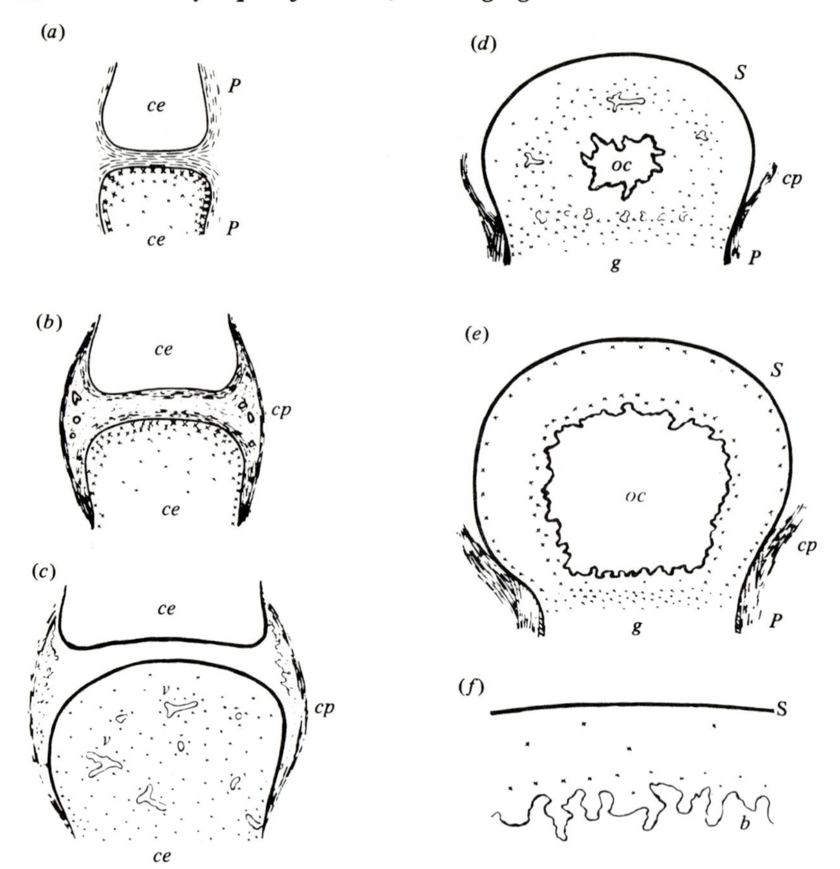

Fig. 100. Growth of the epiphysis and of articular cartilage. Small crosses indicate the distribution of mitoses. (*a*) The early chondro-epiphysis (*ce*) with a unilaminar interzone at the joint site continuous with the perichondrium (*P*) at the sides. Appositional growth is predominant. (*b*) The interzone has become trilaminar and a joint capsule (*cp*) has formed enclosing some vascular mesenchyme. Growth is still mainly appositional. (*c*) Following cavitation of the joint, growth becomes predominantly interstitial except at the periphery of the articular area. Cell proliferation may be more active near the cartilage canals (*v*). (*d*) After formation of the ossification centre (*oc*), interstitial growth continues in a 'mitotic annulus' midway between the articular surface (*S*) and the bony centre. Mitotic activity on the diaphysial aspect serves for epiphysial plate growth (*g*). (*e*) With further growth, the epiphysis becomes hemispherical due to cessation of enlargement on the diaphysial aspect. Two zones of mitoses are found in the cartilage – close to the ossification centre and subjacent to the articular surface. Appositional growth continues at the periphery of the articular area. (*f*) As epiphysial growth nears completion, no definite zone of chondrocyte mitosis is found near the articular surface although proliferation continues near the subchondral bone (*b*).

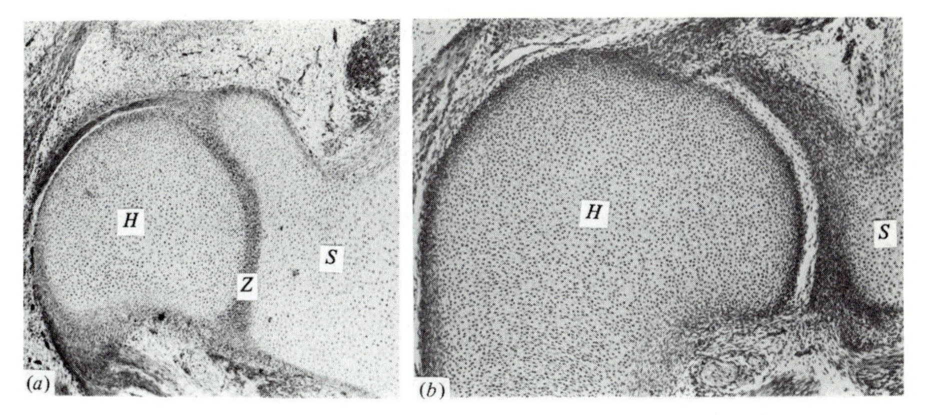

Fig. 101. Human shoulder joint, formalin fixation. (*a*) 20-mm embryo, magnification × 40. (*b*) 25-mm embryo, magnification × 51. In (*a*) there is an avascular interzone (*Z*) between the rudiments of scapula (*S*) and humerus (*H*). In (*b*) the interzone has become trilaminar with a less condensed intermediate layer.

although thereafter articular cartilage growth may be dependent on muscular activity (Murray & Drachman, 1969). Cavitation of the knee and other large synovial joints occurs from 30 mm in the tenth week (Haines, 1947; Gray & Gardner, 1950). Spaces appear in the intermediate layer of the interzone, at first in the peripheral parts of the joint, spreading and coalescing to include the central region. Soon after this the outer layers of the interzone cease to be chondrogenous, the cells becoming rounded and enclosed in matrix, giving this part of the chondroepiphysis a smooth articular contour (Stump, 1925; Gray & Gardner, 1950). Similar changes take place in the vascular perichondrium enclosed by the joint capsule (Haines, 1947). Hence as the joint cavity and surfaces form and extend, the chondro-epiphysis becomes dependent on interstitial growth (Andersen, 1962) and appositional growth can occur only from that part of the surface of the epiphysis still covered by perichondrium (Fig. 100*c*).

There are few data on the location of interstitial mitoses during this stage, what observations there are referring to mitoses in all planes (Dodds, 1930) and throughout the whole depth of the cartilage (Mankin, 1964; Kvinnsland & Kvinnsland, 1975). Lutfi (1970*b*) noted random labelling with ^{3}H-thymidine in the 'cap' zone of the proximal chick tibia. However, there appear to be two locations which may have a constant relationship to anatomical features of the chondro-epiphysis. The first of these is oriented with respect to the articular surface; since most of the observations apply to material where a secondary ossification centre has already formed it is best discussed under stage 3. The other site relates to the cartilage canals when these structures are present (Fig. 100*c*).

Whether the cartilage canals are formed by inclusion or by active ingrowth from the perichondrium, it is nevertheless postulated that they bring the functional properties of the perichondrium into the interior of the cartilage. Thus in addition to a nutritive role, Haines (1933) suggests that cells are added to the cartilage from the canals. According to Lutfi (1970*b*) the results of ^{3}H-thymidine incorporation studies in the chick show that the canals provide new chondrocytes, including stem cells for endochondral ossification distal to the junctional zone of the proximal tibia. Moss-Salentijn (1975) also finds evidence of mitoses next to the canals (Fig. 59*b*) in the human spheno-occipital synchondrosis although in sheep fetal chondro-epiphyses many mitoses are found at some distance from the canals (Stockwell, 1971*a*). In addition, many of the cartilage canals eventually chondrify. This is particularly well seen in the pseudo-epiphysis of the first metacarpal of man (Haines, 1974), perhaps due to its slow growth. However, it is unlikely that infilling of canals is of major importance in the growth process. While the canals may contribute to the cell content of the chondro-epiphysis, it is probable that their most significant function is to ensure adequate nourishment for the interstitial growth in general (Andersen, 1962) as the cartilage mass increases in size.

Despite the elimination of a chondrogenic covering for much of the epiphysial surface, there is significant appositional growth at the margin of the joint space (Figs. 99, 100*c*). It is worth noting that although the interzone initiates the formation of articular cartilage, it is the vascular perichondrium which makes by far the largest contribution (Gray & Gardner, 1950).

STAGE 3: POST-OSSIFICATION

Ossification of the epiphyses is similar to that in most small short bones of the carpus and tarsus (Gardner, 1971). There is no preliminary subperiosteal formation of bone. Groups of cells more or less in the middle of the epiphysis pass through Streeter's phases 3–5, becoming enlarged and degenerate, the matrix near them becoming mineralised. In epiphyses which do not contain cartilage canals, blood vessels from the perichondrium erode and penetrate into the calcified and uncalcified tissue (Kalayjian & Cooper, 1972). Endochondral ossification commences as osteoblasts form and bone is deposited. In epiphyses possessing cartilage canals the vessels furnish the vascular element for endochondral ossification, their expanded ends ('glomeruli') lying within the initial islands of calcification (Fig. 62), as elegantly shown by Wilsman & Van Sickle (1970). However, the precise spatial relationship of the canals to the initial sites of chondrocyte *hypertrophy*, and hence their physiological influence, if any, on the loci of calcification and degeneration, is still in doubt (Chapter 5).

Initial growth of the ossification nucleus is concentric, as well shown by radiology of lines of growth arrest and of phosphorus poisoning (Harris, 1933; Lacroix, 1951) and is much more rapid than that of the cartilaginous element. Later they enlarge at nearly equal rates and growth of the bony centre on the diaphysial side slows down compared with the rest of its circumference.

As described by Kalayjian & Cooper (1972) for small laboratory animals, growth on the diaphysial aspect differs from the usual process of endochondral ossification. Many of the chondrocytes do not degenerate but become enclosed in a fibrous meshwork forming in advance of the vascular buds. Accretion of mineralised matrix eventually replaces vascular invasion. Further growth of the ossification nucleus on other aspects produces a hemispherical bony centre based on the growth plate (where growth is directed toward the diaphysis) and covered by a dome of cartilage continuous at its margin with the growth plate (Fig. 100*d*, *e*).

Once established, the growth of the ossification nucleus is closely related to the location of cell proliferation in the cartilage. Harris & Russell (1933) plotted the positions of mitoses in the chondro-epiphysis of a 72-mm human embryo (Harris, 1933, Figs. 119–121). They describe the now well-known 'mitotic annulus' lying in a broad zone about half way between the centre of the epiphysis and the periphery (Fig. 100*d*). On the diaphysial aspect of the epiphysis the annulus runs through the proliferation zone of the incipient growth plate. Although ossification had evidently not yet commenced in their specimen the central cartilage is hypertrophied and calcified and has acquired a similar spatial organisation to that seen after ossification has begun. Interstitial cell proliferation from this stage onwards becomes set in this pattern. Thus in post-natal laboratory animals, Elliott (1936) describes a sharply defined zone of mitoses with a maximum density close to the articular surface.

Confirming and extending the earlier work, Rigal (1962) and Mankin (1962*a*) describe two zones of mitosis concentric to the ossification centre in 'young' immature animals (Fig. 100*e*). Studying mitotic figures and ^{3}H-thymidine labelling in femoral condyles of 600–800-g rabbits, Mankin describes a superficial region of mitoses in a zone (*A*) lying between 5–20 cell diameters from the articular surface: this is associated with interstitial growth of the articular cartilage. A second deeper zone (*B*) lying at a similar distance from the ossification nucleus provides for the expansion of the epiphysis. Hence as demonstrated by Siegling (1941) the epiphysis grows by its articular surface. The location of the two zones of mitosis is presumably related to nutritional and other factors. In the case of the superficial zone of mitosis, Elliott (1936) postulated a stimulating force proportional to $1/d^2$ (d = distance from articular surface) and an inhibiting force proportional to $1/d^3$. He thought it probable that the stimulatory factor was nutrition, from the synovial fluid. In view of the permeability properties of articular cartilage (Maroudas, 1973) one might speculate that differences in the stimulating and inhibitory factors could reside in molecular weight and electrostatic charge: the stimulatory factor perhaps being of lower molecular weight and relatively more positively charged. It is relevant that somatomedins (growth hormone intermediaries) have a molecular weight of about 7000 and that neither types A or C (acting on cartilage) are acidic peptides (Hall *et al.*, 1975). Somatomedin activity is present in synovial fluid, although in the adult this is lower than in serum (Coates *et al.*, 1977) possibly because somatomedin is bound to high molecular weight protein.

As the animal develops and the epiphysis enlarges, both Rigal and Mankin note that the twin zones become single. In 1700-g rabbits, Mankin (1963*a*) finds that labelled cells occur only in the deeper cartilage within 2–3 cell diameters from the calcified cartilage adjacent to the bone (Fig. 100*f*); cell multiplication is no longer observed in zone *A*. No labelling is found in the cartilage of adult (5-kg) rabbits.

Something is known of the cell kinetics in the epiphysis at this stage. Mankin (1964) gives values for mitotic indices in the distal femur of the rabbit. Cell proliferation declines rapidly during the maturation period. Mitoses are more frequent in the deep zone *B* than in the superficial zone *A* but at two months the index in the epiphysial and articular cartilage combined is only one-twentieth of that in the growth plate. Blackwood (1966) also finds that the articular fibrocartilage (Fig. 98) in the six-week-old rat mandibular condyle is very inactive compared with the intermediate zone which provides for endochondrial ossification. Frommer *et al.* (1968) observe that the labelling index is high in the articular fibrocartilage in the pre-natal mouse condyle but that it declines much more rapidly than that in the intermediate zone after birth. Mankin (1964) observes no diurnal variation and suggests that the duration of mitosis and the cell cycle time are very long in articular chondrocytes. Tonna & Cronkite (1964) confirm this; they find that the thymidine label persists in rat femoral condylar chondrocytes for as long as nine months after administration of the isotope at birth, concluding that no dilution of label by further mitoses has occurred.

Appositional growth continues at the perichondrial margin of the articular cartilage, which in many epiphyses corresponds to the ossification groove. In the lower femur of five-week-old mice, Tonna (1961) observes thymidine-labelled cells migrating into the articular cartilage, concluding that the perichondrial cells serve as a progenitor pool for circumferential growth of the epiphysis as well as the growth plate (Fig. 99).

NON-ARTICULAR PERMANENT CARTILAGES

Classical studies have led to the standard textbook descriptions of growth. Cell proliferation initially occurs throughout the whole cartilage but as the tissue enlarges and is no longer sufficiently plastic to accommodate new cells, this interstitial growth declines. The last of the interstitial mitoses are said to result in the isogenous cell pairs or groups typifying the central regions of the adult tissue. In the later stages of development, the emphasis is placed on the continuing cell multiplication in the subperichondrial region, which permits the cartilage to enlarge by appositional growth.

This established pattern of growth merits one or two comments. The first is that there is very little information based on modern techniques, such as ^{3}H-thymidine incorporation. Secondly, as described earlier, in the growth plate, in the epiphysis and in articular cartilage it is interstitial growth which is predominant in the later stages of development, there being no apparent problems with expansion of the tissue. Lastly, the few thymidine studies indicate that mitoses are more numerous interstitially than in the perichondrium, as in tracheal cartilage of 200-g rats (Messier & Leblond, 1960): FitzGerald & Shtieh (1977*a*) counted 300 consecutive mitoses in post-natal rat ear cartilage, finding that 78% were in the inner three-fifths of the cartilage and only 12% in the perichondrium. Many of the permanent cartilages have an osseochondral junction where interstitial cell proliferation occurs to facilitate endochondral ossification, as with costal (Harris, 1933) and nasal septal cartilage (Scott, 1953). In three-week-old rabbit nasal cartilage, a midline sagittal section shows a high labelling index anteriorly as well as posteriorly at the junction with the ethmoid (Long, Greulich & Sarnat, 1968); this presumably indicates interstitial growth. Regions of high labelling, other than at the posterior osseochondral junction, vary in position in the septal cartilage during development in the rat (Searls, 1976).

REPAIR OF CARTILAGE

In normal adult cartilage, cell proliferation either does not occur or is so infrequent as to be undetectable (Mankin, 1963*a*). On the other hand, once the normal structural and functional organisation of the tissue is impaired, cell multiplication is often observed. Many and varied methods have been used to damage cartilage in studies of repair and tissue reaction; in articular cartilage this has often been in an attempt to reproduce experimentally the pathological lesions of osteoarthrosis. A review of pre-1939 investigations may be found in the monograph by Bennett, Waine & Bauer (1942) and a brief résumé of more recent techniques is given by Bentley (1974). Probably the most gross form of disruption of the tissue consistent with cell survival is the release of chondrocytes by the enzymatic digestion of the matrix (Smith, 1965). Adult cartilage thus treated (Sokoloff, 1976) yields an important tool for research into chondrocyte function, for the cells multiply and synthesise and secrete matrix materials. Teleologically this may be regarded as a form of repair, an attempt by the cell to reconstitute its environment and normal relationships; *in situ* the chondrocyte response to tissue damage is usually multiplication and matrix secretion.

Discussion will largely be confined to the surgically created anatomical defect and its sequelae, since this type of lesion would appear to be that least complicated by extraneous factors. In post-natal animals, age appears to make little difference to the response of the tissue (Bennett & Bauer, 1935; De Palma, McKeever & Subin, 1966). As discussed by Meachim, there are two possible means of repair of surgical defects: (i) extrinsic mechanisms, in

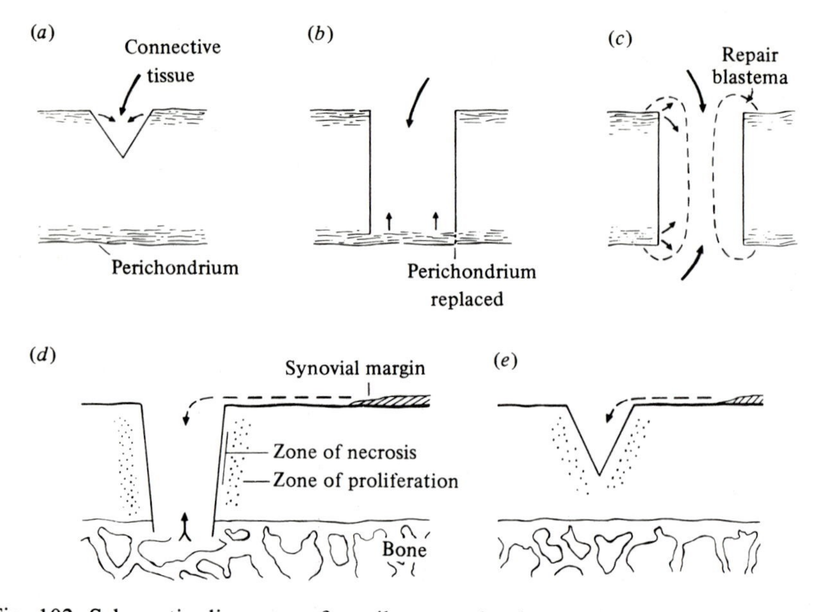

Fig. 102. Schematic diagrams of cartilage repair. Arrows indicate source of repair tissue. (*a*) Small defects with loss of a small part of the perichondrium may be filled by cartilage. (*b*) Large defects with loss of perichondrium are repaired as fibrous tissue unless the perichondrium is resutured into position (lower part of diagram). (*c*) In the rabbit ear, large defects with loss of perichondrium are repaired by cartilage, although repair tissue also comes from the surrounding connective tissue. (*d*) Full thickness defects in articular cartilage are filled by repair tissue, usually from the subchondral bone; this becomes fibrocartilaginous subsequently. A zone of chondrocyte necrosis and adjacent zone of cell proliferation (stippled area) are found in the cartilage at the margins of the defect. (*e*) Partial thickness defects in articular cartilage may not undergo repair, unless new tissue reaches the site from the periphery.

which cells other than or in addition to chondrocytes produce the repair tissue; (ii) intrinsic mechanisms, where the new tissue regenerates from the cartilage alone (Stockwell & Meachim, 1973). Since Redfern published his classic paper in 1851 on the healing of wounds in cartilage, all investigators have shown that extrinsic repair is the most effective of the two mechanisms. Repair is slow and at least in non-articular cartilage a small residual hole may persist (Joseph & Dyson, 1966).

EXTRINSIC MECHANISMS

This form of repair does not, of course, directly involve chondrocyte proliferation. The defect fills with blood clot which later becomes infiltrated with spindle-shaped cells. The repair tissue may come from several sources. In non-articular hyaline and elastic cartilage (Fig. 102) new cells have easy access to the defect from the surrounding connective tissue, although if spared the perichondrium may also act as a source of regenerative cells (Haas, 1914; Joseph, Thomas & Tynen, 1961; Joseph & Dyson, 1966). In articular cartilage, ingrowth is facilitated in 'full-thickness' defects, i.e. where the hole is made right through to the subchondral bone marrow spaces: repair tissue then wells up into the defect (Fig. 102). New tissue may also reach the defect from the synovium at the margin of the articular area (Landells, 1957) or even by 'seeding' from the synovial fluid. These alternative routes may explain the enhanced healing of peripheral as compared with central defects on the articular area (Fisher, 1923; Bennett, Bauer & Maddock, 1932). The synovial route is also available to 'partial thickness' defects, i.e. those limited to the uncalcified layers of the articular cartilage (Fig. 102).

Subsequently the repair tissue becomes vascular, at least at the borders which do not abut on the cartilaginous margins of the defect. New cartilage may or may not be formed by metaplasia after four to eight weeks. In articular cartilage defects it is usually fibrocartilaginous in type (Meachim & Roberts, 1971); although it may sometimes resemble the hyaline variety (Calandruccio & Gilmer, 1962; De Palma *et al.*, 1966; Campbell, 1969), it lacks the zones of the native articular cartilage and contains areas devoid of cells (Ghadially, Fuller & Kirkaldy-Willis, 1971). Similar forms of repair are found in osteoarthrosis (Meachim & Osborne, 1970). In some articular cartilage defects, only fibrous tissue may be produced.

Healing in fibrocartilage also appears to be largely dependent on extrinsic repair tissue. Thus, excision of the menisci of the knee joint can result (though not invariably) in complete regeneration of new fibrous menisci, by tissue proliferating at the synovial periphery of the original cartilages (King, 1936; Walmsley & Bruce, 1938; Smillie, 1943). Incisions of the inner aspect of a meniscus which do not reach the peripheral synovial margin do not heal. In the temporo-mandibular joint, however, no regeneration occurs after total meniscectomy, although minimal damage related to the periphery of the disc may undergo fibrous repair (Sprinz, 1961). Repair in the intervertebral disc can be complicated by the release of nucleus pulposus material into the wound. Tissue proliferation deep to the anterior longitudinal ligament leads to superficial repair (followed by calcification and bony ankylosis) but the deep aspect of the incision does not heal because the repair tissue is blocked by the herniated nucleus pulposus (Smith & Walmsley, 1951).

In non-articular sites, fibrous union occurs in large full-thickness defects where the perichondrium has also been removed (Haas, 1914; Lutfi, 1974; FitzGerald & Shtieh, 1977*b*); although the rabbit ear appears to be an exception, the repair cartilage may later ossify (Joseph *et al.*, 1961).

INTRINSIC MECHANISMS IN NON-ARTICULAR CARTILAGE

In non-articular cartilage, the perichondrium is able to produce cells from its deep surface which differentiate into chondrocytes (Haas, 1914; Amprino & Bairati, 1934; Ohlsen & Sohn, 1976, cited by Skoog & Johansson, 1976). The precise contribution of the perichondrium is difficult to assess. Very small full-thickness holes (Trinick, 1975), small incomplete defects or even where there is a large full-thickness loss of tissue (if the perichondrium is resutured into position) heal with cartilage (Haas, 1914); fibrous repair tissue may be capped by chondroid tissue (FitzGerald & Shtieh, 1977*b*). It is probable, however, that the perichondrium is not the only source of the new tissue; nevertheless the perichondrium and/or the cut surface of the old cartilage plate may have an influence on ingrowth tissue where this is not derived from the perichondrium.

The rabbit's ear may be a useful model on which to study this problem; large full-thickness defects from which the perichondrium is also removed heal with cartilage at this site. In the transparent ear chamber in the rabbit, Clark & Clark (1942) observed that the new cartilage which grows at the margin of the defect, adjacent to the original cartilage plate, is the most persistent form of repair tissue. The regenerating 'blastema' extends forward from the cut end of the old cartilage (Joseph & Dyson, 1966) and at the sides merges with the fibrous layer of the perichondrium (Fig. 102*c*). Grimes (1974*a*) has shown that the original cartilage plate plays an essential role in regeneration since no repair tissue is formed if the old cartilage sheet is extirpated. If the mature cartilage sheet (but not the rest of the tissue of the ear) is X-irradiated, there is ingrowth of new tissue, but no chondrification is observed (Grimes, 1974*b*). Hence in normal repair, the old cartilage or perhaps macromolecules derived from its matrix or cells may be needed to 'stabilise' the formation of new cartilage found in contact with or close to the cut end. It still remains difficult to explain why cartilage regeneration should be so much greater in rabbit than in other mammalian ears, for example the guinea pig (Joseph *et al.*, 1961).

INTRINSIC RESPONSE IN ARTICULAR CARTILAGE

Although it is extrinsic tissue which ultimately bridges surgical defects in articular cartilage, chondrocyte proliferation occurs as part of the intrinsic response of the tissue. Although some have believed that this proliferative response can make a contribution to wound healing, it is normally of little value in repair of experimental defects or of advanced pathological lesions. Nevertheless it is potentially of importance in the case of minute early fissures and irregularities which develop in many parts of the articular surfaces of adult joints.

In general, the changes in the cartilage at the margin of a defect are similar in both articular and non-articular cartilage, immature or adult, but the following description relates to traumatised articular cartilage. In both full- and partial-thickness defects (Key, 1931; Bennett *et al.*, 1932; Carlson, 1957) the cartilage in a zone about 50–100 μm wide bordering the defect undergoes necrosis: the cells die and there is loss of metachromasia. This phase of necrosis starts within one day of operation and is usually complete at two weeks (Mankin, 1962*b*; Calandruccio & Gilmer, 1962; De Palma *et al.*, 1966). After two to three weeks, the region of tissue adjacent to and of similar width to the necrotic zone is characterised by the presence of cell clusters or chondrones. Later on, the degenerate zone often disappears. As regards the infilling of the defect, however, the marginal cartilage makes only a passive contribution: the superficial tissue forms a lip or sloping shoulder at the side of the crater-like defect, attributed to matrix flow (Shands, 1931; Calandruccio & Gilmer, 1962; De Palma *et al.*, 1966; Ghadially *et al.*, 1971), probably caused by the forces set up during joint movement. Scanning electron microscopic studies of this phenomenon demonstrate a corrugated appearance (Ghadially, Ailsby & Oryschak, 1974) reminiscent of smooth waves running up a beach.

Mankin (1962*b*) points out that cartilage is similar to other tissues in its reaction to injury except that being avascular there can be no phase of acute inflammation; thus after trauma there are only the phases of necrosis and of proliferation. It is noteworthy, however, that at least when the whole joint is rendered unstable by severing the ligaments, one of the first reactions is an increased water content of the cartilage (McDevitt & Muir, 1976), which might well be regarded as an oedematous phenomenon. There may be exceptions to the general pattern of necrosis and cell proliferation. Thus hamster articular cartilage does not show cell clusters (Meachim & Illman, 1967) and in fibrocartilage the dual reaction may be absent. In experimental injuries to the menisci of the knee joint of dogs and rabbits (King, 1936; Walmsley & Bruce, 1938), there appears to be no evidence of marginal necrosis or of cell proliferation in the cut edge, although cell clusters are seen in pathologically torn cartilages in man (Noble & Hamblen, 1975). Cartilage cell formation can occur in the stump of the rabbit temporo-mandibular joint meniscus following surgery (Sprinz, 1961) and cell proliferation certainly occurs in the margin of wounds of the anulus fibrosus and in the remains of the nucleus pulposus following experimental incision of the rabbit intervertebral disc (Smith & Walmsley, 1951).

Zone of degeneration

The cause of the marginal necrosis is not known. In view of the elastic properties of cartilage it is unlikely that the incision itself causes chondrocyte death except for the very few cells transected by the knife. Thus one week after

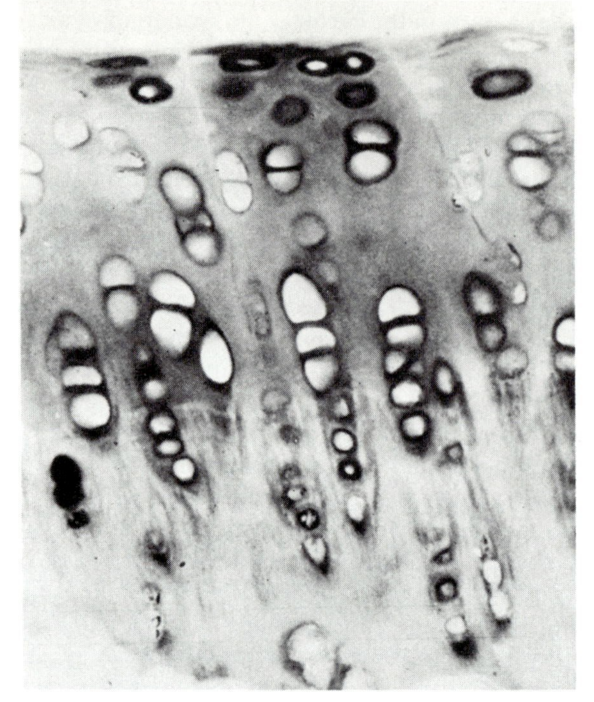

Fig. 103. Adult rabbit femoral condylar articular cartilage. Formolsaline fixation, stained toluidine blue, magnification ×300. Early effectsof scarification (multiple partial thickness incisions) of the articular surface. Seven days after operation, the superficial cells exhibit pericellular 'haloes' of metachromasia but there is little evidence of cell death in the vicinity of the cuts. (From Meachim, 1963.)

rabbit cartilage is scarified by multiple scalpel cuts (Meachim, 1963) as little as 0.1 mm apart, the chondrocytes in the tissue between the incisions show enhanced rims of metachromasia and autoradiographs indicate enhanced sulphate uptake but there is no immediate cell degeneration (Fig. 103). Nor can there be any question of ischaemic necrosis. Increased mechanical stress on the margin of the defect secondary to the disturbance of the normal articular surface contour may be cited but in that case some gradation in the width of the necrotic zone with depth from the articular surface might be expected, yet it often seems to be remarkably uniform in extent. It is also pertinent that intermittent high pressure during experimental clamping of joints results in cell multiplication rather than cell death (Crelin & Southwick, 1964). Again the loss of metachromasia, probably due to leaching out of proteoglycans after disruption of the collagen mesh according to mechanisms proposed by Maroudas (1976), is unlikely to cause the early death of the chondrocytes observed by Mankin (1962*b*). Indeed, depletion of matrix (by papain) leads to elevated sulphate uptake (McElligott & Potter, 1960) and this is also seen after scarification.

The rather constant width of the zone of necrosis suggests some mechanism dependent on diffusion; disturbed nutrition of the cells or inhibitory factors in the synovial fluid (or in the granulation tissue of a full-thickness defect) have been suggested (Calandruccio & Gilmer, 1962; Mankin, 1962*b*). This is an attractive hypothesis but on the other hand scarification does not appear to yield the initial wide zones of cell death seen adjacent to defects involving loss of tissue, although chondrocyte degeneration is eventually observed. However, in the context of an inhibitory hypothesis the absence of a necrotic zone after scarification could result from the poorer access of the synovial fluid to the margins of the incisions in this type of lesion. Whatever the cause of the zone of necrosis, it is possible that its prevention might affect intrinsic repair processes, perhaps enabling cell proliferation to occur earlier and nearer the defect.

Zone of proliferation

In experimental defects it is probable that the cell clusters formed adjacent to the zone of necrosis are due to cell proliferation rather than to redistribution of existing cells; this may be true also of the similar cell groups found in deeply fibrillated cartilage (Meachim & Collins, 1962; Mankin & Lippiello, 1970; Hulth, Lindberg & Telhag, 1972; Rothwell & Bentley, 1973). In experimental work, ^{3}H-thymidine labelling demonstrates uptake by the cells throughout the whole depth of the cartilage near a defect (Mankin, 1962*b*); confirmation of thymidine uptake in the cell clusters both in immature and adult tissue comes from other studies of surgical defects (Dustmann, Puhl & Krempien, 1974) and of papain-induced fibrillation (Bentley, 1974). Negative findings with thymidine (De Palma *et al.*, 1966) may be a consequence of intravenous rather than intra-articular administration of the isotope. Clusters seem to be only an occasional event in the superficial zone (Meachim, 1963) although these cells undergo mitosis in regions adjacent to injury in adult as well as immature tissue (Crelin & Southwick, 1960).

It seems probable that cell proliferation and cluster formation are related to loss of ground substance (Meachim, 1963). Near the margin of the defect as much as 60% of the dry mass of the matrix may be lost (Matthews & Goldstein, 1976). It is possible that proteoglycans in the immediate vicinity of the chondrocyte may exercise a direct control over cell division (Lippmann, 1968). However, it is probable that a more important factor is the indirect effect of matrix changes on diffusion of nutrients and other humoral agents. Not only the quality of the matrix but also the distance traversed by the metabolites through the cartilage are affected by the removal of tissue in the defect. Easier access of nutrients could support a larger number of cells in that region. Thus it has been suggested (Stockwell & Meachim, 1973) that cell cluster formation in experimentally or pathologically fissured cartilage is related to the known normal relationship between cell density and cartilage thickness: fissured cartilage is physiologically, at least, 'thinner', and hence should become more cellular. Similar considerations apply to the zone of proliferation around the crater type of defect. In human fibrillated cartilage the size and number of cells in a cluster is proportional to the distance from the fissure (J. P. Breen, personal communication). It may only be coincidental that the distance of the zone of proliferation from the margin of the defect is similar to that separating the articular surface from the zone of mitosis ('zone *A*') during articular cartilage growth (Mankin, 1962*a*). Elliott's (1936) views on mitotic inhibitory and stimulatory factors and their rate of decline with depth from the articular surface may be relevant here.

However, cellularity is low in the regions around clusters and remnants of dying cells can be seen (Dustmann *et al.*, 1974); this suggests that not all chondrocytes form clusters and that there may be an element of selection in the original chondrocyte population for those cells stimulated to divide. The cell clusters which are formed are highly active in sulphate incorporation and proteoglycan synthesis (Carlson, 1957; McElligott & Collins, 1960; Meachim, 1963). Hence, since there is evidence of chondrocyte heterogeneity in the normal population (Wilsman & Van Sickle, 1971; Mazhuga & Cherkasov, 1974), it is possible that the mother cells selected are of the 'proteoglycan-synthesising' type.

Clusters enlarge at the periphery of the cell group: type I collagen located near cell clusters (Gay *et al.*, 1976) might be associated with repeated mitosis. Initially the cells are highly active as judged by sulphate uptake. The centres of large clusters, which may contain 100 cells or more, show evidence of necrosis and eventually synthetic activity subsides, metachromasia is lost (Meachim, 1963) and the cells degenerate (Dustmann *et al.*, 1974). Such large degenerate masses are unlikely to be beneficial in cartilage repair.

The phase of proliferation is in any case short lived. Although many of the clusters persist, proliferation normally contributes no more to the healing of the wound than some reorganisation of the zone of necrosis (Mankin, 1962*b*).

STIMULATION OF REPAIR IN ARTICULAR CARTILAGE

While the response of cartilage to experimental trauma appears to be only 'reactive' and rather ineffectual as regards repair, it does exhibit cell multiplication and matrix secretion two necessary elements of cartilage repair. It must also be remembered that the defects produced in experiments or indeed the pathological fissures of deeply fibrillated articular cartilage are much larger than the near submicroscopic lacerations and fissures occurring naturally during the life of many areas of the joint surface. Since these minute surface defects are relatively inaccessible to extrinsic repair processes it follows that if they are to be repaired or at least if the process of deterioration is to be kept in abeyance, only intrinsic mechanisms are easily available. At the present time it is not known to what extent, if at all, repair may be effected on this microscopic scale.

Repair may be induced by grafting tissue into defects, either as blocks of tissue or more successfully in the form of pellets of dissociated chondrocytes (Bentley & Greer, 1971). Rapid proliferation and matrix synthesis are needed: implanted epiphysial chondrocytes 'take' while articular chondrocytes are rejected. More recently, grafts of costal cartilage perichondrium (Skoog & Johansson, 1976) have been used, deep surface toward the joint cavity, in articular cartilage regeneration. It may be that repair could also be enhanced if the degradative enzymes present in chondrocytes were inhibited pharmacologically, although according to Barrett (1975) there is still insufficient evidence for their role in the breakdown of living cartilage.

Agents which actively promoted the intrinsic repair mechanism would be of immense value. More is needed than, for example, the inhibition of chondrocyte catheptic activity or the stabilisation of the lysosomal membrane. As it is, increased degradative enzyme activity in chondrocytes of fibrillated cartilage may only reflect increased metabolic activity of the cells (Stockwell & Meachim, 1973), secondary to the anatomical defects arising from mechanical trauma or fatigue processes. Even stimulation of cell proliferation, if this results only in cell clusters, would seem to be insufficient for repair. It may be necessary for the numerous new daughter cells to become separated by matrix secretion before a cluster can form, if an enlarged tissue volume for repair purposes is to be produced. There is also the complex question of the re-formation of the collagen mesh particularly near the articular surface.

Encouraging results have been obtained by Chrisman and his colleagues. Simmons & Chrisman (1965) have demonstrated that salicylates prevent the usual degenerative sequelae of experimental scarification in the rabbit, permitting a considerable degree of incision healing. Used prophylactically in cases of patellar dislocation in man, aspirin again has a beneficial effect (Chrisman, Snook & Wilson, 1972). It is suggested that the drug acts by inhibition of synthesis of prostaglandins (Vane, 1971), which can cause the release of lysosomal enzymes. Results from experiments on the rabbit knee joint support this concept (Teitz & Chrisman, 1975). However, salicylates have no healing effect on advanced lesions and probably act by preventing degenerative processes. They may also have a deleterious effect in the long term since they inhibit glycosaminoglycan synthesis (McKenzie *et al.*, 1976).

Experiments with growth hormone have demonstrated a positive effect on the healing of established cartilage lesions produced by scarification (Chrisman, 1975). Compared with untreated controls, not only are clefts smaller and frequently almost obliterated but often a new tangential collagen layer develops over the surface of a defect. The cell count, DNA content and ^{3}H-thymidine uptake are higher in specimens treated with growth hormone although cell clusters are less prominent. Thus it may be possible to use the hormone, or its intermediaries, in treatment.

Other mechanisms aside, the action of growth hormone, of somatomedins (Hall *et al.*, 1975) and of closely related factors are extremely complex. In addition, pathologically fibrillated cartilage and its clinical manifestation as osteoarthrosis is a condition which has a complex aetiology in which many of the contributory factors are imperfectly understood. Nevertheless, experiments such as those of Chrisman's with growth hormone, and other similar investigations, suggest a biological approach to the treatment of cartilage defects and degenerative joint disease. In an age when we have to resort to the admittedly successful remedies offered by 'spare-part' surgery, it may yet be possible to assist effective healing, mediated by cell activity within the cartilage.

8

CARTILAGE DEGENERATION, CALCIFICATION AND CHONDROCYTE DEATH

Cartilage undergoes changes with age and various forms of degeneration which involve both the cells and the matrix. This is consistent with its avascular nature, low rates of renewal of its constituents, its poor repair capacity and the mechanical stresses acting on the tissue. The origins of certain of the degenerations are of considerable interest but two kinds of problem complicate investigation of their causes. First, it is difficult to segregate those age-related changes which are benign from those which ultimately may have harmful effects. Secondly, it is often not easy to determine whether or not degenerations of the matrix have their origins in cell changes. Inevitably, however, cartilage degenerations at some stage involve chondrocyte death.

CARTILAGE DEGENERATION

The degenerations of the matrix are reviewed by Silberberg & Silberberg (1961). No form of cartilage is immune although the recorded age of onset and prevalence vary in different species and anatomical sites. Contrary to expectation, several of the classical forms of degeneration in man are not confined to adult life but are observed from the beginning of the second decade.

ALBUMINOID DEGENERATION

Finely granular acid- and alkali-fast material appears in the ground substance during the second decade and thereafter calcification may follow.

ASBESTOS FORMATION

Loss of proteoglycan results in unmasking, thickening and disorientation of collagen fibres. It is one of the first degenerations seen. It is observed in the laryngeal cartilages as early as the tenth year, in the costal cartilages during the second decade and in the tracheal rings in the third decade.

GELATINOUS DEGENERATION AND CYST FORMATION

This also is an early phenomenon. It is found in the laryngeal and costal cartilages from the second decade and somewhat later in the trachea. The first sign of degeneration is loss of matrix basophilia followed by loss of cells.

FATTY METAMORPHOSIS

Extracellular fat appears as a consequence of other forms of degeneration. It also occurs in cartilages which are otherwise normal, where it is located in the territorial matrix; in addition, in articular cartilage it is distributed more diffusely in the superficial zone (Chapter 5). It may be present from the second decade and is commonly present in adults (Putschar, 1931; Schallock, 1942; Ghadially *et al.*, 1965; Stockwell, 1965). Many authorities believe it to be a pathological phenomenon although there is no evidence that the superficial lipid of articular cartilage predisposes cartilage to fibrillation (Schallock, 1942; Ghadially *et al.*, 1965).

CALCIFICATION AND OSSIFICATION

Except in the basal layer of articular cartilage, calcification occurs in a random fashion in the permanent cartilages, compared to its orderly progress in the epiphysial growth plate. Mineralisation is observed radiologically from the second decade in the laryngeal (Hately, Evison & Samuel, 1965) and costal cartilages (Elkeles, 1966). Intervertebral discs and even elastic cartilages may also become calcified in the later decades of life (Silberberg & Silberberg, 1961). Commonly it follows and is located at sites of other degenerative change but can occur in their absence. The crystals deposited may be of several related types. Most often the mineral is hydroxylapatite which occurs in all forms of cartilage, but calcium pyrophosphate dihydrate and dicalcium phosphate dihydrate are also found, particularly in fibrocartilage (Dieppe, 1978). In articular cartilage affected by the rare condition of chondrocalcinosis articularis ('pseudogout'), deposition of pyrophosphate crystals (25–50 nm in length) commences near cell lacunae in the middle zone of the tissue (McCarty, 1976; Schumacher, 1976), unlike the more superficial location of urate crystals in gout. Pyrophosphate formation also occurs in the deeper parts of osteoarthrosic cartilage (Howell, Muniz, Pita & Enis, 1976), perhaps associated with the basal remodelling found in this condition.

Ossification usually occurs secondary to calcification or other degenerations, replacing the affected cartilage. Although it can commence in a central part of the tissue, it more commonly starts peripherally and spreads centrally, as in the costal cartilages.

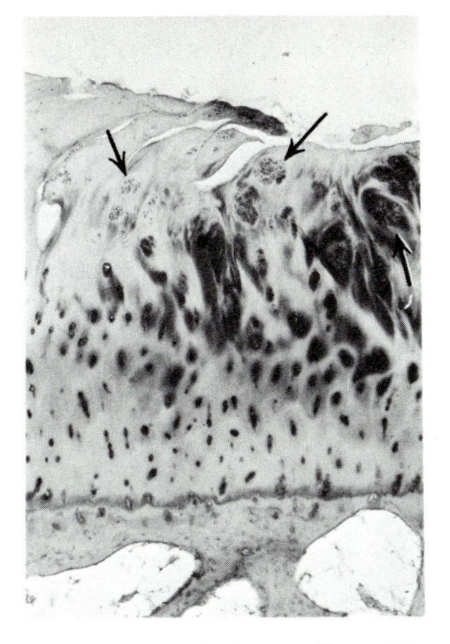

Fig. 104. Human femoral condylar articular cartilage (54 years), formalin fixation, magnification × 32. Early vertical fibrillation. Note the differences in matrix basophilia in relation to the various cell clusters (arrows) found adjacent to the fissures.

ARTICULAR CARTILAGE FIBRILLATION

Like other degenerations this does not affect the whole of the tissue simultaneously but is initially a patchy or focal phenomenon: small isolated areas of the articular surface become broken and frayed. Splitting extends along the general direction of collagen fibre orientation (Collins, 1949). Thus horizontal splitting, or tangential fibrillation, is found while the process is restricted to the superficial zone but radial splitting or vertical fibrillation (Fig. 104) develops when the deeper tissue becomes involved. Fibrillation can extend sideways as well as vertically; hence large areas or even the whole of a joint surface can be affected. Fragments of cartilage become detached from the articular surface, resulting in a variable loss of cartilage and exposure of bone. The incidence and severity of fibrillation, as observed macroscopically, is age-related but varies from joint to joint and in different parts of the same joint (Heine, 1926). Studies of large human joints such as the hip (Byers, Contepomi & Farkas, 1970), knee (Bennett *et al.*, 1942; Meachim & Emery, 1974), shoulder (Meachim & Emery, 1973) and elbow (Goodfellow & Bullough, 1967) show that fibrillation commences as early as the second decade, typically in peripheral regions of the joint surface and in sites such as the para-foveal region of the femoral head and the medial facet of the patella.

Fibrillation is by far the most important degeneration of cartilage because

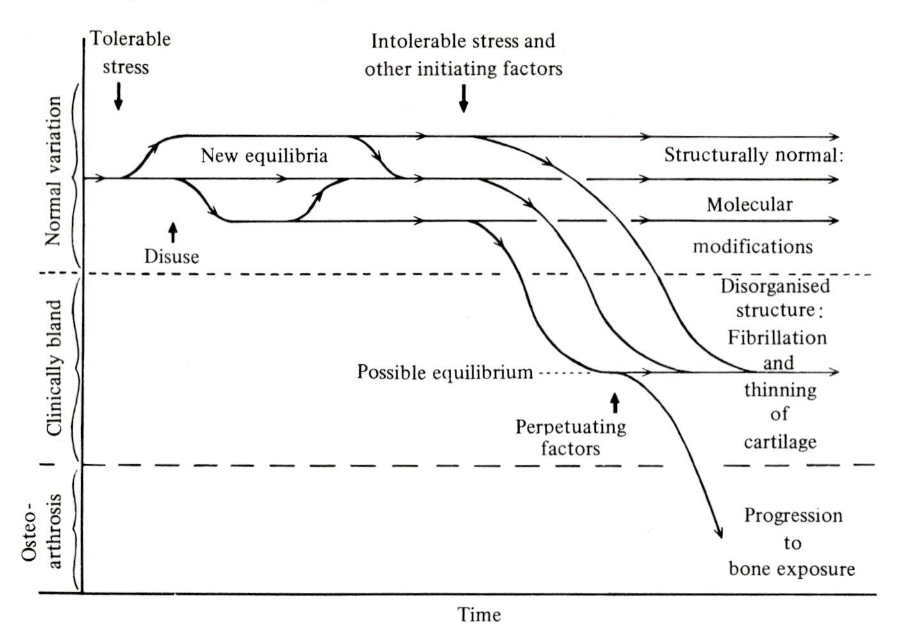

Fig. 105. Scheme indicating the various steady states of articular cartilage which are possible in normal life and in response to adverse factors, with the relationship of fibrillation to the joint failure of osteoarthrosis.

of its relationship to osteoarthrosis (Fig. 105). In this clinical condition, by no means confined to man, the affected joints are painful, with varying degrees of stiffness, deformity and restricted movement. Pathologically there is partial or complete loss of cartilage and exposure of bone together with a number of other features. The cause or causes are unknown in a large number of cases; these cases form a group classified as primary or idiopathic osteoarthrosis. Although some investigators (Radin, Paul & Rose, 1972) suggest that changes in the underlying bone, such as trabecular thickening, may precede cartilage degeneration, many authorities consider that the initial macroscopic lesion in primary osteoarthrosis is cartilage fibrillation (Collins, 1949; Freeman & Meachim, 1973). However, the factors which determine the extent of progression of the lesion may be in the cartilage and/or its environment, for example the character of the underlying bone, the joint surface geometry and the nature of the forces acting on the joint (G. Meachim, personal communication). Thus, in the hip joint, surgically resected osteoarthrosic femoral heads exhibit areas of 'progressive' cartilage destruction and areas showing typical age-related fibrillation, which differ in their location (Byers *et al.*, 1970) and histology (Byers *et al.*, 1977). Hence the lesion which precipitates the surgical condition in the hip probably does not result directly from cartilage destruction associated with age-related fibrillation. In most joints, however, fibrillation may sometimes, though by no means always, progress to osteoarthrosis (Freeman & Meachim, 1973).

In conclusion though not necessarily true of all degenerations there seem to be two general endpoints to cartilage deterioration. Partly depending on anatomical site, the tissue either becomes calcified and may later ossify, or it shows structural disintegration and tissue bulk is lost. Before reviewing calcification and fibrillation in detail it is worth considering changes with age in 'normal' cartilage which may make the tissue more susceptible to these phenomena.

CHANGES IN CARTILAGE WITH AGE

There are difficulties, philosophical, semantic and otherwise, in defining what is meant by ageing and by normality of the tissue. In practice, the criteria adopted tend to be arbitrary and those most convenient for the field of investigation. In the following account of changes in cartilage with the passage of time, the ageing period is taken as that part of the lifespan following cessation of skeletal incremental growth, for example in man after the age of 25–30 years. As regards normality, specimens which are structurally intact both macroscopically and histologically are usually considered suitable for studies of ageing as opposed to pathological investigation.

CELLULARITY

It is well known that aged adult cartilage is much less cellular than immature tissue. Furthermore, chondrocyte death is a well-established phenomenon in growing tissue and occurs in many adult permanent cartilages. These observations have led to the common belief that cell content falls with age. While there is no doubt that a massive reduction in cell density occurs during the growth and maturation period (Figs. 44, 89) it is less certain whether this trend continues after growth has ceased.

There is considerable variation between cartilages. In non-articular cartilage an age-related decrease in cellularity occurs after growth has ceased. This can be as much as 25% in human costal cartilage (Fig. 106) after the third decade (Stockwell, 1967*a*) and a similar decline can be seen in horse nasal cartilage (Fig. 18 in Galjaard, 1962) though post-maturation changes in human tracheal cartilage are minimal (Beneke, 1973). Cell loss also occurs in elastic and fibrocartilage (Silberberg & Silberberg, 1961).

There has been much interest in articular cartilage. In humeral head (25–84 years) and femoral condylar cartilage (31–89 years) in man there is no significant change in cell density of the whole tissue, i.e. of the full thickness of the tissue (Meachim & Collins, 1962; Stockwell, 1967*a*). Similarly, biochemical estimations of cell mass in human articular cartilage indicate that the cell content remains constant with age (Miles & Eichelberger, 1964). However, in human femoral head cartilage, cell content of the full thickness of the tissue decreases by as much as 35% over the period 30–100 years (Vignon *et al.*, 1976). It is possible to reconcile these apparently conflicting

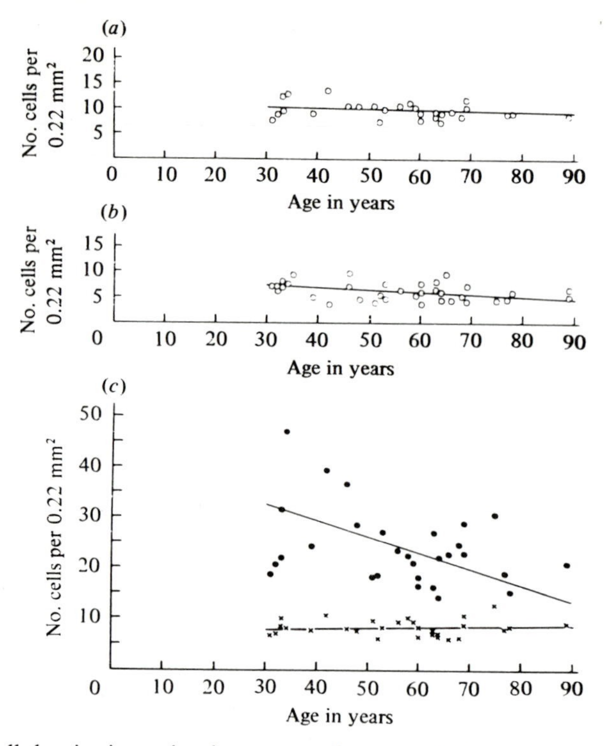

Fig. 106. Cell density in ageing human cartilage. (*a*) Cellularity of the full thickness of femoral condylar articular cartilage: $y = 11\text{–}0.02x$. There is no significant change with age. (*b*) Change in cellularity of the fourth costal cartilage: $y = 8-0.04x$. A significant ($P < 0.05$) loss of cells occurs. (*c*) Zonal changes in cellularity of femoral condylar cartilage during ageing. A significant ($P < 0.01$) loss of cells occurs in the superficial zone (●): $y = 42-0.3x$, which is balanced by a small increase in the cell density of the deeper tissue (×): $y = 7+0.02x$. (From Stockwell, 1967*a*.)

results if the various zones of the cartilage are considered separately. It is then found that a reduction in the cell density of the superficial zone occurs both in femoral head and condylar cartilage (Fig. 106), though not in the humeral head. In femoral head cartilage, cell loss in the superficial zone appears to be sufficient to cause the age-related decline in cellularity of the full thickness of the tissue. However, in femoral condylar cartilage, cell loss in the superficial zone occurs *pari passu* with a slight increase in the deeper tissue; therefore no change occurs as a whole.

It is probable that the changes in the superficial zone are related to the development of fibrillation since cell loss is a feature of overt fibrillation. The differing results in the three joints would then be compatible, since the shoulder joint is much less prone to degeneration than the knee or hip (Meachim & Collins, 1962). It is unlikely that the age-related changes in the superficial zone are merely a dimensional consequence of pathological swelling of the matrix. Thus cell death in the superficial zone has been observed ultrastructurally (Barnett *et al.*, 1963; Roy & Meachim, 1968) as well as by light microscopy (Meachim *et al.*, 1965). Furthermore in the hip joint the decline in cell density of the superficial zone is paralleled by a decrease in the total number of chondrocytes lying deep to a unit area of articular surface (Vignon *et al.*, 1976).

CHANGES IN GLYCOSAMINOGLYCANS AND PROTEOGLYCANS

It is generally assumed that the glycosaminoglycan content of cartilage decreases with age. This is undoubtedly true of the growth period in human articular (Kuhn & Leppelmann, 1958) and costal cartilages (Loewi, 1953; Kaplan & Meyer, 1959) although in femoral head articular cartilage, FCD of the whole tissue (i.e. expressed as a wet weight rather than a dry weight) is lower in early childhood than in the adult (Maroudas *et al.*, 1973). In the non-articular cartilages glycosaminoglycan diminution continues throughout the lifespan, and is more rapid and pronounced in rib than in laryngeal and tracheo-bronchial cartilages (Mathews & Glagov, 1966; Mason & Wusteman, 1970). By contrast, little or no reduction in total glycosaminoglycan content is found in macroscopically normal articular cartilage of the knee (Anderson, Ludowieg, Harper & Engleman, 1964; Bollet & Nance, 1966) or of the femoral head (Maroudas *et al.*, 1973) during adult life.

Changes occur in the types and proportions of glycosaminoglycans. In all human cartilages investigated both the total chondroitin sulphate content and the ratio of chondroitin-4-sulphate to chondroitin-6-sulphate decrease during post-natal maturation. The change in this ratio occurs rapidly in costal cartilage but more slowly in a linear fashion throughout the whole lifespan in femoral condylar cartilage (Greiling & Baumann, 1973). At the same time as the fall in chondroitin-4-sulphate during maturation there is an increase in keratansulphate content (Mathews & Glagov, 1966). Keratansulphate is very low or absent at birth but may form as much as 55% of total glycosaminoglycan in adult life. Changes during adult life are more evident in the non-articular cartilages. In costal cartilage, keratansulphate concentration does not increase further after the third or fourth decades but since chondroitin sulphate continues to fall linearly throughout adult life the keratansulphate/chondroitin sulphate ratio may continue to rise throughout the lifespan (Stidworthy *et al.*, 1958; Kaplan & Meyer, 1959). Similar trends occur in tracheo-bronchial cartilage (Mason & Wusteman, 1970). In the human intervertebral disc Hallen (1962) finds that although the keratansulphate/chondroitin sulphate ratio of the nucleus pulposus increases throughout the whole lifespan it does not change in the anulus fibrosus after the third decade. Small quantities of keratansulphate are found in the elastic cartilage of two-year-old rabbits and steers (Gillard & Wusteman, 1970).

In normal articular cartilage there are no substantial changes in the proportions or concentrations of chondroitin sulphate and keratansulphate of the whole tissue, i.e. full-thickness specimens, during adult life (Bollet & Nance, 1966). A slight fall in the chondroitin sulphate content of femoral condylar cartilage is reported (Kuhn & Leppelmann, 1958; Hjertquist & Lemperg, 1972; Greiling & Baumann, 1973) but in the femoral head no change is found on a wet weight basis (Venn, 1978); there is, however, a small decrease in hydration of femoral head cartilage with age. But in both the knee and the hip, keratansulphate increases slightly during adult life (Kuhn & Leppelmann, 1958; Hjertquist & Lemperg, 1972; Venn, 1978).

Analyses of glycosaminoglycans in full-thickness samples of articular cartilage should be assessed in the light of the zonal heterogeneity of the tissue. Although there appears to be little variation with age in the superficial zone (Venn, 1978), the deep zone shows marked changes. Thus during the third and fourth decades in femoral condylar cartilage the proportion of keratansulphate to total glycosaminoglycan rises significantly from a low level (0.12) in immaturity to a high 'plateau' (0.52) in later life (Stockwell, 1970). This is a real change and is not to be attributed to a misleading comparison of epiphysial with 'true' articular cartilage, since the proportion of keratansulphate is low throughout the entire thickness of immature articular cartilage. In the deep zone, staining due to keratansulphate changes from an interterritorial to a territorial localisation (Fig. 38) over the same period of life. These age changes in adult life probably relate to the reduction of nutrient supply from the subchondral marrow spaces which mostly close at the cessation of growth. A similar increase in keratansulphate during early adult life occurs in the femoral head (Venn, 1978). Histologically, the territorial localisation of keratansulphate extends throughout a greater part of the thickness of the deep zone of femoral head than of femoral condylar cartilage where it is restricted to the region immediately adjacent to the tidemark (Stockwell, 1978*a*).

The decline in chondroitin-4-sulphate during maturation and ageing is not universal. Thus not only in certain amphibia and birds but also in some small mammals, such as the rat and rabbit, chondroitin-4-sulphate becomes the predominant form of this glycosaminoglycan. However, the occurrence and increase of keratansulphate appears to be a remarkably constant feature of ageing, the one exception found being the costal cartilage of the rat (Mathews, 1975).

There is little information about changes in proteoglycans with age. There is a higher proportion of protein and keratansulphate in proteoglycans of aged than of infant human costal cartilage (Rosenberg, Johnson & Schubert, 1965). Similar modifications occur in pig articular cartilage proteoglycans, which also become more difficult to extract; the proportion of residual unextractable proteoglycans associated with collagen also increases with age although much of the change occurs before growth has ceased (Simunek & Muir, 1972). In aged femoral head cartilage, most of the proteoglycans are unable to aggregate with hyaluronic acid. This is probably due to a defect of the hyaluronate-binding site in the core protein (see Chapter 3), although the stabilising 'link' glycoprotein is also absent (Perricone, Palmoski & Brandt, 1977). It is suggested that the changes may have been caused by the surgical condition (fracture of the neck of the femur) rather than ageing.

Chondrocytes of immature cartilage are more active than in the adult as regards oxygen utilisation (Rosenthal *et al.*, 1941), glycolysis (Bywaters, 1937), protein synthesis (Mankin & Baron, 1965) and probably proteoglycan metabolism (Collins & McElligott, 1960; Maroudas, 1975). *In vitro* analysis provides little or no evidence of a continuous decline in sulphate uptake and protein synthesis per cell during ageing (Collins & Meachim, 1961; Mankin & Baron, 1965). Older cartilage nevertheless differs from younger cartilage principally in a higher keratansulphate content of the tissue and of individual molecules. This is probably due to changes in synthesis (Shulman & Meyer, 1968) as well as to the slower turnover of keratansulphate. There may also be a decline in chondrocyte oxygen utilisation (Rosenthal *et al.*, 1941), although as discussed earlier (Chapter 4) this might theoretically be explained by a loss of superficial cells in the specimens investigated.

In conclusion, there is during maturation a decline in tissue metabolic activity and glycosaminoglycan content associated with a massive reduction in cellularity. Similarly, loss of cells in ageing cartilage will result in local cessation of tissue metabolism, which may be the explanation for the diminution in glycosaminoglycan content of ageing non-articular cartilage. In articular cartilage there is little age-related change in overall glycosaminoglycan or cell content, except in the superficial zone. Such changes may render the tissue more prone to degeneration. It is not known at present whether cell loss and possible metabolic changes in ageing cartilage are due to primary factors operating within the chondrocyte, perhaps involving senility of DNA and associated enzymes. Cell loss could be secondary to changes in matrix consistency or in the surrounding tissues which affect the cells by nutritional deprivation or by permitting noxious insult or abnormal mechanical stress.

PIGMENTATION

Aged cartilage has a yellow-brown tinge compared with the pallor of the youthful tissue. In human cartilage, the colour is prominent in the intervertebral disc, the central zone of costal cartilage, in the menisci and in the basal zone of articular cartilage; pigmentation is not confined to human tissues (Silberberg & Silberberg, 1961; van der Korst, Sokoloff & Miller, 1968). In human costal cartilage the amount of pigment is appreciable from the fourth decade. The brown colouration is in general distributed inversely to the degree of tissue basophilia in costal and articular cartilage although the superficial zone of articular cartilage does not conform to this rule. It may be relevant that the superficial region of articular cartilage has fine fibres at all stages, while thicker fibres are found in the regions that become pigmented.

Hass (1943) found that neither lipid nor iron compounds contribute to the pigment. Van der Korst *et al.* (1968) exclude lipofuschins, lipochromes, haemosiderin and melanins and have shown that the pigment is not associated with the glycosaminoglycan fraction of the matrix. Instead it appears to be part of or attached to non-collagenous proteins containing large amounts of acidic amino acid residues. Since proteins of this type may be involved in the collagen–proteoglycan interaction (Chapter 3) it is possible that pigmentation may reflect an increasing redundancy of this type of protein as proteoglycan concentration falls and thicker fibres form in the matrix.

Other forms of pigment occur. For example, the superficial zone of normal articular cartilage is sometimes coloured yellow (Ghadially *et al.*, 1965) as may be the full thickness of degenerate articular cartilage. This appears to be due to carotenoids (Davies, 1967) and bile pigments (van der Korst *et al.*, 1968) in the synovial fluid. Although the black pigment of ochronosis can cause articular fibrillation there is no evidence of an association between knee joint osteoarthrosis and the degree of brown pigmentation of cartilage (van der Korst *et al.*, 1968).

INITIATION OF CARTILAGE CALCIFICATION

Although several mechanisms of calcification may operate at different biological sites it is probable that there are close similarities between provisional calcification in the growth plate and 'dystrophic' calcification in permanent cartilages. Since much is known about provisional calcification and because of its great intrinsic importance it is worth considering first how the growth plate becomes mineralised.

CALCIFICATION IN THE GROWTH PLATE

Whatever the precise mechanism, it appears that some kind of initiating or 'nucleating' agent is required to which calcium salts can attach and from which crystals can grow. Although collagen has been implicated in bone, it does not appear to act as a template in cartilage since the crystals of hydroxylapatite bear no constant spatial relationship to the fibres. Theories of calcification are legion and it is probably fair to say that every constituent of a tissue has been invoked at some time or other as the nucleating agent. There has, however, been increasing evidence during the last decade that lipid has a central role. Although the association of lipid and calcification had long been suspected, it was the work of Irving (1959), showing strong sudanophilia due to extracellular lipid in the hypertrophic zone of growth plate cartilage in normal though not rachitic rats, which suggested that lipid was important.

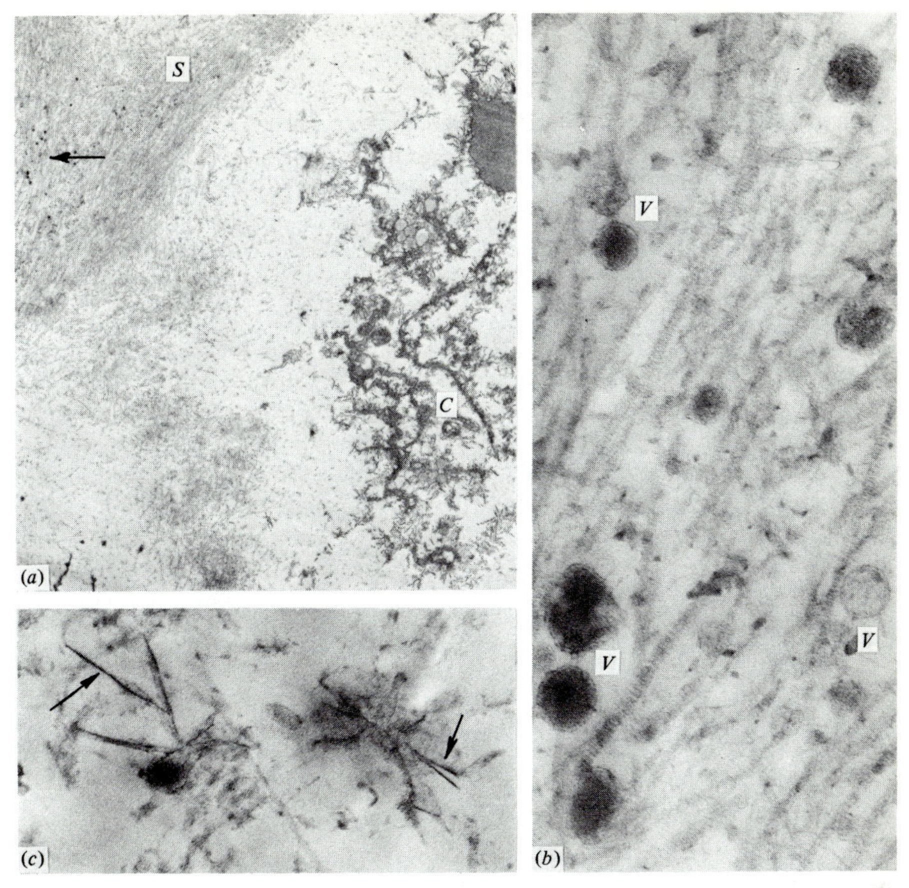

Fig. 107. Rat upper tibial growth plate (four weeks), glutaraldehyde fixation. (*a*) Lower hypertrophic zone showing part of a degenerating chondrocyte (*C*) in its lacuna, and a longitudinal septum (*S*) containing matrix vesicles (arrow). Magnification ×3600. (*b*) Higher magnification of the region indicated by the arrow in (*a*). Both pale and dense matrix vesicles (*V*) are present, about 50–150 nm in diameter. Magnification ×86000. (*c*) Early mineralisation in a longitudinal septum, showing needle-like crystals (arrows) about 150–200 nm long close to matrix vesicles. Magnification ×86000.

Ultrastructural studies have given added impetus since they show that extracellular lipid in the form of membranous vesicles has a close spatial relationship to early calcification.

Matrix vesicles

These structures were first observed by Bonucci (1967) in growth plate cartilage and their morphology is fully described by Anderson (1969) and Bonucci (1970). The membrane-bound vesicles have round or oval profiles and range in size from 30–1000 nm with an average diameter of 250 nm. They are found at all levels in the growth plate and are concentrated in the longitudinal septa from the proliferation as far as the calcified zone. Their content varies in electron density; pale 'matrix vesicles' and dark 'matrix-dense bodies' may be distinguished (Serafini-Fracassini & Smith, 1974) but although there may be more than one class of these structures most authorities group them all together as matrix vesicles. The first needle-like crystals of hydroxylapatite, about 150 × 4 nm (Bonucci, 1969), are associated with these structures in the lower hypertrophic zone (Fig. 107). Hence they are believed to have a primary role in calcification. Although at first a lysosomal function was also considered (Anderson, 1969), histochemical localisation of acid phosphatase activity in the matrix vesicles (Thyberg & Friberg, 1970) may be artefactual due to mineral already *in situ*. Analyses of cell and matrix vesicle fractions isolated following collagenase digestion of the growth plate (Ali *et al.*, 1970) give little evidence of acid phosphatase activity in the vesicles. However, it is possible that in other situations, such as mineralising teeth, matrix vesicles may be heterogeneous, some of them exhibiting acid phosphatase activity (Slavkin *et al.*, 1976). But there is no doubt that the alkaline phosphatase activity of the growth plate is very largely associated with the matrix vesicles (Ali *et al.*, 1970).

In the growth plate the majority of matrix vesicles seems to originate by budding from cell processes. In the reserve zone, vesicles are distributed pericellularly within the lacuna and at about 1–3 μm from the cell. Occasional cell processes can be several microns long and detachment of their bulbous tips results in the pericellular distribution of matrix vesicles (Bonucci, 1970); subsequently they become located in the longitudinal septa. Consistent with this origin, the vesicle membrane has several features characteristic of the plasmalemma. Thus the cholesterol/protein ratio and glycolipid composition are similar (Wuthier, 1975); enzymes such as phospholipase A_1 (Wuthier, 1976); 5-adenosine monophosphatase (Ali *et al.*, 1970) and alkaline phosphatase, located on the external aspect of the vesicle membrane (Matsuzawa & Anderson, 1971), are common to both sets of membranes; at other sites of mineralisation they exhibit the same H2 histocompatibility antigen specificity (Slavkin *et al.*, 1976). Studies of lipid incorporation, using glycerol-2-^{3}H, demonstrate a prompt uptake in the limiting membranes of cells in the resting and proliferation zones and a rapid transfer of the label to the matrix vesicles (Rabinovitch, 1974).

Availability of calcium

Although the first hydroxylapatite crystals are seen in the matrix vesicles, it seems probable that the calcium component reaches them by way of the cells. Counts of mitochondrial granules and localisation of calcium uptake, using ^{45}Ca and histochemical staining procedures (Matthews, Martin, Lynn &

Collins, 1968; Martin & Matthews, 1969; Brighton & Hunt, 1976), indicate a gradient of increasing calcium concentration in the mitochondria from the proliferative to the lower hypertrophic zones. Calcium concentration in the matrix vesicles shows a similar gradient as detected by X-ray emission microanalysis (Ali, 1976) and chemical analysis of isolated vesicles (Wuthier, 1977). Whereas there is evidence that the mitochondria in the calcified zone exhibit loss of calcium, the vesicles show a calcium increase. Time course studies indicate that ^{45}Ca is taken up by the cells prior to its appearance in the matrix vesicles (Matthews *et al.*, 1968).

It is not known in what form calcium leaves the cells. The earliest mineral found in the matrix vesicles is not hydroxylapatite but granules of amorphous calcium phosphate 2 nm in diameter (Bonucci, 1970; Wuthier & Eanes, 1975; Hohling, Steffens, Stamm & Mays, 1976). Lehninger (1970) postulates that 'micropackets' of amorphous calcium phosphate leave the cell complexed to an inhibitor. Wuthier (1977) considers that the matrix vesicle is initially a vector for some form of non-crystalline calcium phosphate possibly stabilised by phosphatidyl-serine during transit (Wuthier & Eanes, 1975). On the other hand, the anionic groups of proteoglycans are known to bind calcium (Boyd & Neuman, 1951; Woodward & Davidson, 1968) and therefore might conceivably act as a carrier (Serafini-Fracassini & Smith, 1974). However, degradation of proteoglycans is probably necessary before the calcium can be released and made available for mineralisation. Degradation occurs only in the lower hypertrophic zone (Campo & Dziewiatkowski, 1962; Matukas & Krikos, 1968; Greer *et al.*, 1968). Hence it is unlikely that this mechanism accounts for the augmentation of calcium seen in the matrix vesicles at higher levels in the plate, although it may be necessary for the massive crystallisation taking place near the metaphysis.

Calcium uptake by matrix vesicles

The mode of uptake is associated with both the enzymes and lipids of the vesicle membrane. Three types of enzyme activity – alkaline phosphatase, pyrophosphatase and adenosine triphosphatase – are located in the vesicle. It is a moot point whether these represent three different enzymes or a single enzyme with a wide substrate specificity (Moss, Eaton, Smith & Whitby, 1967; Majeska & Wuthier, 1975). Inorganic pyrophosphate inhibits calcification of collagen *in vitro* (Fleisch & Bisaz, 1962) and the potential significance of matrix vesicle enzymes in removing such inhibitors has been emphasised by Felix & Fleisch (1976*a*, *b*). ATP is found in growth plate cartilage (Albaum, Hirshfeld & Sobel, 1952; Whitehead & Weidman, 1957) and enhances calcification of cartilage *in vitro* (Cartier & Picard, 1951). It has been postulated that uptake of calcium into matrix vesicles is an active process dependent upon energy provided by ATPase activity (Ali & Evans, 1973*b*; Ali, 1976). According to Goodlad *et al.* (1972) (cited by Serafini-Fracassini &

Smith, 1974), ATP reaches the vicinity of the matrix vesicles by way of the metaphysis, released from erythrocytes as a consequence of their extravasation from ruptured metaphysial capillaries (Fig. 66). As Serafini-Fracassini & Smith (1974) comment, such observations could explain the significance of the metaphysial circulation in calcification (Trueta & Amato, 1960), if indeed ATP is an essential substrate for the matrix vesicles. However, it is doubtful whether ATP has a special role. A large number of substrates, including all the nucleotide triphosphates and indeed the hexosemonophosphates (Ali, 1976), can substitute for ATP in the postulated energy-dependent calcium uptake. In addition, a number of investigators have shown that the matrix vesicle ATPase is atypical in that it is not calcium-dependent; for this and other reasons it is believed that there is no active transport. Therefore the enhanced uptake of calcium is due merely to the generation of phosphate by ATP hydrolysis (Majeska & Wuthier, 1975; Felix & Fleisch, 1976*a*, *b*; Sajdera, Franklin & Fortuna, 1976), either directly or via pyrophosphate. Although now localised to the matrix vesicle, this is reminiscent of Robison's (1923) original concept of the role of alkaline phosphatase and organic phosphates in cartilage mineralisation.

Calcium uptake, instead of occurring via a pump mechanism, may be a passive process dependent on the presence of a high proportion of acid phospholipids in mineralising matrix. The vesicles are lipid-rich and contain more sphingomyelin and phosphatidyl-serine than cell membranes (Peress, Anderson & Sajdera, 1974; Wuthier, 1975). This is thought to result from passive enrichment following separation from the cell, due to a greater degradation by phospholipase of phosphatidyl-choline and phosphatidyl-ethanolamine than of the acidic phospholipids (Wuthier, Majeska & Collins, 1977). Phosphatidyl-serine has a strong affinity for Ca^{2+} (Papahadjopoulos, 1968), inorganic phosphate greatly enhancing binding by the formation of non-crystalline complexes (Cotmore, Nichols & Wuthier, 1971). Hence it is suggested that calcium uptake is an ion-exchange phenomenon (Peress *et al.*, 1974), assisted by the hydrolysis of phosphate esters or of inorganic pyrophosphate in low concentration (Anderson & Reynolds, 1973).

Crystallisation

The transformation of amorphous calcium phosphate into the needle-like crystals of hydroxylapatite is not fully understood. This may occur indirectly via crystals of octacalcium phosphate which have the shape of wrinkled leaflets, the first crystals appearing on the surface of the amorphous phase (Eanes & Meyer, 1977). Formation of hydroxylapatite in the vesicle may require degradation of phospholipid: the presence of degradation products (lyso-phospholipids) and a higher activity of phospholipase toward phosphatidylserine in the calcified zone (Wuthier, 1976, 1977) supports this notion. On the other hand the unaltered phospholipid complex may itself serve as a nucleation site by holding calcium and phosphate ions in the correct orientation for crystal formation (Boskey & Posner, 1977).

Crystal formation is affected by a number of natural and artificial inhibitors including inorganic pyrophosphate, organic phosphonates such as ethane-1-hydroxy-1,1-diphosphonate (Schenk *et al.*, 1973) and nucleotide di- and triphosphates (Termine & Conn, 1976), which stabilise and so prevent crystal growth. The enzyme complement of the matrix vesicles is able to hydrolyse the natural inhibitors, at the same time producing inorganic phosphate to assist mineralisation. However, it appears that proteoglycans, and hyaluronate–proteoglycan aggregates in particular, also act as inhibitors and their disaggregation may be facilitated by lysozyme (Pita *et al.*, 1975) produced by the cells.

CALCIFICATION IN AGEING CARTILAGE

There is good evidence that this also is associated with the presence of extracellular lipid. Studies of human costal cartilage demonstrate that total lipid content increases significantly from 0.4% to 1.2% wet weight over the age range 30–90 years. During the same period, both cell density (Fig. 106) and the mean content of lipid per cell (Fig. 13) decrease. Hence it follows that there must be an age-related increase of extracellular lipid (Stockwell, 1967*c*); this is probably the major component in old age (Stockwell, 1966). Thus both extracellular lipid and the incidence of calcification (Hass, 1943) increase with age. Furthermore, there is a higher lipid content in female than in male costal cartilage at any given age after growth has ceased (Stockwell, 1967*c*) and this correlates with the sex difference in costal cartilage calcification (Elkeles, 1966).

The extracellular lipid, much of it phospholipid, accumulates in the territorial matrix (Fig. 72) at and just external to the lacunar rim (Stockwell, 1966). Calcium salts are also initially deposited in the territorial matrix (Silberberg & Silberberg, 1961), spreading outwards into the inter-territorial matrix and inwards to involve the cell. Early ultrastructural studies of this region of the matrix showed that calcium particles lie close to fat droplets (Zbinden, 1953). It has been demonstrated in articular cartilage (Fig. 37) that extracellular lipid is in the form of electron-dense granules and membrane-bound bodies (Barnett *et al.*, 1963), similar to the matrix vesicles subsequently discovered in the growth plate.

As discussed earlier (Chapter 5) these collections of lipidic material may be either the debris of cell necrosis (Fig. 109) or formed (like the growth plate vesicles) by detachment of the bulbous ends of living cell processes (Ghadially *et al.*, 1965). In ageing human and rat costal cartilage, Bonucci & Dearden (1976) have observed an additional mechanism in which numerous vesicles are simultaneously extruded from the cell via a cytoplasmic vacuole. This process has also been observed in mature rabbit elastic cartilage (Cox & Peacock, 1977). Bonucci & Dearden consider that the most common source of these vesicles in ageing cartilage is from cell necrosis, a view in accord with the age-related decline in cellularity. The membranous bodies so formed are more variable in size and shape than their counterparts in the growth plate but nevertheless the first crystals form inside them and calcification spreads from this origin.

The 'matrix vesicles' of ageing permanent cartilage, unlike those in the growth plate, have not been subjected to exhaustive investigation for enzyme activity and lipid constituents. Although alkaline phosphatase may be present, its activity is too low to be detected histochemically. Disaggregation and degradation of proteoglycan complexes are required to permit calcification to proceed in the growth plate and therefore the diminution in glycosaminoglycan content of ageing permanent non-articular cartilages is conducive to mineralisation. However, such modifications of the ground substance are slow and uncoordinated, probably dependent on the decline in synthesis of proteoglycans and the release of degradative enzymes by degenerating cells. These factors are among those responsible for the slow progression of calcification in ageing cartilage.

That the normally uncalcified region of articular cartilage is relatively unaffected by abnormal mineralisation may be in part due to the lack of age changes in cellularity and glycosaminoglycan content, throughout most of its thickness. A few microcrystals are found in pericellular matrix vesicles in the deeper parts of normal cartilage near the tidemark; such features are much more numerous in osteoarthrosic cartilage which also shows increased levels of alkaline phosphatase activity (Ali & Wisby, 1976). These observations may be accounted for by remodelling taking place at the osseochondral junction. The much larger amounts of lipid in the superficial zone (Fig. 72), which show a small age-related increase (Bonner *et al.*, 1975), do not seem to be associated with mineral deposits even when cell loss and depletion of proteoglycans occur in early regressive change (Meachim *et al.*, 1965). Presumably other conditions for matrix calcification are lacking.

INITIATION OF FIBRILLATION

The factors causing splitting and fraying of the articular surface may be quite distinct from those which cause the pattern of cartilage destruction typical of certain osteoarthrosic joints. However, the possibility that they are related or even identical, in some joints at least, merits discussion of the initial pathogenesis of fibrillation.

The detection of fibrillation depends on the size of the splits and the sensitivity of the method used to examine the articular surface. For practical purposes many investigators regard fibrillation as the presence of surface irregularities visible either with the naked eye and/or with the microscope at magnifications up to about $\times 400$. This represents but one arbitrary point in the development of the lesion, and ideas about its aetiology and pathogenesis stem from observations of earlier and later stages. Thus 'normal' articular surfaces examined with the electron microscope may show minute structural imperfections and the superficial collagen fibres may show an abnormally wide separation (Meachim & Roy, 1969). It is possible that these ultrastructural aberrations could be the forerunners of 'histological' fibrillation. Loss of basophilia and of cells is well documented in advanced fibrillation (Hirsch, 1944; Meachim & Collins, 1962); isolated areas of cell necrosis and loss of basophilia are seen beneath some histologically 'normal' surfaces and it has been suggested that such 'regressive' change may predispose cartilage to fibrillation (Meachim *et al.*, 1965). Living, as well as dead, chondrocytes may be implicated. Thus abnormally high levels of proteolytic enzyme activity associated with clusters of cells (Chrisman, 1967) are found in advanced fibrillation (Ali & Bayliss, 1974), although authorities are divided as to whether or not this is accompanied by increased (Meachim & Collins, 1962; Mankin, 1974) or decreased (Byers *et al.*, 1977) synthesis of glycosaminoglycans.

Such data suggest hypotheses based either on cellular or mechanical causes. Thus various unknown factors might cause the chondrocytes to release proteolytic enzymes, so resulting in ground substance depletion. The loss of resilience reduces the resistance of the cartilage to mechanical stress and so fibrillation develops. On the other hand, splitting at the surface could be due primarily to mechanical forces, with or without the aid of other factors.

TENSILE STRENGTH OF THE SUPERFICIAL CARTILAGE

The superficial cartilage is subject to tangential tensile stresses during joint movement and loading (Zarek & Edwards, 1963). Accordingly the collagen fibres run tangentially, with a preferential direction of orientation within this plane, forming a meshwork rather than continuous bundles from side to side of the joint. Hence the tensile strength of the cartilage depends on the *meshwork*, and splits can only occur if this is broken or its fibres separated. It is not known if or how the fibres of the mesh are bonded together.

Mechanical tests confirm that the tensile strength of strips of articular cartilage depends on collagen fibre orientation and content. The highest tensile strength is found when the strips are tested with their long axis parallel rather than perpendicular to the predominant fibre orientation (Kempson, 1973). Exhaustive digestion of the tissue with enzymes such as trypsin and cathepsins B_1 and D (Kempson, 1973; Kempson *et al.*, 1976) removes as much as 90% of the total uronic acid (chondroitin sulphate) content of the tissue. As might be expected this treatment causes increased deformation in compression. However, neither proteoglycan loss nor the small amounts of collagen released by cathepsin B_1 make any significant difference to the tensile stiffness or fracture stress of the cartilage strips. A small, abnormal increase in initial distension of the strips is found when they are stressed perpendicular to the main orientation of collagen fibres. This probably reflects re-alignment of the collagen fibres when first placed under load (Kempson, 1973). By contrast, treatment with clostridial collagenase reduces the tensile stiffness and the fracture stress considerably.

Since tensile strength is independent of the bulk of the proteoglycan moiety, the integrity of the collagen mesh is unlikely to depend on viscous resistance to movement of the fibres within the proteoglycan gel. A specific interaction between collagen and proteoglycan cannot be excluded altogether since the very small amount of unextracted proteoglycan in the mechanical tests might correspond to the residual collagen-bound proteoglycan found after sequential ionic extraction (see Chapter 3). It seems unlikely that the collagen mesh is one continuous framework of collagen, i.e. that the criss-cross fibres are in continuity or joined by intermolecular cross-links of the collagen type. However, the fracture stress of the cartilage strips is about 15–30 MN m^{-2} in the young adult (Kempson, 1975; Weightman, Chappell & Jenkins, 1978), corresponding to 10% of the tensile strength of collagen, about 15–30 kg mm^{-2} (Harkness, 1968) or approximately 150–300 MN m^{-2}. In the rat, transversely sectioned collagen fibres occupy 25% of the matrix (Palfrey, 1975). Extrapolating to man this suggests that the tensile strength of cartilage strips tested parallel to the long axis of these fibres is only 40% of that which might be expected (38–75 MN m^{-2}) if the fibres stretched the whole length of the test strips. Hence it seems probable that the fibres are discontinuous, possibly linked together by the type of insoluble non-collagenous protein isolated by Shipp & Bowness (1975), itself resistant to collagenase.

MODE OF RUPTURE

Although a single traumatic episode might cause cartilage splitting, in natural fibrillation the breakage probably occurs as a result of a wear process. The phenomenon has been called 'fatigue failure', the failure or breakage occurring after a finite number of applications of a stress lower than that required to produce breakage at a single application (Freeman & Meachim, 1973). Failure is related to the magnitude of the stress as well as the number of applications, the lower the stress the larger the number of applications required. Thus splitting of the articular surface parallel to the direction of predominant fibre orientation has been produced at low load (2 MN m^{-2}) by repetitive compression (Weightman, Freeman & Swanson, 1973). Similarly, strips of superficial cartilage have fractured after repeated tensile stress, showing an age-related diminution in 'fatigue strength' (Weightman *et al.*, 1978), as they do in fracture stress at a single application (Kempson, 1975). In view of its seemingly inert metabolic nature, the collagen rather than the proteoglycan content of the matrix should be more likely to undergo 'fatigue failure'. It is worth noting, however, evidence suggesting that adult articular cartilage collagen continues to be metabolised in the rabbit (Repo & Mitchell, 1971) and the dog (Eyre, McDevitt & Muir, 1975); the few investigations in man (Libby *et al.*, 1964) relate only to 73-year-olds in which the type of cartilage is not specified. Nevertheless since the stresses at the surface are tensile in nature it is the collagen mesh which would be expected to undergo wear or 'fatigue'.

While the collagen meshwork is implicated, it is uncertain whether it is the fibres themselves or the bonds (whatever their nature) between them which undergo 'fatigue'. Nor is the actual mechanism of 'fatigue' specified: possibly there is a succession of stress fractures in the fibres or bonds until the tensile strength of the whole cartilage strip is reduced to an abnormally low level. When this corresponds to the tensile stress applied during the 'fatigue' testing, the strip breaks. Whatever the wear process may be, the collagen mesh would be expected to become more vulnerable if the proteoglycans of the matrix were depleted. What would normally be a tolerable load would then result in increased cartilage deformation and abnormally high tensile stresses in the superficial cartilage. Damage to the mesh would then be more severe at each cycle of loading and 'fatigue' would be more rapid. Theoretically, however, provided that collagen 'turnover' is insufficient, a 'fatigue' process operating on a normal cartilage with normal joint stresses could lead to fibrillation,

CELL CHANGES IN EARLY FIBRILLATION

Fibrillation and rupture of the collagen mesh might occur in two general situations (Fig. 108).

(i) Abnormally high stresses acting on a normal cartilage.

(ii) Normal (or subnormal) stresses acting on a defective cartilage.

A number of clinical conditions known to predispose cartilage to osteoarthrosis could produce one or other of these two states of mechanical imbalance. For example, ligamentous instability, bony misalignments or necrosis of the underlying bone might cause abnormally high stress, while ochronosis and crystal deposition diseases produce a defective cartilage (Mitchell & Cruess, 1977); loss of basophilia beneath some histologically normal articular surfaces (Meachim *et al.*, 1965) has already been mentioned. More subtle mechanisms invoking disuse atrophy have been postulated. Thus during joint usage, part of the articular area might habitually be out of contact with the opposing surface most of the time and therefore would adapt structurally and biochemically to low stress but when infrequently in contact might experience high physiological but damaging stress. This mode of stressing may occur in normal joint usage (Seedhom, Takeda, Wright & Tsubuku, 1978) or in connection with joint remodelling during ageing (Bullough, Goodfellow & O'Connor, 1973).

Hence, there are two aspects of chondrocyte involvement in the initiation of fibrillation to be considered.

(*a*) The effects on the chondrocyte of abnormally high stresses before and after surface disruption.

(*b*) Whether the chondrocyte might contribute to the formation of a defective cartilage.

It is obvious that these are interrelated. The reaction of the cell to abnormally

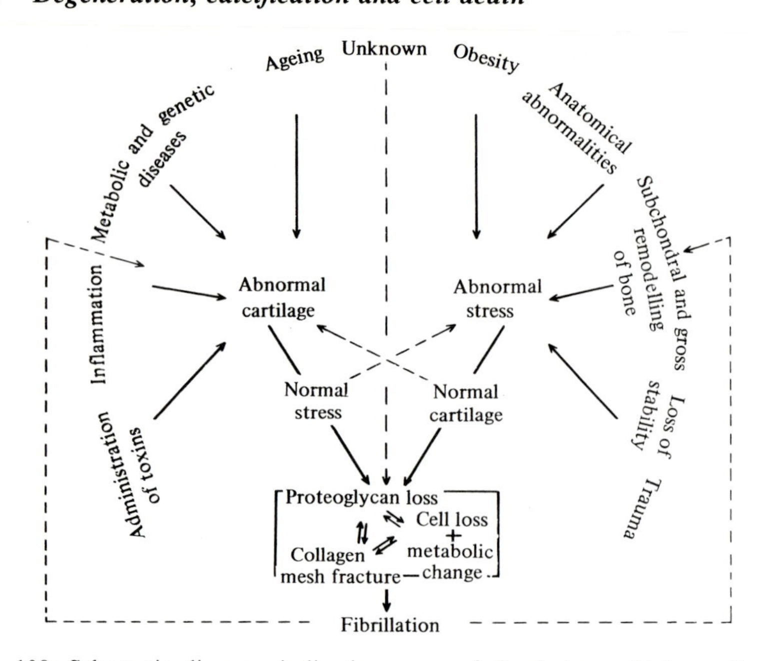

Fig. 108. Schematic diagram indicating some of the factors which predispose to fibrillation and the intracartilaginous defects which may interact during the pathogenesis of the structural disorganisation of articular cartilage. Once established, fibrillation itself constitutes an abnormal cartilage and may lead (dotted lines) to other changes, for example subchondral bone remodelling.

high stresses may affect its metabolism, so causing changes in the matrix. Conversely, if the cell is responsible for the formation of a defective matrix, normal joint forces are likely to cause abnormally high stress on the chondrocyte. In either case a vicious spiral of changes could follow.

If sufficiently great, the stress may produce chondrocyte death. This will reduce the potential for matrix synthesis and turnover, giving a primary deficiency of proteoglycan. The massive chondrocyte necrosis found in pressure immobilisation of joints, probably resulting from nutritional deprivation, certainly leads to cartilage degeneration (Trias, 1961). A lower stress will permit the cells to survive although they are likely to show reactive changes. Thus after normal joint exercise superficial chondrocyte swelling has been observed (Ekholm & Norbäck, 1951). In experiments on sheep where one forelimb is excluded from weight bearing the articular cartilage in the intact forelimb shows an increased synthesis and content of chondroitin sulphate (Lowther & Caterson, 1974). In similar experiments on dogs (Kostenszky & Olah, 1972), the hexosamine content is increased due mostly to changes in glucosamine (probably keratansulphate).

Recent biochemical observations of experiments in which joint instability and eventual arthritis are produced by division of the anterior cruciate ligament in the dog knee have been of considerable interest. Structural damage to the superficial cartilage and changes in proteoglycan content suggestive of altered chondrocyte metabolism are early sequelae. Within three weeks of operation the water content of the whole tissue increases (McDevitt & Muir, 1976). The increase in water (also found in natural fibrillation) implies that the collagen mesh must have expanded, presumably due to damage. This is corroborated by articular surface irregularities observed histologically at one week (McDevitt, Gilbertson & Muir, 1977) and abnormally wide separation of superficial collagen fibres observed ultrastructurally at two weeks after operation (Muir, 1977). However, changes in the proteoglycans are found just as soon after operation. At three weeks the galactosamine/glucosamine ratio of the whole tissue is unusually high and a larger proportion of proteoglycans than normal becomes extractable both at low ionic strength and with 2-M calcium chloride (McDevitt & Muir, 1976). The increase in the hexosamine ratio is even more pronounced in the extractable proteoglycans (Muir, 1977). The number and size of proteoglycan aggregates are reduced in the operated joints as early as one week after operation (Muir, 1977; McDevitt, Billingham & Muir, 1978). However, no consistent change in chondroitin sulphate content of the whole tissue occurs until several months later. These biochemical changes suggest that a considerable proportion of the existing proteoglycans have been replaced by molecules which contain less keratansulphate and which do not participate in hyaluronic acid–proteoglycan aggregates (and perhaps also the 2.7S protein–collagen complex).

Scission of the core protein of existing proteoglycans between the intermediate keratansulphate-rich and the chondroitin sulphate-rich segments (see Chapter 3) would remove the hyaluronate-binding site and explain both the failure to aggregate and the change in hexosamine ratio. This could be mediated by extracellular release of proteolytic enzyme. However, core protein scission may not be the explanation since proteoglycan monomers of normal and operated joint cartilage have similar sedimentation coefficients (McDevitt *et al.*, 1978). Hence either there must be a fault in the tertiary structure of the hyaluronate-binding site or the failure to aggregate resides in the hyaluronic acid itself or in the link glycoprotein. This could be caused by lysozymal or proteolytic activity, respectively. Alternatively, new proteoglycans may be synthesised which are deficient in keratansulphate side chains and also lack hyaluronate-binding capacity. However, degradation of the original proteoglycans would be required if there were to be no overall change in chondroitin sulphate content.

Somewhat similar results – poor aggregation, though in the presence of proteoglycans with intact hyaluronate-binding sites – are found in bovine fibrillated cartilage (Brandt, Palmoski & Perricone, 1976). This suggests a defect or lack of hyaluronic acid or of the stabilising capacity of the link glycoprotein. However, in both ageing (Perricone *et al.*, 1977) and osteoarthrosic (Palmoski & Brandt, 1976) human femoral head cartilage, aggregates do not occur nor do the extracted proteoglycans interact with hyaluronic acid, suggesting that they have a defect in the hyaluronic acid binding site.

In the experiments on dogs a number of results could be explained on the basis of degradative attack (mostly proteolytic) at one or more sites in proteoglycan organisation, very soon after the chondrocyte is subjected to the abnormal stresses resulting from joint instability. The mechanism is probably more subtle than a massive dissolution of proteglycan monomer. Furthermore, since cartilage is known to contain inhibitors to collagenase and other proteolytic enzymes (Kuettner *et al.*, 1976) degradation may be made possible by diminished secretion of enzyme inhibitor or (by analogy with the collagenase-inhibitor hypothesis (Chapter 4) of Reynolds *et al.*, 1977) by proteolytic breakdown of this inhibitor. The biochemical changes occur very soon after operation, i.e. on subjection of the chondrocyte to abnormal stresses. Increased hydration of the tissue, in itself, might be responsible for proteoglycan changes mediated by the chondrocyte. At present, however, there appears to be no evidence that mechanical damage to the collagen mesh and the associated increased water content antedate the biochemical changes in proteoglycan. Indeed it is conceivable, as Brandt *et al.* (1976) suggest in bovine fibrillation, that a proteolytic mechanism causing disaggregation of the proteoglycans might also affect non-collagenous proteins associated with the collagen mesh. It may be relevant that collagenase activity is found in human osteoarthritic cartilage, though it appears to be undetectable in normal tissue (Ehrlich *et al.*, 1977). If released in response to mechanical stress, collagenase activity might also produce weakening of the collagen mesh. However, such speculations as to the early effects of experimental joint instability must await the results of future biochemical investigations.

Various non-mechanical procedures cause changes in articular chondrocytes which secondarily affect the matrix. Thus noxious substances introduced into the joint can cause chondrocyte necrosis (Bennett *et al.*, 1942; Bentley, 1974) and lead to fibrillation. More physiological materials, such as intra-articular steroids, affect chondrocyte metabolism (Mankin *et al.*, 1972) and vitamin A causes potentiation of lysosomal enzymes, resulting in fibrillation (Lenzi *et al.*, 1974); triglyceride injection results in fatty acid uptake by the chondrocyte and this is followed by chondrocyte degeneration and roughening of the articular surface (Ghadially *et al.*, 1970). Such actions of steroids and fats on cartilage may be outweighed by their beneficial effects on the synovial membrane when administered in the form of steroid-loaded liposomes to inflamed joints (Dingle *et al.*, 1978).

In organ culture, degradation of articular cartilage occurs if synovial membrane is grown in contact with it, due to enzymes released by the synovial cells. However, there is also an indirect action mediated through the chondrocytes for which contact is not necessary (Fell & Jubb, 1977). It is not known if similar interactions occur *in vivo*, but the synovial membrane and any subclinical changes therein might indirectly affect the chondrocyte. It seems unlikely that nutritional deprivation could occur on a massive scale although localised effects on the superficial cartilage cell population cannot be excluded. The age-related closure of the subchondral marrow spaces may influence metabolism of the deeper cells and hence the chemical nature and mechanical properties of the deep zone matrix.

As discussed earlier (pp. 245–249), the possibility of changes with age within the chondrocyte cannot be excluded. It is not known if lipid and fibrillar accumulations in the cytoplasm are associated with any change in metabolism.

Over a long period, various adverse factors to which the superficial cells in particular would be maximally exposed might summate to cause changes in cell function, making them more susceptible to mechanical stress and more liable to necrosis. The age-related loss of cells in the superficial zone of the cartilage of the knee and hip may be explained on this basis. This is in a sense 'fatigue' of the cell population though there appears to be no suitable test for the failure point. It is worth commenting that the age-related decline in 'fatigue strength' of the superficial cartilage of the femoral head (Weightman *et al.*, 1978) is paralleled by the decline in its cell density (Vignon *et al.*, 1976). Hence it is possible that the enhanced 'fatigue' of the collagen mesh in ageing cartilage could be a consequence of cell failure.

ULTRASTRUCTURE OF CHONDROCYTE DEGENERATION

In immature rabbit articular cartilage, two forms of cell degeneration have been distinguished, both found in the deep zone still undergoing the last stages of endochondral ossification (Davies *et al.*, 1962). In both forms, the nucleus becomes irregular with an indistinct nuclear envelope. In the first type of degeneration, cytoplasmic organelles are more widely separated by cytosol of reduced electron density. The granular ER is moderately distended but other organelles are less prominent. The cell membrane is indistinct and eventually becomes discontinuous. Davies *et al.* equate the morphology of this type to that of the degenerating hypertrophic cell in the growth plate. Thus although the hypertrophic cell is much larger than the articular chondrocyte, in both cells, the ER rather than the Golgi membrane is a persistent cytoplasmic organelle (Holtrop, 1972). In slightly older though still immature rabbit articular cartilage Palfrey & Davies (1966) describe degenerate cells with dense homogeneous nuclei of an irregular outline. The ER is widely dilated and the remaining cytoplasm in which it is embedded appears compressed, electron dense and of small volume. Mitochondria are still present though they tend to be swollen. This appearance is not uncommon in fetal cartilage (Stockwell, 1971*a*) and has also been described in the superficial zone of human articular cartilage (Roy & Meachim, 1968). As with the hypertrophic cell, in which the mitochondria persist to a remarkably late stage of degeneration (Serafini-Fracassini & Smith, 1974), the morphology suggests that the cells die following intense metabolic and synthetic activity.

The second type of degenerative cell described by Davies *et al.* (1962) has

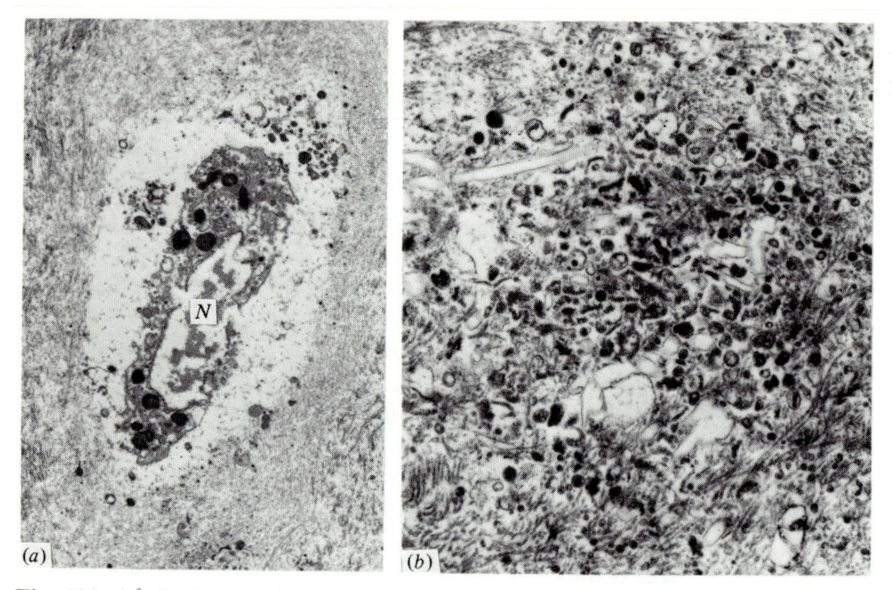

Fig. 109. Adult sheep articular cartilage, glutaraldehyde fixation. (*a*) A necrotic cell in the superficial zone. Nuclear (*N*) chromatin is lost and both the nuclear envelope and the plasmalemma are fragmented. Collections of cell debris are beginning to form in the matrix. Magnification × 7500. (*b*) A later stage in necrosis showing an aggregate of heterogeneous cell debris in the matrix. Magnification × 18000.

a smaller body than normal, with many long radiating processes. The cytoplasm is vacuolated and few organelles can be distinguished. Eventually the cell is reduced to a shrivelled spidery form with little cytoplasm, lying in a fluid-filled lacuna. Although it does not have the various cytoplasmic inclusions found in chondrocytes of ageing cartilage (a feature affecting the amount and nature of the residual debris), this type of cell resembles in its lack of organelles the degenerate cell commonly seen in ageing cartilage. Thus in ageing human costal chondrocytes there is little ER and mitochondria are few and of a bizarre shape, although Golgi vacuoles persist relatively late. Cytoplasmic filaments accumulate together with electron-dense complex bodies and myelin figures; lysosomes may be more frequent (Dearden *et al.*, 1974).

A similar pattern of cell degeneration is found in adult human articular cartilage (Meachim & Roy, 1967). In ageing cartilage, therefore, the morphology suggests that cell death usually follows a gradual decline of cell activity. While in endochondral ossification the degenerate remains of chondrocytes of the growth plate are removed by the erosion front, this cannot occur in the permanent cartilages. Characteristically, fragmentation of the cell membrane results in a heterogeneous aggregate of rounded or irregular membranous bodies, clusters of dense granules and other debris (Fig. 109), the amount of lipidic material partly depending on the number and size of lipid globules formerly contained in the living cell. Collagen fibres of much wider diameter (100–300 nm) than normal are often observed close to degenerating cells and their remains (Barnett *et al.*, 1963; Bonucci *et al.*, 1974). Eventually the site of the lacuna becomes obscured by fibrous matrix and the cell debris disperses.

The overall picture seen in histological section is of a chondrocyte which gradually loses its stainability, the basophilia of the territorial matrix eventually following suit. Both cell and lacuna therefore merge with the surrounding matrix, a process aptly termed 'Verdämmerung' (Schaffer, 1930) or cell 'fading'.

REFERENCES

Abbott, J., Mayne, R. & Holtzer, H. (1972). Inhibition of cartilage development in organ cultures of chick somites by the thymidine analogue, 5-bromo-2'-deoxyuridine. *Devl Biol.*, **28**, 430–42.

Akeson, W. H., Eichelberger, L. & Roma, M. (1958). Biochemical studies of articular cartilage. II. Values following the denervation of an extremity. *J. Bone Jt Surg.*, **40A**, 153–62.

Akeson, W. H., Woo, S. L. Y., Amiel, D., Coutts, R. D. & Daniel, D. (1973). The connective tissue response to immobility: biochemical changes in periarticular connective tissue of the immobilised rabbit knee. *Clin. Orthop.*, **93**, 356–62.

Albaum, H. G., Hirshfeld, A. & Sobel, A. E. (1952). Calcification. VI. Adenosinetriphosphate content of pre-osseous cartilage. *Proc. Soc. exp. Biol. Med.*, **79**, 238–41.

Ali, S. Y. (1976). Analysis of matrix vesicles and their role in the calcification of epiphysial cartilage. *Fedn Proc. Fedn Am. Socs exp. Biol.*, **35**, 135–42.

Ali, S. Y. & Bayliss, M. T. (1974). Enzymic changes in human osteoarthrotic cartilage. In *Normal and osteoarthrotic articular cartilage*, ed. S. Y. Ali, M. W. Elves & D. H. Leaback, pp. 189–205. London: Institute of Orthopaedics.

Ali, S. Y. & Evans, L. (1973*a*). Enzymatic degradation of cartilage in osteoarthritis. *Fedn Proc. Fedn Am. Socs exp. Biol.*, **32**, 1494–8.

Ali, S. Y. & Evans, L. (1973*b*). The uptake of (^{45}Ca) calcium ions by matrix vesicles isolated from calcifying cartilage. *Biochem. J.*, **134**, 647–50.

Ali, S. Y., Sajdera, S. W. & Anderson, H. C. (1970). Isolation and characterization of calcifying matrix vesicles from epiphysial cartilage. *Proc. natn. Acad. Sci. USA*, **67**, 1513–20.

Ali, S. Y. & Wisby, A. (1976). The role of matrix vesicles in human osteoarthrotic cartilage. *Proc. R. microsc. Soc.*, **11** (Suppl.), 62.

Allfrey, V. G., Littau, V. C. & Mirsky, A. E. (1963). On the role of histones in regulating ribonucleic acid synthesis in the cell nucleus. *Proc. natn. Acad. Sci. USA*, **49**, 414–21.

Amprino, R. & Bairati, A. (1933). Studi sulle trasformazioni delle cartilagini del l'uomo nell'accrescimento e nella senescenza. I. Cartilagini jaline. *Z. Zellforsch. mikrosk. Anat.*, **20**, 143–205.

Amprino, R. & Bairati, A. (1934). Studi sulle trasformazioni delle cartilagini de l'uomo nell'accrescimento e nella senescenza. II. Cartilagini elastiche. *Z. Zellforsch. mikrosk. Anat.*, **20**, 489–522.

Andersen, H. (1962). Histochemical studies of the development of the human hip joint. *Acta anat.*, **48**, 258–92.

Andersen, H. & Matthiessen, M. E. (1966). The histiocyte in human foetal tissues. *Z. Zellforsch. mikrosk. Anat.*, **72**, 193–214.

Anderson, B., Hoffman, P. & Meyer, K. (1965). The O-serine linkage in peptides of chondroitin 4- or 6-sulphate. *J. biol. Chem.*, **240**, 156–67.

Anderson, C. E., Ludowieg, J., Harper, H. A. & Engleman, E. P. (1964). The composition of the organic component of human articular cartilage. *J. Bone Jt Surg.*, **46*A***, 1176–83.

Anderson, H. C. (1969). Vesicles associated with calcification in the matrix of epiphyseal cartilage. *J. Cell Biol.*, **41**, 59–72.

Anderson, H. C., Chacko, S., Abbott, J. & Holtzer, H. (1970). The loss of phenotypic traits by differentiated cells *in vitro*. VII. Effects of 5-bromodeoxyuridine and prolonged culturing on fine structure of chondrocytes. *Am. J. Path.*, **60**, 289–312.

Anderson, H. C. & Reynolds, J. J. (1973). Pyrophosphate stimulation of calcium uptake into cultured embryonic bones. Fine structure of matrix vesicles and their role in calcification. *Devl Biol.*, **34**, 211–27.

Anderson, H. C. & Sajdera, S. W. (1971). The fine structure of bovine nasal cartilage. Extraction as a technique to study proteoglycans and collagen in cartilage matrix. *J. Cell Biol.*, **49**, 650–63.

Anson, B. J., Bast, T. H. & Cauldwell, E. W. (1948). The development of the auditory ossicles, the otic capsule and the extracapsular tissues. *Ann. Otol. Rhinol. Lar.*, **57**, 603–32.

Arlet, J., Mole, J. & Barriuso, J. (1958). Études histologiques concernant le cartilage articulaire de bovide à l'etat frais. I: Critère de vitalité des chondrocytes. *Revue Rhum. Mal. ostéo-artic.*, **25**, 565–78.

Arnold, J. (1878). Die Abscheidung des indigschwefelsauren Natrons im Knorpelgewebe. *Virchows Arch. path. Anat. Physiol.*, **73**, 125–46.

Ash, P. & Francis, M. J. O. (1975). Response of isolated rabbit articular and epiphysial chondrocytes to rat liver somatomedin. *J. Endocr.*, **66**, 71–8.

Astbury, W. T. (1938). X-ray adventures among the proteins. *Trans. Faraday Soc.*, **34**, 378–88.

Astrup, T. (1969). Relation between fibrin formation and fibrinolysis. In *Thrombosis*, ed. S. Sherry, K. M. Brinkhaus, E. Genton, & J. M. Stengle, pp. 609–38. Washington D.C.: National Academy of Sciences.

Atkins, E. D. T., Hardingham, T. E., Isaac, D. M. & Muir, H. (1974). X-ray fibre diffraction of cartilage proteoglycan aggregates containing hyaluronic acid. *Biochem. J.*, **141**, 919–21.

Avery, G., Chow, M. & Holtzer, H. (1956). An experimental analysis of the development of the spinal column. V. Reactivity of chick somites. *J. exp. Zool.*, **132**, 409–25.

Axhausen, G. (1924). Über den Abgrenzungsvorgang am epiphysaren Knochen (Osteochondritis dissecans König). *Virchows Arch. path. Anat. Physiol.*, **252**, 458–518.

Bacsich, P. & Wlassics, T. (1931). Die Abhangigkeit des Fettgehaltes der Zellen des elastischen Knorpels vom Ernahrungsgrad des individuums. *Z. mikrosk.-anat. Forsch.*, **24**, 7–16.

Bacsich, P. & Wyburn, G. M. (1947). The significance of the mucoprotein content on the survival of homografts of cartilage and cornea. *Proc. R. Soc. Edinb. B*, **62**, 321–7.

Bailey, A. J. & Peach, C. M. (1971). The chemistry of the collagen cross-links. *Biochem. J.*, **121**, 257–9.

Bailey, A. J. & Robins, S. P. (1973). Role of hydroxylysine in the stabilization of the collagen fibre. In *Biology of the fibroblast*, ed. E. Kulonen & J. Pikkarainen, pp. 385–95. London & New York: Academic Press.

Bailey, A. J., Robins, S. P. & Balian, G. (1974). Biological significance of the intermolecular cross-links of collagen. *Nature, Lond.* **251**, 105–9.

Bailey, A. J., Robins, S. P. & Balian, G. (1975). Function of borohydride reducible components of collagen. In *Protides of biological fluids*, **22**, ed. H. Peeters, pp. 87–90. Oxford: Pergamon.

Baker, J. R., Cifonelli, J. A., Mathews, M. B. & Roden, L. (1969). Mannose-containing glycopeptides from keratosulfate (KS). *Fedn Proc. Fedn Am. Socs exp. Biol.*, **28**, 605.

Baker, J. R., Cifonelli, J. A. & Roden, L. (1975). The linkage of corneal keratansulphate to protein. *Conn. Tissue Res.*, **3**, 149–56.

Baker, J. R., Roden, L. & Stoolmiller, A. C. (1972). Biosynthesis of chondroitin sulphate proteoglycan. Xylosyl transfer to Smith-degraded cartilage proteoglycan and other exogenous acceptors. *J. biol. Chem.*, **247**, 3838–47.

Balazs, E. A., Bloom, G. D. & Swann, D. A. (1966). Fine structure and glycosaminoglycan content of the surface layer of articular cartilage. *Fedn Proc. Fedn Am. Socs exp. Biol.*, **25**, 1813–16.

Balazs, E. A., Davies, J. V., Phillips, G. O. & Young, M. D. (1967). Transient intermediates in the radiolysis of hyaluronic acid. *Radiat. Res.*, **31**, 243–55.

Balduini, C., Brovelli, A. & Castellani, A. A. (1970). Biosyntheses of glycosaminoglycans in bovine cornea. *Biochem. J.*, **120**, 719–23.

Balduini, C., Brovelli, A., De Luca, G., Galligani, L. & Castellani, A. A. (1973). Uridine diphosphate glucose dehydrogenase from cornea and epiphysial plate cartilage. *Biochem. J.*, **133**, 243–9.

Balmain-Oligo, N., Moscofian, A. & Plachot, J. J. (1973). Développement épiphysaire et calcification de la couche basale du cartilage articulaire. *Path. Biol., Paris*, **21**, 611–21.

Balogh, K. & Cohen, R. B. (1961). Histochemical localisation of uridine diphosphoglucose dehydrogenase in cartilage. *Nature, Lond.*, **192**, 1199–200.

Balogh, K., Dudley, H. R. & Cohen, R. B. (1961). Oxidative enzyme activity in skeletal cartilage and bone. *Lab. Invest.*, **10**, 839–45.

Balogh, K. & Kunin, A. S. (1971). The effect of cortisone on the metabolism of epiphyseal cartilage. A histochemical study. *Clin. Orthop.*, **80**, 208–15.

Bancroft, M. & Bellairs, R. (1976). The development of the notochord in the chick embryo, studied by scanning and transmission electron microscopy. *J. Embryol. exp. Morph.*, **35**, 383–401.

Banghoorn, E. S. & Schopf, J. W. (1966). Micro-organisms three billion years old from the Pre-cambrian of South Africa. *Science*, **152**, 758.

Bardos, T. J., Levin, G. M., Herr, R. R. & Gordon, H. L. (1955). Synthesis of compounds related to thymine II. Effect of thymine antagonists on the biosynthesis of DNA. *J. Am. chem. Soc.*, **77**, 4279–86.

Barland, P., Janis, R. & Sandson, J. (1966). Immunofluorescent studies of human articular cartilage. *Ann. rheum. Dis.*, **25**, 156–63.

Barland, P., Novikoff, A. B. & Hamerman, D. (1962). Electron microscopy of the human synovial membrane. *J. Cell Biol.*, **14**, 207–20.

Barland, P., Smith, C. & Hamerman, D. (1968). Localization of hyaluronic acid in synovial cells by radioautography. *J. Cell Biol.*, **37**, 13–26.

Barnett, C. H., Cochrane, W. & Palfrey, A. J. (1963). Age changes in articular cartilage of rabbits. *Ann. rheum. Dis.*, **22**, 389–400.

Barnett, C. H., Davies, D. V. & MacConaill, M. A. (1961). *Synovial joints. Their structure and mechanics.* London: Longmans, Green.

Barrett, A. J. (1975). The enzymatic degradation of cartilage matrix. In *Dynamics of connective tissue macromolecules*, ed. P. M. C. Burleigh & A. R. Poole, pp. 189–215. Amsterdam: North-Holland.

Barrett, A. J., Sledge, C. B. & Dingle, J. T. (1966). Effect of cortisol on the synthesis of chondroitin sulphate by embryonic cartilage. *Nature, Lond.*, **211**, 83–4.

Barrett, A. J. & Starkey, P. M. (1973). The interaction of α_2-macroglobulin with proteinases. *Biochem. J.*, **133**, 709–24.

Bassett, C. A. L. (1971). Biophysical principles affecting bone structure. In *The biochemistry and physiology of bone*, **3**, ed. G. H. Bourne, pp. 1–76. London & New York: Academic Press.

Bassett, C. A. L. & Herrmann, I. (1961). Influence of oxygen concentration and mechanical factors on differentiation of connective tissues *in vitro*. *Nature, Lond.*, **190**, 460–1.

Bauer, W., Ropes, M. W. & Waine, H. (1940). The physiology of articular structures. *Physiol. Rev.* **20**, 272–312.

Baum, H. (1974). Mitochondria and peroxisomes. In *The cell in medical science*, **1**, ed. F. Beck & J. B. Lloyd, pp. 183–272. London: Academic Press.

Beattie, D. S. (1968). Studies on the biogenesis of mitochondrial protein components in rat liver slices. *J. biol. Chem.*, **243**, 4027–33.

Becks, H. & Evans, H. M. (1953). *Atlas of the skeletal development of the rat* (*Long Evans strain*) *normal and hypophysectomised*. San Francisco: Am. Inst. Dent. Med.

Behnke, O. (1975). An outer component of microtubules. *Nature, Lond.*, **257**, 709–10.

Belanger, L. F. & Migicovsky, B. B. (1963). Bone cell formation and survival in H^3-thymidine-labelled chicks under various conditions. *Anat. Rec.* **145**, 385–90.

Bell, E. T. (1909). On the occurrence of fat in the epithelium, cartilage and muscle fibres of the ox. *Am. J. Anat.* **9**, 401–12.

Bellamy, G. & Bornstein, P. (1971). Evidence for procollagen, a biosynthetic precursor of collagen. *Proc. natn. Acad. Sci. USA*, **68**, 1138–42.

Beneke, G. (1973). Cell density in cartilage and DNA content of the cartilage cells dependent on age. In *Connective tissue and ageing*, ed. H. G. Vogel, pp. 91–4. Amsterdam: Excerpta Medica.

Bennett, G. A. & Bauer, W. (1935). Further studies concerning the repair of articular cartilage in dog joints. *J. Bone Jt Surg.*, **17**, 141–50.

Bennett, G. A., Bauer, W. & Maddock, S. J. (1932). A study of the repair of articular cartilage and the reaction of normal joints of adult dogs to surgically created defects of articular cartilage, 'joint mice' and patellar displacement. *Am. J. Path.* **8**, 499–524.

Bennett, G. A., Waine, H. & Bauer, W. (1942). *Changes in the knee joint at various ages*. New York: Commonwealth Fund.

Benninghoff, A. (1925). Form und Bau der Gelenkknorpel in ihren Beziehungen zur Function. II. Der Aufbau des Gelenkknorpels in seinen Beziehungen zur Function. *Z. Zellforsch. mikrosk. Anat.*, **2**, 783–862.

Bentley, G. (1974). Experimental osteoarthrosis. In *Normal and osteoarthrotic articular cartilage*, ed. S. Y. Ali, M. W. Elves & D. H. Leaback, pp. 259–83. London: Institute of Orthopaedics.

Bentley, G. & Greer, R. B. (1971). Homotransplantation of isolated epiphyseal and articular cartilage chondrocytes into joint surfaces of rabbits. *Nature, Lond.*, **230**, 385–8.

Berg, R. A., Kishida, Y., Kobayashi, Y., Inouye, K., Tonelli, A. E., Sakakibara, S. & Prockop, D. J. (1973). A model for the triple-helical structure of $(\text{pro-Hyp-Gly})_{10}$ involving a *cis*-peptide bond and interchain hydrogen bonding to the hydroxyl group of hydroxyproline. *Biochim. biophys. Acta*, **328**, 553–9.

Berg, R. A. & Prockop, D. J. (1973). The thermal transition of a non-hydroxylated form of collagen. Evidence for a role for hydroxyproline in stabilising the triple helix of collagen. *Biochem. biophys. Res. Commun.*, **52**, 115–20.

Bernstein, D. S., Leboeuf, B. & Cahill, G. F. (1961). Studies on glucose metabolism in cartilage *in vitro*. *Proc. Soc. exp. Biol. Med.*, **107**, 458–61.

Bhatnagar, R. S., Kivirikko, K. I. & Prockop, D. J. (1968). Studies on the synthesis and intracellular accumulation of protocollagen in organ culture of embryonic cartilage. *Biochim. biophys. Acta*, **154**, 196–207.

Bhavanandan, V. P. & Meyer, K. (1968). Studies on keratosulfates. Methylation, desulfation and acid hydrolysis studies on old human rib cartilage keratosulfate. *J. biol. Chem.*, **243**, 1052–9.

Bisgard, J. D. & Bisgard, M. E. (1935). Longitudinal growth of long bones. *Archs Surg., Chicago*, **31**, 568–78.

Blackwood, H. J. J. (1965). Vascularization of the condylar cartilage in the human mandible. *J. Anat.*, **99**, 551–62.

Blackwood, H. J. J. (1966). Growth of the mandibular condyle of the rat studied with tritiated thymidine. *Archs oral Biol.*, **11**, 493–500.

Bloom, W. & Bloom, M. A. (1940). Calcification and ossification. Calcification of developing bones in embryonic and newborn rats. *Anat. Rec.*, **78**, 497–524.

Bohmig, R. (1930). Die Blutgefassversorgung der Wirbelbandscheiben das Verhalten des intervertebralen chordasegmente. *Arch. klin. Chir.*, **158**, 374–424.

Bole, G. G. (1962). Synovial fluid lipids in normal individuals and patients with rheumatoid arthritis. *Arthritis Rheum.* **5**, 589–601.

Bole, G. G. (1963). The incorporation of P^{32}-labelled orthophosphate and glycerol-1,3-C^{14} into the lipids of the polyvinyl sponge granuloma. *J. clin. Invest.*, **42**, 787–99.

Bollett, A. J., Bonner, W. M. & Nance, J. L. (1963). The presence of hyaluronidase in various mammalian tissues. *J. biol. Chem.*, **238**, 3522–7.

Bollet, A. J., Goodwin, J. F. & Brown, A. K. (1959). Metabolism of mucopolysaccharide in connective tissues. I. Studies of enzymes involved in glucuronide metabolism. *J. clin. Invest.*, **38**, 451–5.

Bollet, A. J. & Nance, J. L. (1966). Biochemical findings in normal and osteoarthritic articular cartilage. II. Chondroitin sulfate concentration and chain length, water, and ash content. *J. clin. Invest.*, **45**, 1170–7.

Bonner, W. M., Jonsson, H., Malanos, C. & Bryant, M. (1975). Changes in the lipids of human articular cartilage with age. *Arthritis Rheum.*, **18**, 461–73.

Bonucci, E. (1967). Fine structure of early cartilage calcification. *J. Ultrastruct. Res.* **20**, 33–50.

Bonucci, E. (1969). Further investigations on the organic/inorganic relationships in calcifying cartilage. *Calcif. Tissue Res.*, **3**, 38–54.

Bonucci, E. (1970). Fine structure and histochemistry of 'calcifying globules' in epiphyseal cartilage. *Z. Zellforsch. mikrosk. Anat.*, **103**, 192–217.

Bonucci, E., Cuicchio, M. & Dearden, L. C. (1974). Investigations of ageing in costal and tracheal cartilage of rats. *Z. Zellforsch. mikrosk. Anat.*, **147**, 505–27.

Bonucci, E. & Dearden, L. C. (1976). Matrix vesicles in ageing cartilage. *Fedn Proc. Fedn Am. Socs exp. Biol.*, **35**, 163–8.

Bornstein, P. (1974). The biosynthesis of collagen. *A. Rev. Biochem.*, **43**, 567–603.

Bornstein, P. & Ehrlich, H. P. (1973). Intracellular translocation and secretion of collagen. In *Biology of the fibroblast*, ed. E. Kulonen & J. Pikkarainen, pp. 321–38. London & New York: Academic Press.

Bornstein, P., von der Mark, K., Wyke, A. W., Ehrlich, H. P. & Monson, J. M. (1972). Characterization of the pro-α_1 chain of procollagen. *J. biol. Chem.* **247**, 2808–13.

Bornstein, P. & Piez, K. A. (1966). The nature of the intramolecular cross-links in collagen. The separation and characterization of peptides from the cross-link region of rat skin collagen. *Biochemistry*, **5**, 3460–73.

Boskey, A. L. & Posner, A. S. (1977). Extraction of a calcium–phospholipid–phosphate complex from bone. *Calcif. Tissue Res.*, **19**, 273–83.

Bosmann, H. B. (1968). Cellular control of macromolecular synthesis: rates of synthesis of extracellular macromolecules during and after depletion by papain. *Proc. R. Soc., B*, **169**, 399–425.

Boström, H. (1952). On the metabolism of the sulfate group of chondroitin sulphuric acid. *J. biol. Chem.*, **196**, 477–81.

Boström, H. & Mansson, B. (1952). On the enzymatic exchange of the sulfate group of chondroitin sulfuric acid in slices of cartilage. *J. biol. Chem.*, **196**, 483–8.

Boström, H. & Mansson, B. (1953). Factors influencing the exchange of the sulphate group of the chondroitin sulphuric acid of cartilage *in vitro*. *Ark. Kemi*, **6**, 23–37.

Boström, H. & Odeblad, E. (1953). The influence of cortisone upon the sulphate exchange of chondroitin sulphuric acid. *Ark. Kemi*, **6**, 39–42.

Bourrett, L. A. & Rodan, G. A. (1976*a*). The role of calcium in the inhibition of cAMP accumulation in epiphysial cartilage cells exposed to physiological pressure. *J. cell. Physiol.*, **88**, 353–62.

Bourrett, L. A. & Rodan, G. A. (1976*b*). Inhibition of cAMP accumulation in epiphyseal cartilage cells exposed to physiological pressure. *Calcif. Tissue Res.*, **21** (Suppl.), 431–6.

Boyd, E. S. & Neuman, W. F. (1951). The surface chemistry of bone. V. The ion binding properties of cartilage. *J. biol. Chem.* **193**, 243–51.

Boyd, E. S. & Neuman, W. F. (1954). Chondroitin sulfate synthesis and respiration in chick embryonic cartilage. *Archs Biochem.*, **51**, 475–86.

de Brabander, M., Aerts, F., van de Veire, R. & Borgers, M. (1975). Evidence against interconversion of microtobules and filaments. *Nature, Lond.*, **253**, 119–20.

Brandt, K. D. & Muir, H. (1971). Heterogeneity of protein-polysaccharides of porcine articular cartilage. The chondroitin sulphate proteins associated with collagen. *Biochem. J.* **123**, 747–55.

Brandt, K. D., Palmoski, M. J. & Perricone, E. (1976). Aggregation of cartilage proteoglycans. II. Evidence for the presence of a hyaluronate-binding region on proteoglycans from osteoarthritic cartilage. *Arthritis Rheum.*, **19**, 1308–14.

Brash, J. C. (1934). Some problems in the growth and developmental mechanics of bone. *Edinb. med. J.*, **41**, 363–87.

Brashear, H. R. (1963). Epiphyseal avascular necrosis and its relation to longitudinal bone growth. *J. Bone Jt Surg.*, **45*A***, 1423–8.

Bray, B. A., Lieberman, R. & Meyer, K. (1967). Structure of human skeletal keratosulphate. The linkage region. *J. biol. Chem.*, **242**, 3373–80.

Brem, H. & Folkman, J. (1975). Inhibition of tumour angiogenesis mediated by cartilage. *J. exp. Med.*, **141**, 427–39.

Brighton, C. T. & Heppenstall, R. B. (1971). Oxygen tension in zones of the epiphyseal plate, the metaphysis and diaphysis. *J. Bone Jt Surg.*, **53*A***, 719–28.

Brighton, C. T. & Hunt, R. M. (1976). Histochemical localization of calcium in growth plate mitochondria and matrix vesicles. *Fedn Proc. Fedn Am. Socs exp. Biol.*, **35**, 143–7.

Brighton, C. T., Lane, J. M. & Koh, J. K. (1974). *In vitro* rabbit articular cartilage organ model. II. ^{35}S incorporation in various oxygen tensions. *Arthritis Rheum.* **17**, 245–52.

Britten, R. J. & Davidson, E. H. (1969). Gene regulation for higher cells: a theory. *Science*, **165**, 349–57.

Brodin, H. (1955*a*). Longitudinal bone growth: the nutrition of the epiphyseal cartilages and the local blood supply. *Acta orthop. scand.*, Suppl. 20, 1–92.

Brodin, H. (1955*b*). Paths of nutrition in articular cartilage and intervertebral discs. *Acta orthop. scand.*, **24**, 177–83.

Brookes, M. (1958). The vascularization of long bone in the human foetus. *J. Anat.*, **92**, 261–7.

Brookes, M. (1971). *The blood supply of bone*. London: Butterworths.

Brookes, M. & Landon, D. N. (1964). The juxta-epiphyseal vessels in the long bones of foetal rats. *J. Bone Jt Surg.*, **46*B***, 336–45.

Brower, T. D., Akahoshi, Y. & Orlic, P. (1962). Diffusion of dyes through articular cartilage *in vivo*. *J. Bone Jt Surg.*, **44*A***, 456–63.

Bruch, C. (1852). Beiträge zur Entwickelungsgeschichte des Knochensystems. *Neue Denkschr. allg. schweiz. Ges. ges. Naturw.*, **12**, 1–176.

Buddecke, E., Filipovic, I., Beckmann, J. & von Figura, K. (1973). Metabolic processes of energy provision and synthesis in the ox aorta. In *Connective tissue and ageing*, ed. H. G. Vogel, pp. 7–10. Amsterdam: Excerpta Medica.

Bullough, P. & Goodfellow, J. (1968). The significance of the fine structure of articular cartilage. *J. Bone Jt Surg.*, **50*B***, 852–7.

Bullough, P., Goodfellow, J. & O'Connor, J. (1973). The relationship between degenerative changes and load-bearing in the human hip. *J. Bone Jt Surg.*, **55*B***, 746–58.

Burleigh, P. M. C. (1975). Discussion in *Dynamics of connective tissue macromolecules*, ed. P. M. C. Burleigh & A. R. Poole, p. 216. Amsterdam: North-Holland.

Burleigh, M. C., Barrett, A. J. & Lazarus, G. S. (1974). Cathepsin B1. A lysosomal enzyme that degrades native collagen. *Biochem. J.*, **137**, 387–98.

Burnside, B. (1975). The form and arrangement of microtubules: an historical, primarily morphological review. *Ann. NY Acad. Sci.*, **253**, 14–26.

Butler, W. T. & Cunningham, L. W. (1966). Evidence for the linkage of a disaccharide to hydroxylysine in tropocollagen. *J. biol. Chem.*, **241**, 3882–8.

Butler, W. T., Finch, J. E. & Miller, E. J. (1977). The covalent structure of cartilage collagen. Evidence for sequence heterogeneity of bovine α_1(II) chains. *J. biol. Chem.*, **252**, 639–43.

Byers, P. D., Contepomi, C. A. & Farkas, T. A. (1970). A *post mortem* study of the hip joint. *Ann. rheum. Dis.*, **29**, 15–31.

Byers, P. D., Maroudas, A., Oztop, F., Stockwell, R. A. & Venn, M. F. (1977). Histological and biochemical studies on cartilage from osteoarthrotic femoral heads with special reference to surface characteristics. *Conn. Tissue Res.*, **5**, 41–9.

Bywaters, E. G. L. (1937). The metabolism of joint tissues. *J. Path. Bact.*, **44**, 247–68.

Calandruccio, R. A. & Gilmer, W. S. (1962). Proliferation, regeneration and repair of articular cartilage of immature animals. *J. Bone Jt Surg.*, **44*A***, 431–55.

Cameron, D. A. & Robinson, R. A. (1958). Electron microscopy of epiphyseal and articular cartilage matrix in the femur of the newborn infant. *J. Bone Jt Surg.*, **40*A***, 163–70.

Campbell, C. J. (1969). The healing of cartilage defects. *Clin. Orthop.*, **64**, 45–63.

Campbell, P. N. & von der Decken, A. (1974). Cytomembranes and ribosomes. In *The cell in medical science*, **1**, ed. F. Beck & J. B. Lloyd, pp. 143–81. London: Academic Press.

Campo, R. D. & Dziewiatkowski, D. D. (1962). Intracellular synthesis of protein-polysaccharides by slices of bovine costal cartilage. *J. biol. Chem.*, **237**, 2729–35.

Campo, R. D. & Dziewiatkowski, D. D. (1963). Turnover of the organic matrix of cartilage and bone as visualized by autoradiography. *J. Cell Biol.*, **18**, 19–29.

Campo, R. D. & Phillips, S. J. (1973). Electron microscopic visualization of proteoglycans and collagen in bovine costal cartilage. *Calcif. Tissue Res.*, **13**, 83–92.

Caplan, A. I. (1970). Effects of the nicotinamide-sensitive teratogen 3-acetylpyridine on chick limb cells in culture. *Expl Cell Res.*, **62**, 341–55.

Caplan, A. I. & Koutroupas, S. (1973). The control of muscle and cartilage development in the chick limb: the role of differential vascularization. *J. Embryol. exp. Morph.*, **29**, 571–83.

Caputto, R., Barra, H. S. & Cumar, F. A. (1967). Carbohydrate metabolism. *A. Rev. Biochem.*, **36**, 211–46.

Carey, E. J. (1922). The genesis of muscle and bone. *J. Morph.*, **37**, 1–77.

Carlson, H. (1957). Reactions of rabbit patellary cartilage following operative defects. *Acta orthop. scand.*, Suppl. 28, 1–104.

Cartier, P. & Picard, J. (1951). Biochimie de l'ossification: rôle de la phosphatase alcaline dans les premiers stades de l'ossification. *C. r. Séanc. Soc. Biol.*, **145**, 274–6.

Castellani, A. A., de Bernard, B. & Zambotti, V. (1957). Glucuronic acid formation in epiphyseal cartilage homogenate. *Nature, Lond.*, **180**, 859.

Castellani, A. A. & Pedrini, V. (1956). Presenza e distribuzione della succinodeidrogenasi nei diversi strati della cartilagine di coniugazione. *Boll. Soc. ital. Biol. sper.*, **32**, 152.

Castellani, A. A. & Zambotti, V. (1956). Enzymatic formation of hexosamine in epiphyseal cartilage homogenate. *Nature, Lond.*, **178**, 313–14.

Catchpole, H. R., Joseph, N. R. & Engel, M. B. (1966). Thermodynamic relations of polyelectrolytes and inorganic ions of connective tissue. *Fedn Proc. Fedn Am. Socs exp. Biol.*, **25**, 1124–6.

Catto, M. (1965). A histological study of avascular necrosis of the femoral head after transcervical fracture. *J. Bone Jt Surg.*, **47*B***, 749–76.

Chapman, J. A. (1974). The staining pattern of collagen fibrils. I. An analysis of electron micrographs. *Conn. Tissue Res.*, **2**, 137–50.

Chapman, J. A. & Hardcastle, R. A. (1974). The staining pattern of collagen fibrils. II. A comparison with patterns computer-generated from the amino acid sequence. *Conn. Tissue Res.*, **2**, 151–9.

Chesterman, P. J. & Smith, A. U. (1968). Homotransplantation of articular cartilage and isolated chondrocytes. *J. Bone Jt Surg.*, **50*B***, 184–97.

Chrisman, O. D. (1967). Discussion in *Cartilage: degradation and repair*, ed. C. A. L. Bassett, pp. 81–2. Washington D.C.: Natn. Acad. Sci. Nat. Res. Council Pubn.

Chrisman, O. D. (1975). Effect of growth hormone on established cartilage lesions. *Clin. Orthop.*, **107**, 232–8.

Chrisman, O. D., Snook, G. A. & Wilson, T. C. (1972). The protective effect of aspirin against degeneration of human articular cartilage. *Clin. Orthop.*, **84**, 193–6.

Clark, E. R. & Clark, E. L. (1942). Microscopic observations on new formation of cartilage and bone in the living mammal. *Am. J. Anat.*, **70**, 167–200.

Clarke, I. C. (1971*a*). Surface characteristics of human articular cartilage – a scanning electron microscope study. *J. Anat.*, **108**, 23–30.

Clarke, I. C. (1971*b*). Human articular surface contours and related surface depression frequency studies. *Ann. rheum. Dis.*, **30**, 15–23.

Clarke, I. C. (1971*c*). Articular cartilage: a review and scanning electron microscope study. I. The interterritorial fibrillar architecture. *J. Bone Jt Surg.*, **53*B***, 732–50.

Clarke, I. C. (1974). Articular cartilage: a review and scanning electron microscope study. II. The territorial fibrillar architecture. *J. Anat.*, **118**, 261–80.

Coates, C. L., Burwell, R. G., Buttery, P. J., Walker, G. & Woodward, P. M. (1977). Somatomedin activity in synovial fluid. *Ann. rheum. Dis.*, **36**, 50–5.

Collins, D. H. (1949). *The pathology of articular and spinal diseases*. London: Arnold.

Collins, D. H., Ghadially, F. N. & Meachim, G. (1965). Intracellular lipids of cartilage. *Ann rheum. Dis.*, **24**, 123–35.

Collins, D. H. & McElligott, T. F. (1960). Sulphate ($^{35}SO_4$) uptake by chondrocytes in relation to histological changes in osteo-arthritic human articular cartilage. *Ann. rheum. Dis.*, **19**, 318–30.

Collins, D. H. & Meachim, G. (1961). Sulphate ($^{35}SO_4$) fixation by human articular cartilage compared in the knee and shoulder joints. *Ann. rheum. Dis.*, **20**, 117–22.

Cooke, J. (1973). Morphogenesis and regulation in spite of continual mitotic inhibition of *Xenopus* embryos. *Nature, Lond.*, **242**, 55–7.

Coon, H. G. & Cahn, R. D. (1966). Differentiation *in vitro*: effects of sephadex fractions of chick embryo extract. *Science*, **153**, 1116–19.

Cooper, G. W. & Prockop, D. J. (1968). Intracellular accumulation of protocollagen and extrusion of collagen by embryonic cartilage cells. *J. Cell Biol.*, **38**, 523–37.

Cotmore, J. M., Nichols, G. & Wuthier, R. E. (1971). Phospholipid–calcium phosphate complex: enhanced calcium migration in the presence of phosphate. *Science*, **172**, 1339–41.

Cowan, P. M., North, A. C. T. & Randall, J. T. (1953). High-angle X-ray diffraction of collagen fibres. In *Nature and structure of collagen*, ed. J. T. Randall, pp. 241–9. London: Butterworths.

Cox, R. W. & O'Dell, B. L. (1966). High-resolution electron-microscope observations on normal and pathological elastin. *Jl R. microsc. Soc.*, **85**, 401–11.

Cox, R. W. & Peacock, M. A. (1977). The fine structure of developing elastic cartilage. *J. Anat.*, **123**, 283–96.

Craigmyle, M. B. L. (1955). Studies of cartilage autografts and homografts in the rabbit. *Br. J. plast. Surg.*, **8**, 93–100.

Craigmyle, M. B. L. (1958). Antigenicity and survival of cartilage homografts. *Nature, Lond.*, **182**, 1248.

Crelin, E. S. & Southwick, W. O. (1960). Mitosis of chondrocytes induced in the knee joint articular cartilage of adult rabbits. *Yale J. Biol. Med.*, **33**, 243–4.

Crelin, E. S. & Southwick, W. O. (1964). Changes induced by sustained pressure in the knee joint articular cartilage of adult rabbits. *Anat. Rec.*, **149**, 113–34.

Crick, F. (1970). Diffusion in embryogenesis. *Nature, Lond.*, **225**, 420–2.

Curran, R. C. & Gibson, T. (1956). Uptake of labelled sulphate by human cartilage cells and its use as a test for viability. *Proc. R. Soc., B*, **144**, 572–6.

D'Abramo, F. & Lipmann, F. (1957). The formation of adenosine-3-phosphate-5-phosphosulfate in extracts of chick embryo cartilage and its conversion into chondroitin sulfate. *Biochim. biophys. Acta*, **25**, 211–13.

Daems, W. Th. & Van Rijssel, Th. G. (1961). The fine structure of the peribiliary dense bodies in mouse liver tissue. *J. Ultrastruct. Res.*, **5**, 263–90.

Daimon, T. (1977). The presence and distribution of glycogen particles in chondrogenic cells of the tibio-tarsal anlage of developing chick embryos. *Calcif. Tissue Res.*, **23**, 45–51.

Daniel, J. C., Kosher, R. A., Hamos, J. E. & Lash, J. W. (1974). Influence of external potassium on the synthesis and deposition of matrix components by chondrocytes *in vitro*. *J. Cell Biol.*, **63**, 843–54.

D'Aubigne, R. M. (1964). Idiopathic necrosis of the femoral head in adults. *Ann. R. Coll. Surg.*, **34**, 143–60.

Daughaday, W. H., Hall, K., Raben, M. S., Salmon, W. D., Van Den Brande, J. L. & Van Wyk, J. J. (1972). Somatomedin: proposed designation for sulphation factor. *Nature, Lond.*, **235**, 107.

Davidson, E. A. & Meyer, K. (1955). Structural studies on chondroitin sulphuric acid. II. The glucuronidic linkage. *J. Am. chem. Soc.*, **77**, 4796–8.

Davidson, E. A. & Small, W. (1963*a*). Metabolism *in vivo* of connective tissue mucopolysaccharides. I. Chondroitin sulfate C and keratosulfate of nucleus pulposus. *Biochim. biophys. Acta.*, **69**, 445–52.

Davidson, E. A. & Small, W. (1963*b*). Metabolism *in vivo* of connective tissue mucopolysaccharides. III. Chondroitin sulfate and keratosulfate of cartilage. *Biochim. biophys. Acta*, **69**, 459–63.

Davies, D. V. (1967). Properties of synovial fluid. *Proc. Instn mech. Engrs*, **181** (3J), 25–9.

Davies, D. V., Barnett, C. H., Cochrane, W. & Palfrey, A. J. (1962). Electron microscopy of articular cartilage in the young adult rabbit. *Ann. rheum. Dis.*, **21**, 11–22.

Davies, D. V. & Palfrey, A. J. (1966). Electron microscopy of normal synovial membrane. In *Studies on the anatomy and function of bones and joints*, ed. F. Gaynor Evans, pp. 1–16. Berlin: Springer-Verlag.

Davies, D. V. & White, J. E. W. (1961). The structure and weight of synovial fat pads. *J. Anat.*, **95**, 30–7.

Dawson, A. & Kember, N. F. (1974). Compensatory growth in the rat tibia. *Cell. Tissue Kinet.*, **7**, 285–91.

Dawson, A. & Kember, N. F. (1976). Cell kinetics of a chondrosarcoma. *Cell Tissue Kinet.*, **9**, 547–52.

Dearden, L. C. (1974). Enhanced mineralisation of the tibial epiphysial plate in the rat following propylthiouracil treatment: a histochemical light and electron microscopic study. *Anat. Rec.*, **178**, 671–90.

Dearden, L. C., Bonucci, E. & Cuicchio, M. (1974). An investigation of ageing in human costal cartilage. *Cell Tissue Res.*, **152**, 305–37.

Dearden, L. C. & Mosier, H. D. (1972). Long term recovery of chondrocytes in the tibial epiphysial plate in rats after cortisone treatment. *Clin. Orthop.*, **87**, 322–31.

Dehm, P., Jimenez, S. A., Olsen, B. J. & Prockop, D. J. (1972). A transport form of collagen from embryonic tendon: electron microscopic demonstration of an NH_2-terminal extension and evidence suggesting the presence of cystine in the molecule. *Proc. natn. Acad. Sci. USA*, **69**, 60–4.

De Luca, G., Speziale, P., Balduini, C. & Castellani, A. A. (1975). Biosynthesis of glycosaminoglycans: uridine diphosphate glucose-4′-epimerase from cornea and epiphysial plate cartilage. *Conn. Tissue Res.*, **3**, 39–47.

Denison, R. H. (1967). Ordovician vertebrates from western United States. *Fieldiana Geol.*, **16**, 131–92.

De Palma, A. F., Tsaltas, T. T. & Mauler, G. G. (1963). Viability of osteochondral grafts as determined by uptake of S^{35}. *J. Bone Jt Surg.*, **45*A***, 1565–78.

De Palma, A. F., McKeever, C. D. & Subin, D. K. (1966). Process of repair of articular cartilage demonstrated by histology and autoradiography with tritiated thymidine. *Clin. Orthop.*, **48**, 229–42.

Derge, J. G. & Davidson, E. A. (1972). Protein-polysaccharide biosynthesis. Membrane-bound saccharides. *Biochem. J.*, **126**, 217–23.

Deshmukh, K., Kline, W. G. & Sawyer, B. D. (1976). Role of calcium in the phenotypic expression of rabbit articular chondrocytes in culture. *FEBS Lett.* **67**, 48–51.

Deshmukh, K., Kline, W. G. & Sawyer, B. D. (1977). Effects of calcitonin and parathyroid hormone on the metabolism of chondrocytes in culture. *Biochim. biophys. Acta*, **499**, 28–35.

Deshmukh, K. & Nimni, M. E. (1973). Isolation and characterization of cyanogen bromide peptides from the collagen of bovine articular cartilage. *Biochem. J.*, **133**, 615–22.

Deuchar, E. M. (1963). Sites of earliest collagen formation in the chick embryo, as indicated by uptake of tritiated proline. *Expl Cell Res.*, **30**, 528–40.

Dickens, F. & Weil-Malherbe, H. (1936). Metabolism of cartilage. *Nature, Lond.*, **138**, 125–6.

Dickson, G. R. (1977). Ultrastructure of growth cartilage in the frog and salamander. MSc thesis, Queen's University, Belfast.

Diegelman, R. F., Bernstein, L. T. & Peterkovsky, B. (1973). Cell-free collagen synthesis on membrane-bound polysomes of chick embryo connective tissue and the localization of prolyl hydroxylase on the polysome-membrane complex. *J. biol. Chem.*, **248**, 6514–21.

Dienstein, S. R., Biehl, J., Holtzer, S. & Holtzer, H. (1974). Myogenic and chondrogenic lineages in developing limb buds grown *in vitro*. *Devl Biol.*, **39**, 83–95.

Dieppe, P. A. (1978). New knowledge of chondrocalcinosis. *J. clin. Path.*, **31**, Suppl. 12, 214–22.

Digby, K. H. (1916). The measurement of diaphysial growth in proximal and distal directions. *J. Anat. Physiol.*, **50**, 187–8.

Dingle, J. T. (1973). The role of lysosomal enzymes in skeletal tissues. *J. Bone Jt Surg.*, **55*B***, 87–95.

Dingle, J. T. (1975). The secretion of enzymes into the pericellular environment. *Phil. Trans. R. Soc., B*, **271**, 315–24.

Dingle, J. T. & Burleigh, P. M. C. (1974). Connective tissue and its changes in disease. *Trans. ophthal. Soc. UK*, **94**, 696–709.

Dingle, J. T., Fell, H. B. & Lucey, J. A. (1966). Synthesis of connective-tissue components. The effect of retinol and hydrocortisone on cultured limb-bone rudiments. *Biochem. J.*, **98**, 173–81.

Dingle, J. T., Gordon, J. L., Hazleman, B. L., Knight, C. G., Page Thomas, D. P., Phillips, N. C., Shaw, I. H., Fildes, F. J. T., Oliver, J. E., Jones, G., Turner, E. H. & Lowe, J. S. (1978). Novel treatment for joint inflammation. *Nature, Lond.*, **271**, 372–3.

Dingle, J. T., Horsfield, P., Fell, H. B. & Barratt, M. E. J. (1975). Breakdown of proteoglycan and collagen induced in pig articular cartilage in organ culture. *Ann. rheum. Dis.*, **34**, 303–11.

Dixon, B. (1971). Cartilage cell proliferation in the tail-vertebrae of newborn rats. *Cell Tissue Kinet.*, **4**, 21–30.

Dodds, G. S. (1930). Row formation and other types of arrangement of cartilage cells in endochondral ossification. *Anat. Rec.*, **46**, 385–99.

Doyle, B. B., Hukins, D. W. L., Hulmes, D. J. S., Miller, A. & Woodhead-Galloway, J. (1975). Collagen polymorphism: its origin in the aminoacid sequence. *J. molec. Biol.*, **91**, 79–99.

Doyle, B. B., Hulmes, D. J. S., Miller, A., Parry, D. A. D., Piez, K. A. & Woodhead-Galloway, J. (1974*a*). A D-periodic narrow filament in collagen. *Proc. R. Soc., B*, **186**, 67–74.

Doyle, B. B., Hulmes, D. J. S., Miller, A., Parry, D. A. D., Piez, K. A. & Woodhead-Galloway, J. (1974*b*). Axially projected collagen structures and stained bands of SLS accounted for by distribution of charged residues in α1 chain. *Proc. R. Soc., B*, **187**, 37–46.

Drachman, D. B. & Sokoloff, L. (1966). The role of movement in embryonic joint development. *Devl Biol.*, **14**, 401–20.

Drezner, M. K., Eisenbarth, G. S., Neelon, F. A. & Lebovitz, H. E. (1975). Stimulation of cartilage amino acid uptake by growth hormone-dependent factors in serum. *Biochim. biophys. Acta*, **381**, 384–96.

Drezner, M. K., Neelon, F. A. & Lebovitz, H. E. (1976). Stimulation of cartilage macromolecular synthesis by adenosine-3′,5′-monophosphate. *Biochim. biophys. Acta*, **425**, 521–31.

Dubreuil, G. (1913). La croissance des os des mammifères. III. L'accroissement interstitiel n'existe pas dans les os longs. *C.r. Séanc. Soc. Biol.*, **74**, 935–7.

Duncan, D. (1957). Electron microscope study of the embryonic neural tube and notochord. *Tex. Rep. Biol. Med.*, **15**, 367–77.

Dunn, D. B., Smith, J. D., Zamenhoff, S. & Griboff, G. (1954). Incorporation of halogenated pyrimidines into the deoxyribonucleic acids of *Bacterium coli* and its bacteriophages. *Nature, Lond.*, **174**, 305–7.

Dunstone, J. R. (1960). Ion-exchange reactions between cartilage and various cations. *Biochem. J.*, **77**, 164–70.

Dustmann, H. O., Puhl, W. & Krempien, B. (1974). Das Phänomen der Cluster im Arthroseknorpel. *Arch. orthop. Unfallchir.*, **79**, 321–33.

de Duve, C. & Wattiaux, R. (1966). Functions of lysosomes. *A. Rev. Physiol*, **28**, 435–92.

Dziewiatkowski, D. D. (1964). Effect of hormones on the turnover of polysaccharides in connective tissues. *Biophys. J.*, **4** (Suppl.), 215–38.

Eanes, E. D. & Meyer, J. L. (1977). The maturation of crystalline calcium phosphates in aqueous suspension at physiologic pH. *Calcif. Tissue Res.*, **23**, 259–69.

Ede, D. A. & Agerbak, G. S. (1968). Cell adhesion and movement in relation to the developing limb pattern in normal and talpid[3] mutant chick embryos. *J. Embryol. exp. Morph.*, **20**, 81–100.

Ede, D. A. & Flint, O. P. (1972). Patterns of cell division, cell death and chondrogenesis in cultured aggregates of normal and talpid mutant chick limb mesenchyme cells. *J. Embryol. exp. Morph.*, **27**, 245–60.

Ehrlich, M. G., Mankin, H. J., Jones, H., Wright, R., Crispen, C. & Vigliani, G. (1977). Collagenase and collagenase inhibitors in osteoarthritic and normal human cartilage. *J. clin. Invest.*, **59**, 226–33.

Eichelberger, L., Miles, J. S. & Roma, M. (1952). The histochemical characterization of articular cartilages of poliomyelitis patients. *J. Lab. clin. Med.*, **40**, 284–96.

Eichelberger, L., Roma, M. & Moulder, P. V. (1959). Biochemical studies of articular cartilage. III. Values following the immobilization of an extremity. *J. Bone Jt Surg.*, **41*A***, 1127–42.

Eilberg, R. G., Zuckerberg, D. A. & Person, P. (1975). Mineralization of invertebrate cartilage. *Calcif. Tissue Res.*, **19**, 85–90.

Eisenstein, R., Kuettner, K. E., Neapolitan, C., Soble, L. W. & Sorgente, N. (1975). The resistance of certain tissues to invasion. III. Cartilage extracts inhibit the growth of fibroblasts and endothelial cells in culture. *Am. J. Path.*, **81**, 337–47.

Eisenstein, R., Sorgente, N., Soble, L. W., Miller, A. & Kuttner, K. E. (1973). The resistance of certain tissues to invasion: penetrability of explanted tissue by vascularised mesenchyme. *Am. J. Path.* **73**, 765–74.

Ekholm, R. (1951). Articular cartilage nutrition. *Acta anat.*, **11**, Suppl. 15–2, 1–76.

Ekholm, R. (1955). Nutrition of articular cartilage. *Acta anat.*, **24**, 329–38.

Ekholm, R. & Norbäck, B. (1951). On the relationship between articular changes and function. *Acta orthop. scand.*, **21**, 81–98.

Elkeles, A. (1966). Sex differences in the calcification of the costal cartilages. *J. Am. Geriat. Soc.*, **14**, 456–62.

Elliott, H. C. (1936). Studies on articular cartilage. I. Growth mechanisms. *Am. J. Anat.*, **58**, 127–45.

Ellison, M. L., Ambrose, E. J. & Easty, G. C. (1969). Chondrogenesis in chick embryo somites *in vitro*. *J. Embryol. exp. Morph.*, **21**, 331–40.

Elves, M. W. (1974). A study of the transplantation antigens on chondrocytes from articular cartilage. *J. Bone Jt Surg.* **56*B***, 178–85.

Elves, M. W. (1976). Newer knowledge of the immunology of bone and cartilage. *Clin. Orthop.*, **120**, 232–59.

Engfeldt, B. & Hjertquist, S.-O. (1968). Studies on the epiphysial growth zone. I. The preservation of acid glycosaminoglycans in tissues in some histochemical procedures for electron microscopy. *Virchows Arch. Abt. B. Zellpath.*, **1**, 222–9.

Enneking, W. F. & Harrington, P. (1969). Pathological changes in scoliosis. *J. Bone Jt Surg.*, **51*A***, 165–84.

Evans, E. B., Eggers, G. W. N., Butler, J. K. & Blumel, J. (1960). Experimental immobilisation and remobilisation of rat knee joints. *J. Bone Jt Surg.*, **42*A***, 737–58.

Eyre, D. R., McDevitt, C. A. & Muir, H. (1975). Experimentally-induced osteoarthrosis in the dog. *Ann. rheum. Dis.*, **34,** Suppl. 2, 138–40.

Eyre, D. R. & Muir, H. (1975). The distribution of different molecular species of collagen in fibrous, elastic and hyaline cartilage of the pig. *Biochem. J.*, **151**, 595–602.

Fahmy, A., Lee, S. & Johnson, P. (1971). Ultrastructural effects of testosterone on epiphysial cartilage. *Calcif. Tissue Res.*, **7**, 12–22.

Falchak, K. H., Goetzl, E. J. & Kulka, J. P. (1970). Respiratory gases of synovial fluids. *Am. J. Med.*, **49**, 223–31.

Fawns, H. T. & Landells, J. W. (1953). Histological studies of rheumatic conditions. I. Observations on the fine structure of the matrix of normal bone and cartilage. *Ann. rheum. Dis.*, **12**, 105–13.

Feingold, D. S., Neufeld, E. F. & Hassid, W. Z. (1960). The 4-epimerization and decarboxylation of uridine diphosphate D-glucuronic acid by extracts from *Phaseolus aureus* seedlings. *J. biol. Chem.*, **235**, 910–13.

Felix, R. & Fleisch, H. (1976*a*). Pyrophosphatase and ATPase of isolated cartilage matrix vesicles. *Calcif. Tissue Res.*, **22**, 1–7.

Felix, R. & Fleisch, H. (1976*b*). Role of matrix vesicles in calcification. *Fedn Proc. Fedn Am. Socs exp. Biol.*, **35**, 169–71.

Fell, H. B. (1925). The histogenesis of cartilage and bone in the long bones of the embryonic fowl. *J. Morph. Physiol.*, **40**, 417–60.

Fell, H. B. (1969). Role of biological membranes in some skeletal reactions. *Ann. rheum. Dis.*, **28**, 213–27.

Fell, H. B. & Barratt, M. E. J. (1973). The role of the soft connective tissue in the breakdown of pig articular cartilage cultivated in the presence of complement-sufficient antiserum to pig erythrocytes. I. Histological changes. *Int. Archs Allergy appl. Immun.*, **44**, 441–68.

Fell, H. B. & Canti, R. G. (1934). Experiments on the development *in vitro* of the avian knee joint. *Proc. R. Soc., B*, **116**, 316–51.

Fell, H. B. & Dingle, J. T. (1963). Studies on the mode of action of excess of vitamin A. 6. Lysosomal protease and the degradation of cartilage matrix. *Biochem. J.* **87**, 403–8.

Fell, H. B. & Jubb, R. W. (1977). The effect of synovial tissue on the breakdown of articular cartilage in organ culture. *Arthritis Rheum.*, **20**, 1359–71.

Fell, H. B. & Mellanby, E. (1952). Effect of hypervitaminosis A on embryonic limb bones cultivated *in vitro*. *J. Physiol., Lond.*, **116**, 320–49.

Fessler, J. H. (1960). A structural function of mucopolysaccharide in connective tissue. *Biochem. J.*, **76**, 124–32.

Fessler, J. H. & Fessler, L. I. (1966). Electron microscopic visualization of the polysaccharide hyaluronic acid. *Proc. natn. Acad. Sci. USA*, **56**, 141–47.

Filopovic, I. & Buddecke, E. (1971). Increased fatty acid synthesis of arterial tissue in hypoxia. *Eur. J. Biochem.*, **20**, 587–92.

Fischer, G. & Boedecker, C. (1861). Kunstliche Bildung von Zucker aus Knorpel (Chondrogen), und über die Umsetzung des genossenen Knorpels in menschlicher Korper. *Justus Liebigs Annln Chem.*, **117**, 111–18.

Fisher, A. G. T. (1923). Physiological principles underlying the treatment of injuries and diseases of the articulations. *Lancet*, **1**, 541–8.

Fitton Jackson, S. (1970). Environmental control of macromolecular synthesis in cartilage and bone; morphogenetic response to hyaluronidase. *Proc. R. Soc., B*, **175**, 405–53.

FitzGerald, M. J. T. & Shtieh, M. M. (1977*a*). Interstitial versus appositional growth in developing non-articular cartilage. *J. Anat.*, **124**, 503–4.

FitzGerald, M. J. T. & Shtieh, M. M. (1977*b*). Repair of whole thickness defects in articular and non-articular cartilages in the rat. *J. Anat.*, **124**, 504.

Fleisch, H. & Bisaz, S. (1962). Le rôle inhibiteur du pyrophosphate dans la calcification. *J. Physiol., Paris*, **54**, 340–1.

Flickinger, R. A. (1975). The effect of 5-bromodeoxyuridine on chick embryo limb bud mesenchyme in organ culture. *Cell Differ.*, **4**, 295–304.

Foster, L. N., Kelly, R. P. & Watts, W. M. (1951). Experimental infarction of bone and bone marrow. *J. Bone Jt Surg.*, **33*A***, 396–406.

Franco-Browder, S., de Rydt, J. & Dorfman, A. (1963). The identification of a sulfated mucopolysaccharide in chick embryos, stages 11–23. *Proc. natn. Acad. Sci. USA*, **49**, 643–7.

Frazer, R. D. B., Miller, A. & Parry, D. A. D. (1974). Packing of microfibrils in collagen. *J. molec. Biol.*, **83**, 281–3.

Freeman, M. A. R. (1975). Discussion in *Dynamics of connective tissue macromolecules*, ed. P. M. C. Burleigh & A. R. Poole, p. 99. Amsterdam: North-Holland.

Freeman, M. A. R. & Kempson, G. E. (1973). Load carriage. In *Adult articular cartilage*, ed. M. A. R. Freeman, pp. 228–46. London: Pitman Medical.

Freeman, M. A. R. & Meachim, G. (1973). Ageing, degeneration and remodelling of articular cartilage. In *Adult articular cartilage*, ed. M. A. R. Freeman, pp. 287–329. London: Pitman Medical.

Froesch, E. R., Zapf, J., Audhya, T. K., Ben-Porath, E., Segan, B. J. & Gibson, K. D. (1976). Non-suppressible insulin-like activity and thyroid hormone: major pituitary-dependent sulfation factors for chick embryo cartilage. *Proc. natn. Acad. Sci. USA*, **73**, 2904–8.

Frommer, J., Monroe, C. W., Morehead, J. R. & Belt, W. D. (1968). Autoradiographic study of cellular proliferation during early development of the mandibular condyle in mice. *J. dent. Res.*, **47**, 816–19.

Fry, H. & Robertson, W. Van B. (1967). Interlocked stresses in cartilage. *Nature, Lond.*, **215**, 53–4.

Fuller, J. A. & Ghadially, F. N. (1972). Ultrastructural observations in surgically produced partial-thickness defects in articular cartilage. *Clin. Orthop.*, **86**, 193–205.

Fyfe, F. W. (1964). Predominance of epiphyseal over metaphyseal blood supply in nourishment of epiphyseal cartilage, demonstrated by ^{35}S radioautography and vascular ablation. *J. Anat.*, **98**, 471–2.

Galjaard, H. (1962). Histochemisch en interferometrisch onderzoek van hyalinen kraakbeen. Doctoral Thesis, University of Leiden.

Gallie, W. E. (1956). Avascular necrosis involving articular surfaces. *J. Bone Jt Surg.*, **38*A*** 732–8.

Gardner, D. L. (1972). The influence of microscopic technology on knowledge of cartilage surface structure. *Ann. rheum. Dis.*, **31**, 235–58.

Gardner, D. L. & McGillivray, D. C. (1971). Living articular cartilage is not smooth. The structure of mammalian and avian joint surfaces demonstrated *in vivo* by immersion incident light microscopy. *Ann. rheum. Dis.*, **30**, 3–14.

Gardner, E. (1971). Osteogenesis in the human embryo and fetus. In *The biochemistry and physiology of bone*, **3**, 2nd edn, ed. G. H. Bourne, pp. 77–144. New York & London: Academic Press.

Gardner, E., Gray, D. J. & O'Rahilly, R. (1959). The pre-natal development of the skeleton and joints of the human foot. *J. Bone Jt Surg.*, **41*A***, 847–76.

Gay, S., Muller, P. K., Lemmen, C., Remberger, K., Matzen, K. & Kuhn, K. (1976). Immunohistological study on collagen in cartilage–bone metamorphosis and degenerative osteoarthrosis. *Klin. Wschr.*, **54**, 969–76.

Gerber, B. R. & Schubert, M. (1964). The exclusion of large solutes by cartilage protein polysaccharide. *Biopolymers*, **2**, 259–73.

Gersh, I. & Catchpole, H. R. (1960). The nature of ground substance of connective tissue. *Perspect. Biol. Med.*, **3**, 282–319.

Ghadially, F. N., Ailsby, R. L. & Oryschak, A. F. (1974). Scanning electron microscopy of superficial defects in articular cartilage. *Ann. rheum. Dis.*, **33**, 327–32.

Ghadially, F. N., Fuller, J. A. & Kirkaldy-Willis, W. H. (1971). Ultrastructure of full thickness defects in articular cartilage. *Archs Path.*, **92**, 356–69.

Ghadially, F. N., Ghadially, J. A., Oryschak, A. F. & Yong, N. K. (1976). Experimental production of ridges on rabbit articular cartilage: a scanning electron microscope study. *J. Anat.*, **121**, 119–32.

Ghadially, F. N., Ghadially, J. A., Oryschak, A. F. & Yong, N. K. (1977). The surface of dog articular cartilage: a scanning electron microscope study. *J. Anat.*, **123**, 527–36.

Ghadially, F. N., Meachim, G. & Collins, D. H. (1965). Extracellular lipids in the matrix of human articular cartilage. *Ann. rheum. Dis.* **24**, 136–46.

Ghadially, F. N., Mehta, P. N. & Kirkaldy-Willis, W. H. (1970). Ultrastructure of articular cartilage in experimentally produced lipoarthrosis. *J. Bone Jt Surg.*, **52*A***, 1147–58.

Ghadially, F. N., Moshurchak, E. M. & Ghadially, J. A. (1978). A maturation change in the surface of cat articular cartilage detected by the scanning electron microscope. *J. Anat.*, **125**, 349–60.

Ghosh, S., Blumenthal, H. J., Davidson, E. & Roseman, S. (1960). Glucosamine metabolism. V. Enzymatic synthesis of glucosamine-6-phosphate. *J. biol. Chem.*, **235**, 1265–73.

Gibson, T., Curran, R. C. & Davis, W. B. (1957). The survival of living homograft cartilage in man. *Transplantn Bull.*, **4**, 105–6.

Gibson, T. & Davis, W. B. (1958). The distortion of autogenous cartilage grafts: its cause and prevention. *Br. J. plast. Surg.*, **10**, 257–74.

Gill, G. G. & Abbott, L. C. (1942). Practical method of predicting the growth of the femur and tibia in the child. *Archs Surg.*, **45**, 286–315.

Gillard, G. C. & Wusteman, F. S. (1970). Acid glycosaminoglycans of elastic cartilage. *Biochem. J.*, **118**, 25*P*.

Gilmour, R. S., Windass, J. D., Affara, N. & Paul, J. (1975). Control of transcription of the globin gene. *J. cell. Physiol.*, **85**, 449–58.

Ginsberg, J. M., Eyring, E. J. & Curtiss, P. H. (1969). Continuous compression of rabbit articular cartilage producing loss of hydroxyproline before loss of hexosamine. *J. Bone Jt Surg.*, **51*A***, 467–74.

Glaser, L. (1959). Uridine diphosphate-*N*-acetyl glucosamine-4-epimerase from *Bacillus subtilis*. *Biochim. biophys. Acta*, **31**, 575–6.

Glucksmann, A. (1939). Studies on bone mechanics *in vitro*. II. The role of tension and pressure in chondrogenesis. *Anat. Rec.*, **73**, 39–56.

Glucksmann, A. (1951). Cell death in normal vertebrate ontogeny. *Biol. Rev.*, **26**, 59–86.

Godman, G. C. & Lane, N. (1964). On the site of sulfation in the chondrocyte. *J. Cell biol.*, **21**, 353–66.

Godman, G. C. & Porter, K. R. (1960). Chondrogenesis studied with the electron microscope. *J. biochem. biophys. Cytol.*, **8**, 719–60.

Goel, S. C. (1970). Electron microscopic studies on developing cartilage. I. The membrane system related to the synthesis and secretion of extracellular material. *J. Embryol. exp. Morph.*, **23**, 169–84.

Goetinck, P. F., Pennypacker, J. P. & Royal, P. D. (1974). Proteo-chondroitin sulfate synthesis and chondrogenic expression. *Expl Cell Res.*, **87**, 241–8.

Goodfellow, J. W. & Bullough, P. G. (1967). The pattern of ageing of the articular cartilage of the elbow joint. *J. Bone Jt Surg.*, **49*B***, 175–81.

Goodsir, J. (1857). On the morphological relations of the nervous system in the annulose and vertebrate types of organization. *Edinb. New Phil. J.* (N.S.), **5**. See *Anatomical memoirs of John Goodsir*, **2** (1868), pp. 78–87. Edinburgh: Black.

Gotte, L., Giro, M. G., Volpin, D. & Horne, R. W. (1974). The ultrastructural organization of elastin. *J. Ultrastruct. Res.*, **46**, 23–33.

Gotte, L. & Volpin, D. (1975). Ultrastructural organization of elastin. In *Protides of biological fluids*, **22**, ed. H. Peeters, pp. 137–44. Oxford: Pergamon.

Gould, R. P., Day, A. & Wolpert, L. (1972). Mesenchymal condensation and cell contact in early morphogenesis of the chick limb. *Expl Cell Res.*, **72**, 325–36.

Gould, R. P., Selwood, L., Day, A. & Wolpert, L. (1974). The mechanism of cellular orientation during early cartilage formation in the chick limb and regenerating amphibian limb. *Expl Cell Res.*, **83**, 287–96.

Gray, D. J. & Gardner, E. (1950). Prenatal development of the human knee and superior tibio-fibular joints. *Am. J. Anat.*, **86**, 235–87.
Gray, D. J. & Gardner, E. (1969). Prenatal development of the human humerus. *Am. J. Anat.*, **124**, 431–46.
Gray, D. J., Gardner, E. & O'Rahilly, R. (1957). The prenatal development of the skeleton and joints of the human hand. *Am. J. Anat.*, **101**, 169–224.
Gray, L. H. & Scott, O. C. A. (1964). Oxygen tension and the radio-sensitivity of tumours. In *Oxygen in the animal organism*, ed. F. Dickens & E. Neil, pp. 537–48. Oxford: Pergamon.
Green, W. T., Martin, G. N., Eanes, E. D. & Sokoloff, L. (1970). Microradiographic study of the calcified layer of articular cartilage. *Archs. Path.*, **90**, 151–8.
Greenlee, T. K. & Ross, R. (1967). The development of the rat flexor digital tendon, a fine structure study. *J. Ultrastruct. Res.*, **18**, 354–76.
Greenlee, T. K., Ross, R. & Hartman, J. L. (1966). The fine structure of elastic fibres. *J. Cell Biol.*, **30**, 59–71.
Greenwald, A. S. & Haynes, D. W. (1969). A pathway for nutrients from the medullary cavity to the articular cartilage of the human femoral head. *J. Bone Jt Surg.*, **51*B***, 747–53.
Greenwald, R. A., Schwartz, C. E. & Cantor, J. O. (1975). Interaction of cartilage proteoglycans with collagen-substituted agarose gels. *Biochem. J.* **145**, 601–5.
Greer, R. B., Janicke, G. H. & Mankin, H. J. (1968). Protein-polysaccharide synthesis at three levels of the normal growth plate. *Calcif. Tissue Res.*, **2**, 157–64.
Gregory, J. D. (1973). Multiple aggregation factors in cartilage proteoglycan. *Biochem. J.*, **133**, 383–6.
Gregory, J. D., Laurent, T. C. & Roden, L. (1964). Enzymatic degradation of chondromucoprotein. *J. biol. Chem.*, **239**, 3312–20.
Greiling, H. & Baumann, G. (1973). Age dependent changes of non-sulfated disaccharide groups in the proteoglycans of knee joint cartilage. In *Connective tissue and ageing*, ed. H. G. Vogel, pp. 160–2. Amsterdam: Excerpta Medica.
Grimes, L. N. (1974*a*). The effect of supernumerary cartilaginous implants upon rabbit ear regeneration. *Am. J. Anat.*, **141**, 447–51.
Grimes, L. N. (1974*b*). Selective X-irradiation of the cartilage at the regenerating margin of rabbit ear holes. *J. exp. Zool.*, **190**, 237–42.
Grobstein, C. & Holtzer, H. (1955). *In vitro* studies of cartilage induction in mouse mesoderm. *J. exp. Zool.*, **128**, 333–59.
Grobstein, C. & Parker, G. (1954). *In vitro* induction of cartilage in mouse somite mesoderm by embryonic spinal cord. *Proc. Soc. exp. Biol. Med.*, **85**, 477–81.
Gross, J., Highberger, J. H. & Schmitt, F. O. (1954). Collagen structures considered as states of aggregation of a kinetic unit. The tropocollagen particle. *Proc. natn. Acad. Sci. USA*, **40**, 679–88.
Gross, J. & Lapière, Ch. M. (1962). Collagenolytic activity in amphibian tissues. *Proc. natn. Acad. Sci. USA*, **48**, 1014–22.
Gross, J. & Schmitt, F. O. (1948). The structure of human skin collagen as studied with the electron microscope. *J. exp. Med.*, **88**, 555–68.
Gross, J. I., Mathews, M. B. & Dorfman, A. (1960). Sodium chondroitin sulphate–protein complexes of cartilage. II. Metabolism. *J. biol. Chem.*, **235**, 2889–92.
Gubisch, W. & Schlager, F. (1961). Fermente im Knochen- und Knorpel-gewebe. III. Mitteillung: β-D-Glucuronidase. *Acta histochem.*, **12**, 69–74.
Gustafsson, P.-O., Kasstrom, H., Lindberg, L. T. & Olsson, S.-E. (1975). Growth and mitotic rate of the proximal tibial epiphysial plate in hypophysectomised rats given estradiol and human growth hormone. *Acta radiol.*, Suppl. 344, 69–74.
Gustavson, K. H. (1955). The function of hydroxyproline in collagens. *Nature, Lond.*, **175**, 70–4.

Haas, S. L. (1914). Regeneration of cartilage and bone with a special study of these processes as they occur at the chondro-costal junction. *Surgery Gynec. Obstet.*, **19**, 604–17.

Haas, S. L. (1917). The relation of the blood supply to the longitudinal growth of bone. *Am. J. orthop. Surg.*, **15**, 157–71.

Haba, G. & Holtzer, H. (1965). Chondroitin sulfate: inhibition of synthesis by puromycin. *Science*, **149**, 1263–5.

Haines, R. W. (1933). Cartilage canals. *J. Anat.*, **68**, 45–64.

Haines, R. W. (1937). Growth of cartilage canals in the patella. *J. Anat.*, **71**, 471–9.

Haines, R. W. (1942). The evolution of epiphyses and of endochondral bone. *Biol. Rev.*, **17**, 267–92.

Haines, R. W. (1947). The development of joints. *J. Anat.*, **81**, 33–55.

Haines, R. W. (1953). The early developments of the femoro-tibial and tibio-fibular joints. *J. Anat.*, **87**, 192–206.

Haines, R. W. (1974). The pseudo-epiphysis of the first metacarpal of man. *J. Anat.*, **117**, 145–58.

Haines, R. W. (1975). The histology of epiphysial union in mammals. *J. Anat.*, **120**, 1–25.

Haines, R. W. (1976). Destruction of hyaline cartilage in the sigmoid notch of the human ulna. *J. Anat.*, **122**, 331–4.

Haines, R. W. & Mohuiddin, A. (1968). Metaplastic bone. *J. Anat.*, **103**, 527–38.

Hales, S. (1727). *Statical essays*. London: Innys.

Hall, B. K. (1970). Cellular differentiation in skeletal tissues. *Biol. Rev.*, **45**, 455–84.

Hall, K., Takano, K., Fryklund, L. & Sievertsson, H. (1975). Somatomedins. *Adv. Metab. Disord.*, **8**, 19–46.

Hall-Craggs, E. C. B. (1969). Influence of epiphyses on the regulation of bone growth. *Nature, Lond.*, **221**, 1245.

Hall-Craggs, E. C. B. & Lawrence, C. A. (1969). The effect of epiphysial stapling on growth in length of the rabbit tibia and femur. *J. Bone Jt Surg.*, **51*B***, 359–65.

Hallen, A. (1962). The collagen and ground substance of human intervertebral disc at different ages. *Acta chem. scand.*, **16**, 705–10.

Halstead, L. B. (1969). Calcified tissues in the earliest vertebrates. *Calcif. Tissue Res.*, **3**, 107–24.

Ham, A. W. (1932). Cartilage and bone. In *Special cytology*, **2**, 2nd edn, ed. E. V. Cowdry, pp. 981–1051. New York: Hoeber.

Hamburger, V. & Hamilton, H. L. (1951). A series of normal stages in the development of the chick embryo. *J. Morph.*, **88**, 49–92.

Hamerman, D., Rojkind, M. & Sandson, J. (1966). Protein bound to hyaluronate: chemical and immunological studies. *Fedn Proc. Fedn Am. Socs exp. Biol.*, **25**, 1040–5.

Hamerman, D., Rosenberg, L. C. & Schubert, M. (1970). Diarthrodial joints revisited. *J. Bone Jt Surg.*, **52*A***, 725–74.

Hamerman, D. & Schubert, M. (1962). Diarthrodial joints, an essay. *Am. J. Med.*, **33**, 555–90.

Hamerman, D. & Schuster, H. (1958). Hyaluronate in normal human synovial fluid. *J. clin. Invest.*, **37**, 57–64.

Hamilton, W. J., Boyd, J. D. & Mossman, H. W. (1972). *Human embryology*, 4th edn. London: Macmillan.

Handley, C. J., Bateman, J. F., Oakes, B. W. & Lowther, D. (1975). Characterization of the collagen synthesised by cultured cartilage cells. *Biochim. biophys. Acta*, **386**, 444–50.

Handley, C. J. & Lowther, D. A. (1976). Inhibition of proteoglycan biosynthesis by hyaluronic acid in chondrocytes in cell culture. *Biochim. biophys. Acta*, **444**, 69–74.

Handley, C. J. & Phelps, C. F. (1972*a*). The biosynthesis *in vitro* of chondroitin sulphate in neonatal rat epiphysial cartilage. *Biochem. J.*, **126**, 417–32.

Handley, C. J. & Phelps, C. F. (1972*b*). The biosynthesis *in vitro* of keratan sulphate in bovine cornea. *Biochem. J.*, **128**, 205–13.

Haraldsson, S. (1962). The vascular pattern of a growing and full-grown human epiphysis. *Acta anat.*, **48**, 156–67.

Hardingham, T. E., Ewins, R. J. F. & Muir, H. (1976). Cartilage proteoglycans. Structure and heterogeneity of the protein core and the effects of specific protein modifications on the binding to hyaluronate. *Biochem. J.*, **157**, 127–43.

Hardingham, T. E. & Muir, H. (1972*a*). The specific interaction of hyaluronic acid with cartilage proteoglycans. *Biochim. biophys. Acta*, **279**, 401–5.

Hardingham, T. E. & Muir, H. (1972*b*). Biosynthesis of proteoglycans in cartilage slices. *Biochem. J.*, **126**, 791–803.

Hardingham, T. E. & Muir, H. (1973*a*). Binding of oligosaccharides of hyaluronic acid to proteoglycans. *Biochem. J.*, **135**, 905–8.

Hardingham, T. E. & Muir, H. (1973*b*). Hyaluronic acid in cartilage. *Biochem. Soc. Trans.*, **1**, 282–4.

Hardingham, T. E. & Muir, H. (1974*a*). Hyaluronic acid in cartilage and proteoglycan aggregation. *Biochem. J.*, **139**, 565–81.

Hardingham, T. E. & Muir, H. (1974*b*). The function of hyaluronic acid in proteoglycan aggregation. In *Normal and osteoarthrotic articular cartilage*, ed. S. Y. Ali, M. W. Elves & D. H. Leaback, pp. 51–63. London: Institute of Orthopaedics.

Harkness, R. D. (1968). Mechanical properties of collagen tissues. In *Biology of collagen*, **2**, ed. B. S. Gould, pp. 247–310. London & New York: Academic Press.

Harpuder, K. (1926). Physikalisch-chemische Untersuchungen am normalen Knorpel. *Biochem. Z.*, **169**, 308–19.

Harris, H. A. (1933). *Bone growth in health and disease*. London: Oxford Medical Pubns.

Harris, H. A. & Russell, A. E. (1933). The atypical growth in cartilage as the fundamental factor in dwarfism and achondroplasia. *Proc. R. Soc. Med.*, **26**, 779–87.

Harwood, R., Grant, M. E. & Jackson, D. S. (1975). Post-translational processing of procollagen polypeptides. In *Protides of biological fluids*, **22** (ed. H. Peeters, pp. 83–5. Oxford: Pergamon.

Harwood, R., Grant, M. E. & Jackson, D. S. (1976). The route of secretion of procollagen. *Biochem. J.*, **156**, 81–90.

Hascall, V. C. & Heinegard, D. (1974*a*). Aggregation of cartilage proteoglycans. I. Role of hyaluronic acid. *J. biol. Chem.*, **249**, 4232–41.

Hascall, V. C. & Heinegard, D. (1974*b*). Aggregation of cartilage proteoglycans. II. Oligosaccharide competition of the proteoglycan–hyaluronic acid interaction. *J. biol. Chem.*, **249**, 4242–9.

Hascall, V. C. & Sajdera, S. W. (1969). Proteinpolysaccharide complex from bovine nasal cartilage. The function of glycoprotein in the formation of aggregates. *J. biol. Chem.*, **244**, 2384–96.

Hascall, V. C. & Sajdera, S. W. (1970). Physical properties and polydispersity of proteoglycans from bovine nasal cartilage. *J. biol. chem.*, **245**, 4920–30.

Hass, G. M. (1943). Studies of cartilage: a morphologic and chemical analysis of aging human costal cartilage. *Archs Path.*, **35**, 275–84.

Hately, W., Evison, G. & Samuel, E. (1965). The pattern of ossification in the laryngeal cartilages: a radiological study. *Br. J. Radiol.*, **38**, 585–91.

Havers, C. (1691). *Osteologia Nova*. London: Smith.

Hay, E. D. (1968). Organization and fine structure of epithelium and mesenchyme in the developing chick embryo. In *Epithelial–mesenchymal interactions*, ed. R. Fleischmajer & R. E. Billingham, pp. 31–55. Baltimore: Williams & Wilkins.

Hay, E. D. & Meier, S. (1974). Glycosaminoglycan synthesis by embryonic inductors: neural tube, notochord and lens. *J. Cell Biol.*, **62**, 889–98.

Hayflick, L. (1965). The limited *in vitro* lifetime of human diploid cell strains. *Expl Cell Res.*, **37**, 614–36.

Heine, J. (1926). Über die Arthritis deformans. *Virchows Arch. path. Anat. Physiol.*, **260**, 521–663.

Heingard, D. (1972). Extraction, fractionation and characterization of proteoglycans from bovine tracheal cartilage. *Biochem. biophys. Acta*, **285**, 193–207.

Heinegard, D. (1977). Polydispersity of cartilage proteoglycans. Structural variations with size and buoyant density of the molecules. *J. biol. Chem.*, **252**, 1980–9.

Heinegard, D. & Axensson, I. (1977). Distribution of keratansulfate in cartilage proteoglycans. *J. biol. Chem.*, **252**, 1971–9.

Heinegard, D. & Hascall, V. C. (1974*a*). Aggregation of cartilage proteoglycans. III. Characteristics of the proteins isolated from trypsin digests of aggregates. *J. biol. Chem.*, **249**, 4250–6.

Heinegard, D. & Hascall, V. C. (1974*b*). Characterization of chondroitin sulfate isolated from trypsin–chymotrypsin digests of cartilage proteoglycan. *Archs Biochem. Biophys.*, **165**, 427–41.

Heins, J. N., Garland, J. T. & Daughaday, W. H. (1970). Incorporation of ^{35}S-sulphate into rat cartilage explants *in vitro*. Effects of aging on responsiveness to stimulation by sulphation factor. *Endocrinology*, **87**, 688–92.

Helting, T. (1971). Further characterization of the galactosyltransferases in chick cartilage. *Biochim. biophys. Acta*, **227**, 42–55.

Helting, T. & Roden, L. (1968). The carbohydrate–protein linkage region of chondroitin-6-sulfate. *Biochim. biophys. Acta*, **170**, 301–8.

Helting, T. & Roden, L. (1969*a*). Biosynthesis of chondritin sulfate. I. Galactosyl transfer in the formation of the carbohydrate–protein linkage region. *J. biol. Chem.*, **244**, 2790–8.

Helting, T. & Roden, L. (1969*b*). Biosynthesis of chondroitin sulfate. II. Glucuronosyl transfer in the formation of the carbohydrate–protein linkage region. *J. biol. Chem.*, **244**, 2799–805.

Herzog, R. O. & Gonnell, H. W. (1925). Über Kollagen. *Ber. dt. chem. Ges.*, **58**, 2228–30.

Heyner, S. (1973). The antigenicity of cartilage grafts. *Surgery Gynec. Obstet.*, **136**, 298–305.

Highberger, J. H., Gross, J. & Schmitt, F. O. (1950). Electron microscope observations of certain fibrous structures obtained from connective tissue extracts. *J. Am. chem. Soc.*, **72**, 3321–2.

Hills, G. M. (1940). The metabolism of articular cartilage. *Biochem. J.*, **34**, 1070–7.

Hintzche, E. (1928). Untersuchungen an Stützgeweben. II. Über Knochenbildungsfaktoren, insbesondere über den Anteil der Blutgefasse an der Ossifikation. *Z. mikrosk.-anat. Forsch.*, **14**, 373–440.

Hirano, S. & Meyer, K. (1971). Enzymatic degradation of corneal and cartilaginous keratosulfates. *Biochem. biophys. Res. Commun.*, **44**, 1371–5.

Hirsch, C. (1944). A contribution to the pathogenesis of chondromalacia of the patella. *Acta chir. scand.*, **90**, Suppl. 83, 1–106.

Hirschman, A. & Dziewiatkowski, D. D. (1966). Protein-polysaccharide loss during endochondral ossification: immunochemical evidence. *Science*, **154**, 393–5.

Hjertquist, S.-O. & Lemperg, R. (1972). Identification and concentration of the glycosaminoglycans of human articular cartilage in relation to age and osteoarthritis. *Calcif. Tissue Res.*, **10**, 223–37.

Hodge, A. J. & Petruska, J. A. (1963). Recent studies with the electron microscope on ordered aggregates of the tropocollagen macromolecule. In *Aspects of protein structure*, ed. G. N. Ramachandran, pp. 289–300. London & New York: Academic Press.

Hodge, A. J. & Schmitt, F. O. (1960). The charge profile of the tropocollagen macromolecule and the packing arrangement in native-type collagen fibrils. *Proc. natn. Acad. Sci. USA*, **46**, 186–97.

Hodge, J. A. & McKibbin, B. (1969). The nutrition of matrix and immature cartilage in rabbits. *J. Bone Jt Surg.*, **51*B***, 140–7.

Hohling, H. J., Steffens, H., Stamm, G. & Mays, U. (1976). Transmission microscopy of freeze dried, unstained epiphyseal cartilage of the guinea pig. *Cell Tissue Res.*, **167**, 243–63.

Holmdahl, D. E. & Ingelmark, B. E. (1950). The contact between the articular cartilage and the medullary cavities of the bone. *Acta orthop. scand.*, **20**, 156–65.

Holtfreter, J. (1968). Mesenchyme and epithelium in induction and morphogenetic processes. In *Epithelial–mesenchymal interactions*, ed. R. Fleischmajer & R. E. Billingham, pp. 1–30. Baltimore: Williams & Wilkins.

Holtrop, M. E. (1972). The ultrastructure of the epiphysial plate. II. The hypertrophic chondrocyte. *Calcif. Tissue Res.*, **9**, 140–51.

Holtzer, H., Chacko, S., Abbott, J., Holtzer, S. & Anderson, H. (1970). Variable behaviour of chondrocytes *in vitro*. In *Chemistry and molecular biology of the intercellular matrix*, **3**, ed. E. A. Balazs, pp. 1471–84. London & New York: Academic Press.

Holtzer, H. & Detwiler, S. R. (1953). An experimental analysis of the development of the spinal column. III. Induction of skeletogenous cells. *J. exp. Zool.*, **123**, 335–69.

Holtzer, H. & Matheson, D. W. (1970). Induction of chondrogenesis in the embryo. In *Chemistry and molecular biology of the intercellular matrix*, **3**, ed. E. A. Balazs, pp. 1753–69. London & New York: Academic Press.

Holtzer, H. Weintraub, H., Mayne, R. & Mochan, B. (1972). The cell cycle, cell lineages and cell differentiation. *Curr. Top. Devl Biol.*, **7**, 229–56.

Honner, R. & Thompson, R. C. (1971). The nutritional pathways of articular cartilage. *J. Bone Jt Surg.*, **53*A***, 742–8.

Hopfinger, A. J. & Walton, A. G. (1970). Theoretical structure of the polar regions of the tropocollagen molecule. *Biopolymers*, **9**, 433–44.

Hopwood, J. J. & Robinson, H. C. (1974*a*). The alkali-labile linkage between keratansulphate and protein. *Biochem. J.*, **141**, 57–69.

Hopwood, J. J. & Robinson, H. C. (1974*b*). The structure and composition of cartilage keratansulphate. *Biochem. J.*, **141**, 517–26.

Horstadius, S. (1950). *The neural crest*. London: Oxford University Press.

Horwitz, A. L. & Dorfman, A. (1968). Subcellular sites for synthesis of chondromucoprotein of cartilage. *J. Cell Biol.*, **38**, 358–68.

Hoseman, R., Dreissig, W. & Nemetschek, Th. (1974). Schachtelhalm-structure of the octafibrils in collagen. *J. molec. Biol.*, **83**, 275–80.

Howell, D. S., Muniz, O., Pita, J. C. & Enis, J. E. (1976). Pyrophosphate release by osteoarthritis cartilage incubates. *Arthritis Rheum.*, **19** (Suppl.), 488–94.

Hoyle, F. & Wickramasinghe, N. C. (1977). Polysaccharides and infra-red spectra of galactic sources. *Nature, Lond.*, **268**, 610–12.

Huang, D. (1977). Extracellular matrix–cell interactions and chondrogenesis. *Clin. Orthop.*, **123**, 169–76.

Hulmes, D. J. S., Miller, A., Parry, D. A. D., Piez, K. A. & Woodhead-Galloway, J. (1973). Analysis of the primary structure of collagen for the origins of molecular packing. *J. molec. Biol.*, **79**, 137–48.

Hulth, A., Lindberg, L. & Telhag, H. (1972). Mitosis in human osteoarthritic cartilage. *Clin. Orthop.*, **84**, 197–9.

Hunt, S. & Oakes, K. (1977). Intracelluar cationic counterion composition of an acid mucopolysaccharide. *Nature, Lond.*, **268**, 370–2.

Hunter, W. (1743). Of the structure and diseases of articulating cartilages. *Phil. Trans. R. Soc.*, **42**, 514–21.

Hurrell, D. J. (1934). The vascularization of cartilage. *J. Anat.*, **69**, 47–61.

Ingelmark, B. E. & Ekholm, R. (1948). A study on variations in the thickness of articular cartilage in association with rest and periodic load. *Uppsala LäkFor. Förh.*, **53**, 61–74.

Ingelmark, B. E. & Saaf, J. (1948). Über die Ernahrung des Gelenknorpels und die Bildung der Gelenkflussigheit unter verscheidenen funktionellen Verhaltnissen. *Acta orthop. scand.*, **17**, 303–57.

Inman, V. T. & Saunders, J. B. de C. M. (1957). Anatomico-physiological aspects of injury to the intervertebral disc. *J. Bone Jt Surg.*, **29**, 461–8.

Irving, J. T. (1959). Histochemical changes in the epiphyseal cartilage during rickets. *Nature, Lond.*, **183**, 1734–5.

Irving, M. H. (1964). The blood supply of the growth cartilage in young rats. *J. Anat.*, **98**, 631–9.

Ishido, B. (1923). Gelenkuntersuchungen. *Virchows Arch. path. Anat. Physiol.*, **244**, 424–8.

Ishikawa, H., Bischoff, R. & Holtzer, H. (1969). Formation of arrowhead complexes with heavy meromyosin in a variety of cell types. *J. Cell Biol.*, **43**, 312–28.

Jacob, F. & Monod, J. (1963). Genetic regulatory mechanisms in the synthesis of proteins. *J. molec. Biol.*, **3**, 318–56.

Jacobs, M., Bennett, P. M. & Dickens, M. J. (1975). Duplex microtubule is a new form of tubulin assembly induced by polycations. *Nature, Lond.*, **257**, 707–9.

Jacobson, W. & Fell, H. B. (1941). Developmental mechanics and potencies of undifferentiated mesenchyme. *Q. Jl microsc. Sci.*, **82**, 563–86.

Janoff, A. (1970). Mediators of tissue damage in leucocyte lysosomes. X. Further studies on human granulocyte elastase. *Lab. Invest.*, **22**, 228–36.

Jasin, H. E., Cooke, T. D., Hurd, E. R., Smiley, J. D. & Ziff, M. (1973). Immunologic models used for the study of rheumatoid arthritis. *Fedn Proc. Fedn Am. Socs exp. Biol.*, **32**, 147–52.

Jeffrey, A. K. (1975). Osteophytes and the osteoarthrotic femoral head. *J. Bone Jt Surg.*, **57*B***, 314–24.

Jibril, A. O. (1967). Proteolytic degradation of ossifying cartilage matrix and the removal of acid muco-polysaccharides prior to bone formation. *Biochim. biophys. Acta.* **136**, 162–5.

Johnson, A. H. & Baker, J. R. (1973). Amino acid sequences of peptides from the linkage region of chondroitin sulphate. *Biochem. Soc. Trans.*, **1**, 277–9.

Johnston, P. M. & Comer, C. L. (1957). Autoradiographic studies of the utilization of S^{35}-sulphate by the chick embryo. *J. biochem. biophys. Cytol.*, **3**, 231–8.

Johnston, R. A. (1972). Relation of growth cartilage thickness to bone growth in the rat tibia. *J. Anat.*, **111**, 339–40.

Jones, K. L. & Addison, J. (1975). Pituitary fibroblast growth factor as a stimulator of growth in cultured rabbit articular chondrocytes. *Endocrinology*, **97**, 359–65.

Joseph, J. & Dyson, M. (1966). Tissue replacement in the rabbit's ear. *Br. J. Surg.*, **53**, 372–80.

Joseph, J., Thomas, G. A. & Tynen, J. (1961). The reaction of the ear cartilage of the rabbit and guinea-pig to trauma. *J. Anat.*, **95**, 564–8.

Jurand, A. (1965). Ultrastructural aspects of early development of the fore-limb buds in the chick and the mouse. *Proc. R. Soc., B*, **162**, 387–405.

Juva, K., Prockop, D. J., Cooper, G. W. & Lash, J. W. (1966). Hydroxylation of proline and the intracellular accumulation of a polypeptide precursor of collagen. *Science*, **152**, 92–4.

Kadar, A. & Robert, B. (1975). Electron microscopical studies on the role of microfibrils in elastogenesis. In *Protides of biological fluids*, **22**, ed. H. Peeters, pp. 149–55. Oxford: Pergamon.

Kalayjian, D. B. & Cooper, R. R. (1972). Osteogenesis of the epiphysis. *Clin. Orthop.*, **85**, 242–56.

Kaminski, M., Kaminski, G., Jakobisiak, M. & Brzezinski, W. (1977). Inhibition of lymphocyte-induced angiogenesis by isolated chondrocytes. *Nature, Lond.*, **268**, 238–40.

Kantor, T. G. & Schubert, M. (1957). Difference in permeability of cartilage to cationic and anionic dyes. *J. Histochem. Cytochem.*, **5**, 28–32.

Kaplan, D. & Fisher, B. (1964). The effect of methyl prednisolone on mucopolysaccharides of rabbit vitreous humour and costal cartilage. *Biochim. biophys. Acta*, **83**, 102–12.

Kaplan, D. & Meyer, K. (1959). Ageing of human cartilage. *Nature, Lond.*, **183**, 1267–8.

Kaplan, N. O., Ciotti, M. M. & Stolzenbach, F. E. (1956). Reaction of pyridine nucleotide analogues with dehydrogenases. *J. biol. Chem.*, **221**, 833–44.

Kawiak, J., Moskalewski, S. & Darzynkiewicz, Z. (1965). Isolation of chondrocytes from calf cartilage. *Expl Cell Res.*, **39**, 59–68.

Keates, R. A. B. & Hall, R. H. (1975). Tubulin requires an accessory protein for self assembly into microtubules. *Nature, Lond.*, **257**, 418–20.

Keith, A. (1919). *Menders of the maimed.* London: Oxford Medical Pubns.

Keith, A. (1920). Studies on the anatomical changes which accompany certain growth disorders of the human body. *J. Anat.*, **54**, 101–15.

Kelley, R. O. & Fallon, J. F. (1976). Ultrastructural analysis of the apical ectodermal ridge during vertebrate limb morphogenesis. 1. The human forelimb with special reference to gap junctions. *Devl Biol.*, **51**, 241–56.

Kelley, R. O. & Palmer, G. C. (1976). Regulation of mesenchymal cell growth during human limb morphogenesis through glycosaminoglycan-adenylate cyclase interaction at the cell surface. In *Embryogenesis in mammals*, ed. K. Elliott & M. O'Connor, pp. 275–90. Amsterdam: Elsevier. CIBA Foundation Symposium 40 (N.S.).

Kember, N. F. (1960). Cell division in endochondral ossification. *J. Bone Jt Surg.*, ***42B***, 824–39.

Kember, N. F. (1971). Cell population kinetics of bone growth: the first ten years of autoradiographic studies with tritiated thymidine. *Clin. Orthop.*, **76**, 213–30.

Kember, N. F. (1972). Comparative patterns of cell division in epiphysial cartilage plates in the rat. *J. Anat.*, **111**, 137–42.

Kember, N. F. (1973). Aspects of the maturation process in growth cartilage in the rat tibia. *Clin. Orthop.*, **95**, 288–94.

Kember, N. F. & Sissons, H. A. (1976). Quantitative histology of the human growth plate. *J. Bone Jt Surg.*, ***58B***, 426–35.

Kempson, G. E. (1973). Mechanical properties of articular cartilage. In *Adult articular cartilage*, ed. M. A. R. Freeman, pp. 171–227. London: Pitman Medical.

Kempson, G. E. (1975). Mechanical properties of articular cartilage and their relationship to matrix degradation and age. *Ann. rheum. Dis.*, **34**, Suppl. 2, 111–13.

Kempson, G. E., Tuke, M. A., Dingle, J. T., Barrett, A. J. & Horsfield, P. H. (1976). The effects of proteolytic enzymes on the mechanical properties of adult human articular cartilage. *Biochim. biophys. Acta*, **428**, 741–60.

Kennedy, J. F. (1976). Chemical and biochemical aspects of the glycosaminoglycans and proteoglycans in health and disease. *Adv. clin. Chem.*, **18**, 1–101.

Key, A. (1931). Experimental arthritis: the changes in joints produced by creating defects in the articular cartilage. *J. Bone Jt Surg.*, **13**, 725–39.

Key, J. A. (1932). The synovial membrane of joints and bursae. In *Special cytology*, **2**, 2nd edn, ed. E. V. Cowdry, pp. 1055–76. New York: Hoeber.

Kincaid, S. A., Van Sickle, D. C. & Wilsman, N. J. (1972). Histochemical evidence of a functional heterogeneity of the chondrocytes of adult canine articular cartilage. *Histochem. J.*, **4**, 237–43.

King, D. (1936). The healing of semi-lunar cartilages. *J. Bone Jt Surg.*, **18**, 333–42.

Kistler, G. D. (1936). Effects of circulatory disturbances on the structure and healing of bone. *Arch Surg.*, **33**, 225–47.

Kleinsmith, L. J. (1975). Phosphorylation of non-histone proteins in the regulation of chromosome structure and function. *J. cell. Physiol.*, **85**, 459–75.

Kornfeld, S., Kornfeld, R., Neufeld, E. & O'Brien, P. J. (1964). The feedback control of sugar nucleotide biosynthesis in liver. *Proc. natn. Acad. Sci. USA*, **52**, 371–9.

van der Korst, J. K., Sokoloff, L. & Miller, E. J. (1968). Senescent pigmentation of cartilage and degenerative joint disease. *Archs Path.*, **86**, 40–7.

Kosher, R. A. (1976). Inhibition of spontaneous, notochord-induced and collagen-induced *in vitro* somite chondrogenesis by cyclic AMP derivatives and theophylline. *Devl Biol.*, **53**, 265–76.

Kosher, R. A. & Church, R. L. (1975). Stimulation of *in vitro* somite chondrogenesis by procollagen and collagen. *Nature, Lond.*, **258**, 327–30.

Kosher, R. A. & Lash, J. W. (1975). Notochordal stimulation of *in vitro* somite chondrogenesis before and after enzymatic removal of perinotochordal materials. *Devl Biol.*, **42**, 362–78.

Kosher, R. A., Lash, J. W. & Minor, R. R. (1973). Environmental enhancement of *in vitro* chondrogenesis. IV. Stimulation of somite chondrogenesis by exogenous chondromucoprotein. *Devl Biol.*, **35**, 210–20.

Kostenszky, K. S. & Olah, E. H. (1972). Effect of increased functional demand on the glucosaminoglycan (mucopolysaccharide) content of the articular cartilage. *Acta biol. hung.*, **23**, 75–82.

Kostovic-Knezevic, Lj. & Svalger, A. (1975). Cytoplasmic filaments associated with lipid droplets in chondrocytes of the rat auricular cartilage. *Experientia*, **31**, 581–2.

Krane, S. M. (1975). Degradation of collagen in connective tissue diseases. Rheumatoid arthritis. In *Dynamics of connective tissue macromolecules*, ed. P. M. C. Burleigh & A. R Poole, pp. 309–26. Amsterdam: North-Holland.

Krishnan, A. & Hsu, D. (1969). Observations on the association of helical polyribosomes and filaments with vincristine-induced crystals in Earle's L-cell fibroblasts. *J. Cell Biol.*, **43**, 553–63.

Krishnan, A. & Hsu, D. (1971). Binding of colchicine-^{3}H to vinblastine- and vincristine-induced crystals in mammalian tissue culture cells. *J. Cell Biol.*, **48**, 407–10.

Krukenberg, C. F. W. (1884). Die chemischen Bestandtheile des Knorpels. *Z. Biol.*, **20**, 307–26.

Kuettner, K. E., Hiti, J., Eisenstein, R. & Harper, E. (1976). Collagenase inhibition by cationic proteins derived from cartilage and aorta. *Biochem. biophys. Res. Commun.*, **72**, 40–6.

Kuhlman, R. E. (1960). A microchemical study of the developing epiphyseal plate. *J. Bone Jt Surg.*, **42A**, 457–66.

Kuhn, K., Grassmann, W. & Hofmann, U. (1960). Über den Aufbau der Kollagenfibrille ans Tropokollagenmolekeln. *Naturwissenschaften*, **47**, 258–9.

Kuhn, R. & Leppelmann, H. J. (1958). Galaktosamin und Glucosamin im Knorpel in Abhangigkeit vom Lebensalter. *Justus Liebigs Annln Chem.*, **611**, 254–8.

Kurki, P., Linder, E., Virtanen, I. & Stenman, S. (1977). Human smooth muscle auto-antibodies reacting with intermediate (100 Å) filaments. *Nature, Lond.*, **268**, 240–1.

Kvinnsland, S. & Kvinnsland, S. (1975). Growth in cranio-facial cartilages studied by ^{3}H-thymidine incorporation. *Growth*, **39**, 305–14.

Kvist, T. N. & Finnegan, C. V. (1970). The distribution of glycosaminoglycan in the axial region of the developing chick embryo. II. Biochemical analysis. *J. exp. Zool.*, **175**, 241–58.

Lacroix, P. (1951). *The organization of bones*, translated by S. Gilder. London: J. & A. Churchill.

Landells, J. W. (1957). The reaction of injured human articular cartilage. *J. Bone Jt Surg.* **39*B***, 548–62.

Lane, J. M., Brighton, C. T., Menkowitz, B., Cochran, W. & Robinson, M. A. (1976). Aerobic vs. anaerobic metabolism of articular cartilage. American Orthopedics Research Society, 22nd meeting, p. 99. New Orleans. (Abstract.)

Lane, L. B., Villacin, A. & Bullough, P. G. (1977). The vascularity and remodelling of subchondral bone and calcified cartilage in adult human femoral and humeral heads. *J. Bone Jt Surg.*, **59*B***, 272–8.

Langer, R., Brem, H., Falterman, K., Klein, M. & Folkman, J. (1976). Isolation of a cartilage factor that inhibits tumour neo-vascularization. *Science*, **193**, 70–2.

Lapière, Ch. M., Lenaers, A. & Pierard, G. (1973). Procollagen and procollagen peptidase in skin as a functional system. In *Biology of the fibroblast*, ed. E. Kulonen & J. Pikkarainen, pp. 379–84. New York: Academic Press.

Lapière, Ch. M., Nusgens, B., Pierard, G. & Hermanns, J. F. (1975). The involvement of procollagen in spatially arranged fibrogenesis. In *Dynamics of connective tissue macromolecules*, ed. P. M. C. Burleigh & A. R. Poole, pp. 33–50. Amsterdam: North-Holland.

Larsson, S.-E. (1976). The metabolic heterogeneity of glycosaminoglycans of the different zones of the epiphysial growth plate and the effect of ethane-1-hydroxy-1,1,-diphosphonate (EHDP) upon glycosaminoglycan synthesis *in vivo. Calcif. Tissue Res.*, **21**, 67–82.

Larsson, S.-E. & Lemperg, R. K. (1975). Concentration, distribution and metabolism of glycosaminoglycans in different layers of normal human articular cartilage. *Ann. rheum. Dis.*, **34**, Suppl. 2, 42–3.

Larsson, S.-E., Ray, R. D. & Kuettner, K. E. (1973). Microchemical studies on acid glycosaminoglycans of the epiphyseal zones during endochondral calcification. *Calcif. Tissue Res.*, **13**, 271–85.

Lash, J. W. (1968). Chondrogenesis: genotypic and phenotypic expression. *J. Cell. Physiol.*, **72**, Suppl. 1, 35–46.

Lash, J. W., Holtzer, S. & Holtzer, H. (1957). An experimental analysis of the development of the spinal column. VI. Aspects of cartilage induction. *Expl Cell Res.*, **13**, 292–303.

Laskin, D. M., Sarnat, B. G. & Bain, J. A. (1952). Respiration and anaerobic glycolysis of transplanted cartilage. *Proc. Soc. exp. Biol. Med.*, **79**, 474–6.

Laurent, T. C. (1964). The interaction between polysaccharides and other macromolecules. 9. The exclusion of molecules from hyaluronic acid gels and solutions. *Biochem. J.*, **93**, 106–12.

Laurent, T. C. (1968). The exclusion of macromolecules from polysaccharide media. In *The chemical physiology of mucopolysaccharides*, ed. G. Quintarelli, pp. 153–68. Boston: Little Brown.

Laurent, T. C. (1970). Structure of hyaluronic acid. In *Chemistry and molecular biology of the intercellular Matrix*, **2**, ed. E. A. Balazs, pp. 703–32. London & New York: Academic Press.

Laurent, T. C. & Scott, J. E. (1964). Molecular weight fractionation of polyanions by cetylpyridinium chloride in salt solutions. *Nature, Lond.*, **202**, 661–2.

Lavietes, B. B. (1970). Cellular interaction and chondrogenesis *in vitro*. *Devl Biol.*, **21**, 584–610.

Lawrence, P. A., Crick, F. H. C. & Munro, M. (1972). A gradient of positional information in an insect, *Rhodnius*. *J. Cell Sci.*, **11**, 815–53.

Layman, D. L., McGoodwin, E. B. & Martin, G. R. (1971). The nature of the collagen synthesised by cultured human fibroblasts. *Proc. natn. Acad. Sci. USA*, **68**, 454–8.

Layman, D. L. & Ross, R. (1973). The production and secretion of procollagen peptidase by human fibroblasts in culture. *Archs Biochem. Biophys.*, **147**, 451–6.

Layman, D. L., Sokoloff, L. & Miller, E. J. (1972). Collagen synthesis by articular chondrocytes in monolayer culture. *Expl Cell Res.*, **73**, 107–12.

Layton, L. L. (1951). Effect of cortisone upon chondroitin sulfate synthesis by animal tissues. *Proc. Soc. exp. Biol. Med.*, **76**, 596–8.

Leaback, D. H. (1974). Studies on some glycosidases from the chondrocytes of articular cartilage and from certain other cells and tissues. In *Normal and osteoarthrotic articular cartilage*, ed. S. Y. Ali, M. W. Elves & D. H. Leaback, pp. 73–83. London: Institute of Orthopaedics.

Lee-Own, V. & Anderson, J. C. (1975). The isolation of collagen-associated proteoglycan from bovine nasal cartilage and its preferential interaction with α_2 chains of Type I collagen. *Biochem. J.*, **149**, 57–63.

Lee-Own, V. & Anderson, J. C. (1976). Interaction between proteoglycan subunits and Type II collagen from bovine nasal cartilage and the preferential binding of proteoglycan subunit to Type I collagen. *Biochem. J.*, **153**, 259–64.

Leeson, T. S. & Leeson, C. R. (1958). Observations on the histochemistry and fine structure of the notochord in rabbit embryos. *J. Anat.*, **92**, 278–85.

Lehninger, A. L. (1970). Mitochondria and calcium ion transport. *Biochem. J.*, **119**, 129–38.

Leidy, J. (1849). On the intimate structure and history of the articular cartilages. *Am. J. med. Sci.* (N.S.), **17**, 277–94.

Leloir, L. F. (1951). The enzymatic transformation of uridine diphosphate glucose into a galactose derivative. *Archs Biochem. Biophys.*, **33**, 186–90.

Lemperg, R. (1971*a*). The subchondral bone plate of the femoral head in adult rabbits. I. Spontaneous remodelling studied by microradiography and tetracycline labelling. *Virchows Arch. Abt. A. Path. Anat.*, **352**, 1–13.

Lemperg, R. (1971*b*). The subchondral bone plate of the femoral head in adult rabbits. II. Changes induced by intracartilaginous defects studied by microradiography and tetracycline labelling. *Virchows Arch. Abt. A. Path. Anat.*, **352**, 14–25.

Lemperg, R., Larsson, S.-E. & Hjertquist, S.-O. (1974). The glycosaminoglycans of bovine articular cartilage. I. Concentration and distribution in different layers in relation to age. *Calcif. Tissue Res.*, **15**, 237–51.

Lenaers, A., Ansay, M., Nusgens, B. V. & Lapière, Ch. M. (1971). Collagen made of extended α-chains, procollagen, in genetically-defective dermatosparaxic calves. *Eur. J. Biochem.* **23**, 533–43.

Lenzi, L., Berlanda, P., Flora, A., Aureli, G., Rizzotti, M., Balduini, G. & Boni, M. (1974). Vitamin A induced osteoarthritis in rabbits: an experimental model for the study of the human disease. In *Normal and osteoarthrotic articular cartilage*, ed. S. Y. Ali, M. W. Elves & D. H. Leaback, pp. 243–57. London: Institute of Orthopaedics.

Levene, C. (1964). The patterns of cartilage canals. *J. Anat.*, **98**, 515–38.

Levene, C. I. & Bates, C. J. (1973). Ascorbic acid and collagen. In *Biology of the fibroblast*, ed. E. Kulonen & J. Pikkarainen, pp. 396–410. London & New York: Academic Press.

Levene, C. I. & Murray, J. C. (1977). The aetiological role of maternal vitamin B6 deficiency in the development of atherosclerosis. *Lancet*, **1**, 628–30.

Levene, P. A. & La Forge, F. B. (1914). On chondroitin sulphuric acid. *J. biol. Chem.*, **18**, 123–30.

Levenson, G. E. (1969). The effect of ascorbic acid on monolayer cultures of three types of chondrocytes. *Expl Cell Res.*, **55**, 225–8.

Levitt, D. & Dorfman, A. (1973). The differentiation of cartilage. In *Biology of the fibroblast*, ed. E. Kulonen & J. Pikkarainen, pp. 79–91. London & New York: Academic Press.

Levitt, D. & Dorfman, A. (1974). Concepts and mechanisms of cartilage differentiation. *Curr. Top. Devl Biol.*, **8**, 103–50.

Levitt, D., Ho, P. L. & Dorfman, A. (1975). Effect of 5-bromodeoxyuridine on the ultrastructure of developing limb bud cells *in vitro*. *Devl Biol.*, **43**, 75–90.

Libby, W. F., Berger, R., Mead, J. F., Alexander, G. V. & Ross, J. F. (1964). Replacement rates for human tissue from atmospheric radiocarbon. *Science*, **146**, 1170–2.

Lin, S.-Y., Lin, D. & Riggs, A. D. (1976). Histones bind more tightly to bromodeoxyuridine-substituted DNA than to normal DNA. *Nucleic Acids Res.*, **3**, 2183–91.

Lin, S.-Y. & Riggs, A. D. (1972). *Lac*-operator analogues: bromodeoxyuridine substitution in the *lac*-operator affects the rate of dissociation of the *lac*-repressor. *Proc. natn. Acad. Sci. USA*, **69**, 2574–6.

Lindahl, O. (1948). Über den Wassergehalt des Knorpels. *Acta orthop. scand.*, **17**, 134–6.

Linn, F. C. & Sokoloff, L. (1965). Movement and composition of interstitial fluid of cartilage. *Arthritis Rheum.*, **8**, 481–94.

Linsenmayer, T. F., Toole, B. P. & Trelstad, R. L. (1973*a*). Temporal and spatial transitions in collagen types during embryonic chick limb development. *Devl Biol.*, **35**, 232–9.

Linsenmayer, T. F., Trelstad, R. L. & Gross, J. (1973*b*). The collagen of chick embryonic notochord. *Biochim. biophys. Res. Commun.*, **53**, 39–45.

Lippmann, M. (1968). Glycosaminoglycans and cell division. In *Epithelial–mesenchymal interactions*, ed. R. Fleischmajer & R. E. Billingham, pp. 208–29, Baltimore: Williams & Wilkins.

Little, K., Pimm, L. H. & Trueta, J. (1958). Osteoarthritis of the hip: an electron microscope study. *J. Bone Jt Surg.*, **40*B***, 123–31.

Lloyd, J. B. & Beck, F. (1974). Lysosomes. In *The cell in medical science*, **1**, ed. F. Beck & J. B. Lloyd, pp. 273–313. London: Academic Press.

Loeb, L. (1926). Auto-transplantation and homo-transplantation of cartilage in the guinea-pig. *Am. J. Path.*, **2**, 111–22.

Loewi, G. (1953). Changes in the ground substance of ageing cartilage. *J. Path. Bact.*, **65**, 381–8.

Loewi, G. (1965). Localization of chondromucoprotein in cartilage. *Ann. rheum. Dis.*, **24**, 528–35.

Lohmander, S., Antonopoulos, C. A. & Friberg, U. (1973). Chemical and metabolic heterogeneity of chondroitin sulfate and keratansulfate in guinea pig cartilage and nucleus pulposus. *Biochim. biophys. Acta*, **304**, 430–48.

Lohmander, S. & Hjerpe, A. (1975). Proteoglycans of mineralising rib and epiphyseal cartilage. *Biochim. biophys. Acta*, **404**, 93–109.

Long, R., Greulich, R. C. & Sarnat, B. G. (1968). Regional variations in chondrocyte proliferation in the cartilaginous nasal septum of the growing rabbit. *J. dent. Res.*, **47**, 505.

Longmore, R. B. & Gardner, D. L. (1975). Development with age of human articular cartilage surface structure. A survey by interference microscopy of the lateral femoral condyle. *Ann. rheum. Dis.*, **34**, 26–37.

Lowther, D. A. & Caterson, B. (1974). Stress induced changes in the biosynthesis of proteoglycans in articular cartilage. Paper read to a Symposium on Articular Cartilage, Imperial College of Science, London, 2–6 September 1974.

Lowther, D. A. & Natargan, M. (1972). The influence of glycoproteins on collagen fibril formation in the presence of chondroitin sulphate proteoglycan. *Biochem. J.*, **127**, 607-8.

Lowther, D. A., Toole, B. P. & Hetherington, A. C. (1970). Interactions of proteoglycans with tropocollagen. In *Chemistry and molecular biology of the intercellular matrix*, **2**, ed. E. A. Balazs, pp. 1135–53. London & New York: Academic Press.

Lucy, J. A. & Dingle, J. T. (1964). Fat soluble vitamins and biological membranes. *Nature, Lond.* **204**, 156–60.

Lund-Oleson, K. (1970). Oxygen tensions in synovial fluids. *Arthritis Rheum.*, **13**, 769–76.

Luschka, H. von (1858). *Die Halbegelenke des Menschlichen Korpers*. Berlin: Reimer.

Luscombe, M. & Phelps, C. F. (1967). The composition and physico-chemical properties of bovine nasal-septa protein–polysaccharide complex. *Biochem. J.*, **102**, 110–19.

Lutfi, A. M. (1970*a*). Mode of growth, fate and functions of the cartilage canals. *J. Anat.*, **106**, 135–45.

Lutfi, A. M. (1970*b*). Study of cell multiplication in the cartilaginous upper end of the tibia of the domestic fowl by tritiated thymidine autoradiography. *Acta anat.*, **76**, 454–63.

Lutfi, A. M. (1974*a*). Fate of auricular transplants and repair of auricular gaps in the grivet monkey (*Cercopithecus aethiops aethiops*). *J. Anat.*, **117**, 81–8.

Lutfi, A. M. (1974*b*). The role of cartilage in long bone growth: a reappraisal. *J. Anat.*, **117**, 413–17.

Lutwak-Mann, C. (1940). Enzymes in articular cartilage. *Biochem. J.* **34**, 517–27.

Luxembourger, M. M., Malkani, K. & Rebel, A. (1974). Etude au microscope électronique de la region de transition entre le périchondre et la cartilage de la plaque épiphysaire chez le fetus de cobaye. *Archs Anat. microsc.*, **63**, 117–32.

MacCabe, J. A. & Parker, B. W. (1976). Evidence for a gradient of a morphogenetic substance in the developing limb. *Devl Biol.*, **54**, 297–303.

McCall, J. G. (1968). Scanning electron microscopy of articular surfaces. *Lancet*, **2**, 1194.

McCall, J. G. (1969). Load-deformation response of the micro-structure of articular cartilage. In *Lubrication and wear in joints*, ed. V. Wright, pp. 39–48. London: Sector.

MacConaill, M. A. (1951). The movements of bones and joints. 4: The mechanical structure of articulating cartilage. *J. Bone Jt Surg.*, **33*B***, 251–7.

McCarty, D. J. (1976). Calcium pyrophosphate dihydrate crystal deposition disease – 1975. *Arthritis Rheum.*, **19** (Suppl.), 275–85.

McClain, P. E. & Wiley, E. R. (1972). Differential scanning calorimeter studies of the thermal transitions of collagen. *J. biol. Chem.*, **247**, 692–7.

MacLean, N. & Hilder, V. A. (1977). Mechanisms of chromatin activation and repression. *Int. Rev. Cytol.*, **48**, 1–54.

McConaghey, P. (1972). The production of sulphation factor by rat liver. *J. Endocr.*, **52**, 1–9.

McCutchen, C. W. (1962). Frictional properties of animal joints. *Wear*, **5**, 1–17.

McCutchen, C. W. (1965). A note upon tensile stresses in the collagen fibers of articular cartilage. *Med. Electron. Biol. Engng*, **3**, 447–8.

McDevitt, C. A., Billingham, M. E. J. & Muir, H. (1978). A defect in proteoglycan aggregates in early experimental osteoarthrosis. *Ann. rheum. Dis.*, **37**, 484.

McDevitt, C. A., Gilbertson, E. & Muir, H. (1977). An experimental model of osteoarthritis; early morphological and biochemical changes. *J. Bone Jt Surg.*, **59*B***, 24–35.

McDevitt, C. A. & Muir, H. (1976). Biochemical changes in the cartilage of the knee in experimental and natural osteoarthritis in the dog. *J. Bone Jt Surg.*, **58*B***, 94–101.

McElligott, T. F. & Collins, D. H. (1960). Chondrocyte function of human articular and costal cartilage compared by measuring the *in vitro* uptake of labelled ^{35}S sulphate. *Ann. rheum. Dis.*, **19**, 31–41.

McElligott, T. F. & Potter, J. L. (1960). Increased fixation of sulphur^{35} by cartilage *in vitro* following depletion of the matrix by intravenous papain. *J. exp. Med.* **112**, 743–50.

McKenzie, L. S., Horsburgh, B. A., Ghosh, P. & Taylor, T. K. F. (1976). Effect of anti-inflammatory drugs on sulphated glycosaminoglycan synthesis in aged human articular cartilage. *Ann. rheum. Dis.*, **35**, 487–97.

McKibbin, B. (1973). Nutrition. In *Adult articular cartilage*, ed. M. A. R. Freeman, pp. 277–86. London: Pitman Medical.

McKibbin, B. & Holdsworth, F. W. (1966). The nutrition of immature joint cartilage in the lamb. *J. Bone Jt Surg.*, **48*B***, 793–803.

McKibbin, B. & Holdsworth, F. (1968). The source of nutrition of articular cartilage. *J. Bone Jt Surg.*, **50*B***, 876–7.

Majeska, R. J. & Wuthier, R. E. (1975). Studies on matrix vesicles isolated from chick epiphyseal cartilage. *Biochim. biophys. Acta*, **391**, 51–60.

Makowsky, L. (1948). Studien über den Wasserhaushalt des Kniegelenkknorpels. *Helv. chir. Acta*, **15**, 44–59.

Malseed, Z. M. & Heyner, S. (1976). Antigenic profile of the rat chondrocyte. *Arthritis Rheum.*, **19**, 223–31.

Mankin, H. J. (1962*a*). Localization of tritiated thymidine in articular cartilage of rabbits. I. Growth in immature cartilage. *J. Bone Jt Surg.*, **44*A***, 682–8.

Mankin, H. J. (1962*b*). Localization of tritiated thymidine in articular cartilage of rabbits. II. Repair in immature cartilage. *J. Bone Jt Surg.*, **44*A***, 688–98.

Mankin, H. J. (1963*a*). Localization of tritiated thymidine in articular cartilage of rabbits. III. Mature articular cartilage. *J. Bone Jt Surg.*, **45*A***, 529–40.

Mankin, H. J. (1963*b*). Localization of tritiated cytidine in articular cartilage of immature and adult rabbits after intra-articular injection. *Lab. Invest.*, **12**, 543–8.

Mankin, H. J. (1964). Mitoses in articular cartilage of immature rabbits. *Clin. Orthop.*, **34**, 170–83.

Mankin, H. J. (1974). Biochemical abnormalities in articular cartilage in osteoarthritis. In *Normal and osteoarthrotic articular cartilage*, ed. S. Y. Ali, M. W. Elves & D. H. Leaback, pp. 153–72. London: Institute of Orthopaedics.

Mankin, H. J. (1975*a*). Discussion in *Dynamics of connective tissue macromolecules*, ed. P. M. C. Burleigh & A. R. Poole, p. 101. Amsterdam: North-Holland.

Mankin, H. J. (1975*b*). The metabolism of articular cartilage in health and disease. In *Dynamics of connective tissue macromolecules*, ed. P. M. C. Burleigh & A. R. Poole, pp. 327–53. Amsterdam: North-Holland.

Mankin, H. J. & Baron, P. A. (1965). The effect of ageing on protein synthesis in articular cartilage of rabbits. *Lab. Invest.*, **14**, 658–64.

Mankin, H. J. & Lippiello, L. (1969*a*). Nucleic acid and protein synthesis in epiphysial plates of rachitic rats. *J. Bone Jt Surg.*, **51*A***, 862–74.

Mankin, H. J. & Lippiello, L. (1969*b*). The turn-over of adult rabbit articular cartilage. *J. Bone Jt Surg.*, **51*A***, 1591–600.
Mankin, H. J. & Lippiello, L. (1970). Biochemical and metabolic abnormalities in articular cartilage from osteoarthritic human hips. *J. Bone Jt Surg.*, **52*A***, 424–34.
Mankin, H. J., Zarins, A. & Jaffe, W. L. (1972). The effect of systemic corticosteroids on rabbit articular cartilage. *Arthritis Rheum.*, **15**, 593–99.
Marcus, R. E. (1973). The effect of low oxygen concentration on growth, glycolysis and sulphate incorporation by articular chondrocytes in monolayer culture. *Arthritis Rheum.*, **16**, 646–56.
von der Mark, K. & Bornstein, P. (1973). Characterization of the pro-α_1 chain of procollagen. *J. biol. Chem.*, **248**, 2285–9.
von der Mark, K., von der Mark, H. & Gray, S. (1976). Study of differential collagen synthesis during development of the chick embryo by immunofluorescence. II. Localisation of Type I and Type II collagen during long bone development. *Devl Biol.*, **53**, 153–70.
von der Mark, K., Wendt, P., Rexrodt, F. & Kuhn, K. (1970). Direct evidence for a correlation between amino acid sequence and cross striation pattern of collagen. *FEBS Lett.* **11**, 105–8.
Maroudas, A. (1968). Physicochemical properties of cartilage in the light of ion-exchange theory. *Biophys. J.*, **8**, 575–95.
Maroudas, A. (1970). Distribution and diffusion of solutes in articular cartilage. *Biophys. J.*, **10**, 365–79.
Maroudas, A. (1972). X-ray microprobe analysis of articular cartilage. *Conn. Tissue Res.*, **1**, 153–63.
Maroudas, A. (1973). Physico-chemical properties of articular cartilage. In *Adult articular cartilage*, ed. M. A. R. Freeman, pp. 131–70. London: Pitman Medical.
Maroudas, A. (1975). Glycosaminoglycan turn-over in articular cartilage. *Phil. Trans. R. Soc., B*, **271**, 293–313.
Maroudas, A. (1976). Transport of solutes through cartilage: permeability to large molecules. *J. Anat.*, **122**, 335–47.
Maroudas, A., Bullough, P. G., Swanson, S. A. V. & Freeman, M. A. R. (1968). The permeability of articular cartilage. *J. Bone Jt. Surg.*, **50*B***, 166–77.
Maroudas, A. & Evans, H. (1974). Sulphate diffusion and incorporation into human articular cartilage. *Biochim. biophys. Acta*, **338**, 265–79.
Maroudas, A., Evans, H. & Almeida, L. (1973). Cartilage of the hip joint. *Ann. rheum. Dis.*, **32**, 1–9.
Maroudas, A., Muir, H. & Wingham, J. (1969). The correlation of fixed negative charge with glycosaminoglycan content of human articular cartilage. *Biochim. biophys. Acta*, **177**, 492–500.
Maroudas, A., Stockwell, R. A., Nachemson, A. & Urban, J. (1975). Factors involved in the nutrition of the human lumbar intervertebral disc: cellularity and diffusion of glucose *in vitro*. *J. Anat.*, **120**, 113–30.
Maroudas, A. & Thomas, H. (1970). A simple physicochemical micromethod for determining fixed anionic groups in connective tissue. *Biochim. biophys. Acta*, **215**, 214–16.
Maroudas, A. & Venn, M. (1977). Chemical composition and swelling of normal and osteoarthrotic femoral head cartilage. II. Swelling. *Ann. rheum. Dis.*, **36**, 399–406.
Martin, G. R., Byers, P. H. & Smith, B. D. (1973). On the nature of the polypeptide precursors of collagen. In *Biology of the fibroblast*, ed. E. Kulonen & J. Pikkarainen, pp. 339–47. London & New York: Academic Press.
Martin, J. H. & Matthews, J. L. (1969). Mitochondrial granules in chondrocytes. *Calcif. Tissue Res.*, **3**, 184–93.

Martoja, R. & Bassot, J.-M. (1965). Existence d'un tissu conjunctif de type cartilagineux chez certains insectes orthoptères. *C. r. Séanc. Acad. Sci. Biol.*, **261**, 2954–7.

Mason, R. M. (1971). Observations on the glycosaminoglycans of aging bronchial cartilage studied with alcian blue. *Histochem. J.*, **3**, 421–34.

Mason, R. M. & Wusteman, F. S. (1970). The glycosaminoglycans of human tracheobronchial cartilage. *Biochem. J.*, **120**, 777–85.

Matchinsky, F. M. (1971). An improved catalytic assay for DPN. *Meth. Enzym.*, **18*B***, 3–11.

Mathews, M. B. (1958). Isomeric chondroitin sulphates. *Nature, Lond.*, **181**, 421–2.

Mathews, M. B. (1965). The interaction of collagen and acid mucopolysaccharides. A model for connective tissue. *Biochem. J.*, **96**, 710–16.

Mathews, M. B. (1971). Comparative biochemistry of chondroitin sulphate-proteins of cartilage and notochord. *Biochem. J.*, **125**, 37–46.

Mathews, M. B. (1975). Connective tissue. Macromolecular structure and evolution. *Molec. Biol. Biochem. Biophys.*, **19**, 1–318.

Mathews, M. B. & Cifonelli, J. A. (1965). Comparative biochemistry of keratosulfates. *J. biol. Chem.*, **240**, 4140–5.

Mathews, M. B. & Decker, L. (1968). The effect of acid mucopolysaccharides and acid mucopolysaccharide-proteins on fibril formation from collagen solutions. *Biochem. J.*, **109**, 517–26.

Mathews, M. B. & Glagov, S. (1966). Acid mucopolysaccharide patterns in aging human cartilage. *J. clin. Invest.*, **45**, 1103–11.

Mathews, M. B. & Lozaityte, I. (1958). Sodium chondroitin sulphate-protein complexes of cartilage. I. Molecular weight and shape. *Archs Biochem. Biophys.*, **74**, 158–74.

Matsuzawa, T. & Anderson, H. C. (1971). Phosphatases of epiphyseal cartilage studied by electron microscopic cytochemical methods. *J. Histochem. Cytochem.*, **19**, 801–8.

Matthews, B. F. (1953). Composition of articular cartilage in osteoarthritis. Changes in collagen/chondroitin sulphate ratio. *Br. med. J.* **2**, 660–1.

Matthews, J. G. & Goldstein, D. J. (1976). Remodelling in immature cartilage: local matrix dry mass changes. *J. Bone Jt Surg.*, **58*B***, 373–4.

Matthews, J. L., Martin, J. H., Lynn, J. A. & Collins, E. J. (1968). Calcium incorporation in the developing cartilaginous epiphysis. *Calcif. Tissue Res.*, **1**, 330–6.

Matukas, V. J. & Krikos, G. A. (1968). Evidence for changes in protein polysaccharide associated with the onset of calcification in cartilage. *J. Cell Biol.*, **39**, 43–8.

Matukas, V. J., Panner, B. J. & Orbison, J. L. (1967). Studies on ultrastructural identification and distribution of proteinpolysaccharides in cartilage matrix. *J. Cell Biol.*, **32**, 365–77.

Maxwell, E. (1957). The enzymic interconversion of uridine diphosphogalactose and uridine diphosphoglucose. *J. biol. Chem.*, **229**, 139–51.

Mayes, R. W., Mason, R. M. & Griffin, D. C. (1973). The composition of cartilage proteoglycans. An investigation using high- and low-ionic-strength extraction procedures. *Biochem J.*, **131**, 541–53.

Mayne, R., Abbott, J. & Holtzer, H. (1973*a*). Requirement for cell proliferation for the effects of 5-bromo-2′-deoxyuridine on cultures of chick chondrocytes. *Expl Cell Res.*, **77**, 255–63.

Mayne, R., Schiltz, J. R. & Holtzer, H. (1973*b*). Some overt and covert properties of chondrogenic cells. In *Biology of the fibroblast*, ed. E. Kulonen & J. Pikkarainen, pp. 61–78. London & New York: Academic Press.

Mayne, R., Vail, M. S. & Miller, E. J. (1975). Analysis of changes in collagen biosynthesis that occur when chick chondrocytes are grown in 5-bromo-2′-deoxyuridine. *Proc. natn. Acad. Sci. USA*, **72**, 4511–15.

Mayne, R., Vail, M. S., Mayne, P. M. & Miller, E. J. (1976*a*). Changes in type of collagen synthesised as clones of chick chondrocytes grow and eventually lose division capacity. *Proc. natn. Acad. Sci. USA*, **73**, 1674–8.

Mayne, R., Vail, M. S. & Miller, E. J. (1976*b*). The effect of embryo extract on the types of collagen synthesised by cultured chick chondrocytes. *Devl Biol.*, **54**, 230–40.

Mazhuga, P. M. & Cherkasov, V. V. (1974). Adaptive distribution of specific biosyntheses in a homogeneous population of the articular cartilage chondrocytes. *Z. mikrosk.-anat. Forsch.*, **88**, 364–74.

Meachim, G. (1963). The effect of scarification on articular cartilage in the rabbit. *J. Bone Jt Surg.*, **45*B***, 150–61.

Meachim, G. (1967). The histology and ultrastructure of cartilage. In *Cartilage: degradation and repair*, ed. C. A. L. Bassett, p. 3. Washington D.C.: Natn. Acad. Sci. Nat. Res. Council Pubn.

Meachim, G. (1972). Meshwork patterns in the ground substance of articular cartilage and nucleus pulposus. *J. Anat.*, **111**, 219–27.

Meachim, G. (1975). Articular cartilage lesions in the Liverpool population. *Ann. rheum. Dis.*, **34**, Suppl. 2, 122–4.

Meachim, G. & Collins, D. H. (1962). Cell counts of normal and osteoarthritic articular cartilage in relation to the uptake of sulphate ($^{35}SO_4$) *in vitro*. *Ann. rheum. Dis.*, **21**, 45–50.

Meachim, G., Denham, D., Emery, I. H. & Wilkinson, P. H. (1974). Collagen alignments and artificial splits at the surface of human articular cartilage. *J. Anat.*, **118**, 101–18.

Meachim, G. & Emery, I. H. (1973). Cartilage fibrillation in shoulder and hip joints in Liverpool necropsies. *J. Anat.*, **16**, 161–79.

Meachim, G. & Emery, I. H. (1974). Quantitative aspects of patello-femoral cartilage fibrillation in Liverpool necropsies. *Ann. rheum. Dis.*, **33**, 39–47.

Meachim, G., Ghadially, F. N. & Collins, D. H. (1965). Regressive changes in the superficial layer of human articular cartilage. *Ann. rheum. Dis.*, **24**, 23–30.

Meachim, G. & Illman, O. (1967). Articular cartilage degeneration in hamsters and in pigs. *Z. Versuchstierk.*, **9**, 33–46.

Meachim, G. & Osborne, G. V. (1970). Repair at the femoral articular surface in osteo-arthritis of the hip. *J. Path.* **102**, 1–8.

Meachim, G. & Roberts, C. (1971). Repair of the joint surface from subarticular tissue in the rabbit knee. *J. Anat.*, **109**, 317–27.

Meachim, G. & Roy, S. (1967). Intracytoplasmic filaments in the cells of adult human articular cartilage. *Ann. rheum. Dis.*, **26**, 50–8.

Meachim, G. & Roy, S. (1969). Surface ultrastructure of mature adult human articular cartilage. *J. Bone Jt Surg.*, **51*B***, 529–39.

Meachim, G. & Stockwell, R. A. (1973). The matrix. In *Adult articular cartilage*, ed. M. A. R. Freeman, pp. 1–50. London: Pitman Medical.

Mechanic, G. L., Gallop, P. M. & Tanzer, M. L. (1971). The nature of crosslinking in collagens from mineralized tissues. *Biochem. biophys. Res. Commun.*, **45**, 644–53.

Medoff, J. (1967). Enzymatic events during cartilage differentiation in the chick embryonic limb bud. *Devl Biol.*, **16**, 118–43.

Messier, B. & Leblond, C. P. (1960). Cell proliferation and migration as revealed by radioautography after injection of thymidine-H^3 into male rats and mice. *Am. J. Anat.*, **106**, 247–85.

Meyer, K. (1958). Chemical structure of hyaluronic acid. *Fedn Proc. Fedn Am. Socs exp. Biol.*, **17**, 1075–7.

Meyer, K. (1970). Reflections on 'mucopolysaccharides' and their protein complexes. In *Chemistry and molecular biology of the intercellular matrix*, **1**, ed. E. A. Balazs, pp. 5–24. London & New York: Academic Press.

Meyer, K., Hoffman, P. & Linker, A. (1958). Mucopolysaccharides of costal cartilage. *Science*, **128**, 896.

Meyer, K., Linker, A., Davidson, E. A. & Weissmann, B. (1953). The mucopolysaccharides of bovine cornea. *J. biol. Chem.*, **205**, 611–16.

Meyer, K. & Palmer, J. W. (1934). The polysaccharide of the vitreous humor. *J. biol. chem.*, **107**, 629–34.

Meyer, K., Smyth, E. M. & Dawson, M. H. (1939). The isolation of a mucopolysaccharide from synovial fluid. *J. biol. Chem.*, **128**, 319–27.

Miles, J. S. & Eichelberger, L. (1964). Biochemical studies of human cartilage during the aging process. *J. Am. geriat. Soc.*, **12**, 1–20.

Miller, A. & Parry, D. A. D. (1973). Structure and packing of microfibrils in collagen. *J. molec. Biol.*, **75**, 441–7.

Miller, A. & Wray, J. S. (1971). Molecular packing in collagen. *Nature, Lond.*, **230**, 437–9.

Miller, E. J. (1971*a*). Isolation and characterization of a collagen from chick cartilage containing three identical α-chains. *Biochemistry*, **10**, 1652–9.

Miller, E. J. (1971*b*). Isolation and characterization of the cyanogen bromide peptides from the α_1(II) chain of chick cartilage collagen. *Biochemistry*, **10**, 3030–5.

Miller, E. J. (1972). Structural studies on cartilage collagen employing limited cleavage and solubilization with pepsin. *Biochemistry*, **11**, 4903–9.

Miller, E. J., van der Korst, J. K. & Sokoloff, L. (1969). Collagen of human articular and costal cartilage. *Arthritis Rheum.*, **12**, 21–9.

Miller, E. J. & Matukas, V. J. (1969). Chick cartilage collagen: a new type of α1 chain not present in bone or skin of the species. *Proc. natn. Acad. Sci. USA*, **64**, 1264–8.

Miller, E. J. & Matukas, V. J. (1974). Biosynthesis of collagen. The biochemist's view. *Fedn Proc. Fedn Am. Socs exp. Biol.*, **33**, 1197–204.

Miller, E. J., Woodall, D. L. & Vail, M. S. (1973). Biosynthesis of cartilage collagen. *J. biol. Chem.*, **248**, 1666–71.

Millington, P. F. & Clarke, I. C. (1973). Bioengineering orientated studies on the structure of human hip joint cartilage. In *Perspectives in biomedical engineering*, ed. R. M. Kenedi, pp. 147–52. London: Macmillan.

Millroy, S. J. & Poole, A. R. (1974). Pig articular cartilage in organ culture. Effect of enzymatic depletion of the matrix on response of chondrocytes to complement-sufficient antiserum against pig erythrocytes. *Ann. rheum. Dis.*, **33**, 500–8.

Minns, R. J. & Steven, F. S. (1977). The collagen fibril organization in human articular cartilage. *J. Anat.*, **123**, 437–57.

Minor, R. R. (1973). Somite chondrogenesis. A structural analysis. *J. Cell Biol.*, **56**, 27–50.

Mital, M. A. & Millington, P. F. (1970). Osseous pathway of nutrition to articular cartilage of the human femoral head. *Lancet*, **1**, 842.

Mitchell, N. S. & Cruess, R. L. (1977). Classification of degenerative arthritis. *Can. med. Ass. J.*, **117**, 763–5.

Montagna, W. (1949). Glycogen and lipids in human cartilage with some cytochemical observations on the cartilage of the dog, cat and rabbit. *Anat. Rec.*, **103**, 77–92.

Moores, G. R. & Partridge, T. A. (1974). The cell surface. In *The cell in medical science*, **1**, ed. F. Beck & J. B. Lloyd, pp. 75–104. London: Academic Press.

Morrison, R. I. G., Barrett, A. J., Dingle, J. T. & Prior, D. (1973). Cathepsins B1 and D. Action on human cartilage proteoglycans. *Biochim. biophys. Acta*, **302**, 411–19.

Moskalewski, S. & Kawiak, J. (1965). Cartilage formation after transplantation of isolated chondrocytes. *Transplantation*, **3**, 737–47.

Moss, D. W., Eaton, R. H., Smith, J. K. & Whitby, L. G. (1967). Association of inorganic-pyrophosphatase activity with human alkaline-phosphatase preparations. *Biochem. J.*, **102**, 53–7.

Moss-Salentijn, L. (1975). Cartilage canals in the human spheno-occipital synchondrosis during fetal life. *Acta anat.*, **92**, 595–606.

Mottet, N. K. (1967). Activity of aminopeptidases (amino naphthylamidases) during early limb bud differentiation in the chick embryo. *J. exp. Zool.*, **165**, 279–92.

Muir, H. (1958). The nature of the link between protein and carbohydrate of a chondroitin sulphate complex from hyaline cartilage. *Biochem. J.*, **69**, 195–204.

Muir, H. (1973). Biochemistry. In *Adult articular cartilage*, ed. M. A. R. Freeman, pp. 100–30. London: Pitman Medical.

Muir, H. (1977). Molecular approach to the understanding of osteoarthrosis. *Ann. rheum. Dis.*, **36**, 199–208.

Muir, H., Bullough, P. & Maroudas, A. (1970). The distribution of collagen in human articular cartilage with some of its physiological implications. *J. Bone Jt Surg.*, **52*B***, 554–63.

Mulder, G. J. (1838). Zusammensetzung des Chondrins. *Justus Liebigs Annln Chem.*, **28**, 328–30.

Mulholland, R. (1974). Lateral hydraulic permeability and morphology of articular cartilage. In *Normal and osteoarthrotic articular cartilage*, ed. S. Y. Ali, M. W. Elves & D. H. Leaback, pp. 85–101. London: Institute of Orthopaedics.

Muller, H. (1858). Ueber die Entwickelung der Knochensubstanz nebst Bemarkungen über den Bau rachitischer Knochen. *Z. wiss. Zool.*, **9**, 147–233.

Müller, J. (1837). Ueber Knorpel und Knochen. *Justus Liebigs Annln Chem.*, **21**, 277–82.

Muller, W. (1929). *Biologie der Gelenke*, p. 28. Leipzig: Barth.

Murray, P. D. F. & Huxley, J. S. (1925). Self-differentiation in the grafted limb-bud of the chick. *J. Anat.*, **59**, 379–84.

Murray, P. D. F. & Drachman, D. B. (1969). The role of movement in the development of joints and related structures: the head and neck in the chick embryo. *J. Embryol. exp. Morph.*, **22**, 349–71.

Myers, D. B. (1976). Electron microscopic autoradiography of $^{35}SO_4$-labelled material closely associated with collagen fibrils in mammalian synovium and ear cartilage. *Histochem. J.*, **8**, 191–9.

Myers, D. B., Highton, T. C. & Rayns, D. G. (1973). Ruthenium red-positive elements interconnecting collagen filaments. *J. Ultrastruct. Res.*, **42**, 87–92.

Nachemson, A. (1969). Intradiscal measurements of pH in patients with lumbar rhizopathies. *Acta orthop. scand.*, **40**, 23–42.

Nachemson, A., Lewin, T., Maroudas, A. & Freeman, M. A. R. (1970). *In vitro* diffusion of dye through through the end plates and the annulus fibrosus of human lumbar intervertebral discs. *Acta orthop. scand.*, **41**, 589–607.

Nakao, K. & Bashey, R. I. (1972). Fine structure of collagen fibrils as revealed by ruthenium red. *Expl molec. Path.*, **17**, 6–13.

Nameroff, M., Trotter, J. A., Keller, J. M. & Muner, E. (1973). Inhibition of cellular differentiation by phospholipase C. I. Effects of the enzyme on myogenesis and chondrogenesis *in vitro*. *J. Cell Biol.*, **58**, 107–18.

Nesbitt, R. (1736). *Human osteogeny*. London: Wood. Cited by Haines, R. W. (1974).

Neufeld, E. F. & Hall, C. W. (1965). The inhibition of UDP-D-glucose dehydrogenase by UDP-D-xylose. A possible regulatory mechanism. *Biochem. biophys. Res. Commun.*, **19**, 456–61.

Neutra, M. & Leblond, C. P. (1966). Radioautographic comparison of uptake of galactose-H^3 and glucose-H^3 in the Golgi region of various cells secreting glycoproteins or mucopolysaccharides. *J. Cell Biol.*, **30**, 137–50.

Nevo, Z. & Dorfman, A. (1972). Stimulation of chondromucoprotein synthesis in chondrocytes by extracellular chondromucoprotein. *Proc. natn. Acad. Sci. USA*, **69**, 2069–72.

Nevo, Z., Horwitz, A. L. & Dorfman, A. (1972). Synthesis of chondromucoprotein by chondrocytes in suspension culture. *Devl Biol.*, **28**, 219–28.

Newcombe, D. S., Thanassi, N. M. & Ciosek, C. P. (1974). Cartilage cyclic nucleotide phosphodiesterase: inhibition by anti-inflammatory agents. *Life Sci.*, **14**, 505–19.

Newsome, D. A. (1976). *In vitro* stimulation of cartilage in embryonic chick neural crest cells by products of retinal pigmented epithelium. *Devl Biol.*, **49**, 496–507.

Nimni, M. E. & Deshmukh, K. (1973). Differences in collagen metabolism between normal and osteoarthritic human articular cartilage. *Science*, **181**, 751–2.

Nist, C., von der Mark, K., Hay, E. D., Olsen, B. R., Bornstein, P., Ross, R. & Dehra, P. (1975). Location of procollagen in chick corneal and tendon fibroblasts with ferritin-conjugated antibodies. *J. Cell Biol.*, **65**, 75–87.

Noble, J. & Hamblen, D. L. (1975). The pathology of the degenerate meniscus lesion. *J. Bone Jt Surg.*, ***57B***, 180–6.

North, A. C. T., Cowan, P. M. & Randall, J. T. (1954). Structural units in collagen fibrils. *Nature, Lond.*, **174**, 1142–3.

Norton, L. A., Rodan, G. A. & Bourrett, L. A. (1977). Epiphyseal cartilage cAMP changes produced by electrical and mechanical perturbations. *Clin. Orthop.*, **124**, 59–68.

Novikoff, P. M., Novikoff, A. B., Quintana, N. & Hann, J. J. (1971). Golgi apparatus, Gerl and lysosomes of neurones in rat dorsal root ganglia studied by thick section and thin section cytochemistry. *J. Cell Biol.*, **50**, 859–86.

Nowikoff, M. (1908). Beobachtungen über die Vermehrung der Knorpelzellen nebst einigen Bemerkungen über die Struktur der 'hyalinen' Knorpelgrundsubstanz. *Z. wiss. Zool.*, **90**, 205–57.

Nussbaum, A. (1923). Ueber Osteochondritis coxae juvenilis – Calve-Legg-Perthes. *Dt. med. Wschr.*, **49**, 849–50.

Obrink, B. (1973*a*). A study of the interactions between monomeric tropocollagen and glycosaminoglycans. *Eur. J. Biochem.*, **33**, 387–400.

Obrink, B. (1973*b*). The influence of glycosaminoglycans on the formation of fibres from monomeric tropocollagen *in vitro*. *Eur. J. Biochem.*, **34**, 129–37.

Oegama, T. R., Laidlaw, J., Hascall, V. C. & Dziewiatkowski, D. D. (1975). The effect of proteoglycans on the formation of fibrils from collagen solutions. *Archs Biochem. Biophys.*, **170**, 698–709.

Ogston, A. (1875). On articular cartilage. *J. Anat. Physiol.*, **10**, 49–74.

Ogston, A. G. (1958). The spaces in a uniform random suspension of fibres. *Trans. Faraday Soc.*, **54**, 1754–7.

Ogston, A. G. (1970). The biological functions of the glycosaminoglycans. In *Chemistry and molecular biology of the intercellular matrix*, **3**, ed. E. A. Balazs, pp. 1231–40. London & New York: Academic Press.

Ogston, A. G. & Phelps, C. F. (1961). The partition of solutes between buffer solutions and solutions containing hyaluronic acid. *Biochem. J.*, **78**, 827–33.

O'Hare, M. J. (1972*a*). Differentiation of chick embryo somites in chorioallantoic culture. *J. Embryol. exp. Morph.*, **27**, 215–28.

O'Hare, M. J. (1972*b*). Chondrogenesis in chick embryo somites grafted with adjacent and heterologous tissues. *J. Embryol. exp. Morph.*, **27**, 229–34.

O'Hare, M. J. (1972*c*). Aspects of spinal cord indication of chondrogenesis in chick embryo somites. *J. Embryol. exp. Morph.*, **27**, 235–43.

O'Hare, M. J. (1973). A histochemical study of sulphated glycosaminoglycans associated with the somites of the chick embryo. *J. Embryol. exp. Morph.*, **29**, 197–208.

Ohlsson, K. & Delshammar, M. (1975). Interactions between granulocyte elastase and collagenase and the plasma proteinase inhibitors *in vitro* and *in vivo*. In *Dynamics of connective tissue macromolecules*, ed. P. M. C. Burleigh & A. R. Poole, pp. 259–75. Amsterdam: North-Holland.

Ohnsorge, J., Schutt, G. & Holm, R. (1970). Rasterelektronenmikroskopische Untersuchungen des gesunden und des arthrotischen Gelenkknorpels. *Z. Orthop.*, **108**, 268–77.

Okayama, M., Pacifici, M. & Holtzer, H. (1976). Differences among sulfated proteoglycans synthesised in non-chondrogenic cells, presumptive chondroblasts and chondroblasts. *Proc. natn. Acad. Sci., USA*, **73**, 3224–8.

Olah, E. H. & Kostenszky, K. S. (1976). Effect of loading and prednisolone treatment on the glycosaminoglycan content of articular cartilage in dogs. *Scand. J. Rheumatol.*, **5**, 49–52.

Olsen, B. R. (1963). Electron microscope studies on collagen. II. Mechanism of linear polymerization of tropocollagen molecules. *Z. Zellforsch. mikrosk. Anat.*, **59**, 199–213.

Olsen, B. R., Berg, R. A., Kishida, Y. & Prockop, D. J. (1973). Collagen synthesis: localization of prolyl hydroxylase in tendon cells detected with ferritin-labelled antibodies. *Science*, **182**, 825–7.

Olsen, B. R., Berg, R. A., Kishida, Y. & Prockop, D. J. (1975). Further characterization of embryonic tendon fibroblasts and the use of immunoferritin techniques to study collagen synthesis. *J. Cell Biol.*, **64**, 340–55.

Olsen, B. R., Hoffman, H. P. & Prockop, D. J. (1976). Interchain disulfide bonds at the COOH-terminal end of procollagen synthesised by matrix free cells from chick embryonic tendon and cartilage. *Archs Biochem. Biophys.*, **175**, 341–50.

Olson, M. D. & Low, F. N. (1971). The fine structure of developing cartilage in the chick embryo. *Am. J. Anat.*, **131**, 197–216.

Olsson, I. (1972). Subcellular sites for synthesis of glycosaminoglycans (mucopolysaccharides) of rabbit bone marrow cells. *Expl Cell Res.*, **70**, 173–84.

Oryschak, A. F., Ghadially, F. N. & Bhatnagar, R. (1974). Nuclear fibrous lamina in the chondrocytes of articular cartilage. *J. Anat.*, **118**, 511–15.

Owen, M., Jowsey, J. & Vaughan, J. (1955). Investigation of the growth and structure of the tibia of the rabbit by microradiographic and autoradiographic techniques. *J. Bone Jt Surg.*, **37*B***, 324–42.

Palfrey, A. J. (1975). Matrix structure in articular cartilage. *Ann. rheum. Dis.*, **34**, Suppl. 2, 20.

Palfrey, A. J. & Davies, D. V. (1966). The fine structure of chondrocytes. *J. Anat.*, **100**, 213–26.

Palmer, J. F. (1835). Experiments and observations on the growth of bones. In *The works of John Hunter, F.R.S. with notes*, **4**, pp. 315–18. London: Longman, Rees, Orme, Brown, Green & Longman.

Palmoski, M. & Brandt, K. (1976). Hyaluronate-binding by proteoglycans. Comparison of mildly and severely osteoarthritic regions of human femoral cartilage. *Clinica chim. Acta*, **70**, 87–95.

Papahadjopoulos, D. (1968). Surface properties of acidic phospholipids: interaction of monolayers and hydrated liquid crystals with uni- and bi-valent metal ions. *Biochim. biophys. Acta*, **163**, 240–54.

Parsons, F. G. (1905). On pressure epiphyses. *J. Anat. Physiol.*, **39**, 402–12.

Partridge, S. M. (1966). Biosynthesis and nature of elastin structures. *Fedn Proc. Fedn Am. Socs. exp. Biol.*, **25**, 1023–9.

Partridge, S. M. (1973). Elastin biosynthesis and the influence of some extracellular factors on macromolecular organization. In *Biology of the fibroblast*, ed. E. Kulonen & J. Pikkarainen, pp. 13–39. London & New York: Academic Press.

Partridge, S. M., Davis, H. F. & Adair, G. S. (1961). The chemistry of connective tissues. 6. The constitution of the chondroitin sulphate-protein complex in cartilage. *Biochem J.*, **79**, 15–26.

Partridge, S. M., Elsden, D. F. & Thomas, J. (1963). Constitution of the cross-linkages in elastin. *Nature, Lond.*, **197**, 1297–8.
Pastan, I. & Perlman, R. C. (1971). Cyclic AMP in metabolism. *Nature New Biol.*, **229**, 5–7.
Patnaik, B. K. (1967). Effect of age on the oxygen consumption and glucose uptake by the elastic cartilage of rat. *Gerontologia*, **13**, 173–6.
Paul, J. (1972). General theory of chromosome structure and gene activation in eukaryotes. *Nature*, **238**, 444–6.
Pawalek, J. M. (1969). Effects of thyroxine and low oxygen tension on chondrogenic expression in cell culture. *Devl Biol.*, **19**, 52–72.
Payton, C. G. (1932). The growth in length of the long bones in the madder-fed pig. *J. Anat.*, **66**, 414–25.
Payton, C. G. (1934). The position of the nutrient foramen and direction of the nutrient canal in the long bones of the madder-fed pig. *J. Anat.*, **68**, 500–10.
Peacock, A. (1951). Observations on the pre-natal development of the intervertebral disc in man. *J. Anat.*, **85**, 260–74.
Peacock, E. E., Weeks, P. M. & Petty, J. M. (1960). Some studies on the antigenicity of cartilage. *Ann. NY Acad. Sci.*, **87**, 175–83.
Peress, N. S., Anderson, H. C. & Sajdera, S. W. (1974). The lipids of matrix vesicles from bovine fetal epiphyseal cartilage. *Calcif. Tissue Res.*, **14**, 275–81.
Perez-Tamayo, R. (1973). Collagen degradation and resorption: physiology and pathology. In *Molecular pathology of connective tissues*, ed. R. Perez-Tamayo & M. Rojkind, p. 229–322. New York: Dekker.
Perez-Tamayo, R. & Rojkind, M. (1974). *Molecular pathology of connective tissues.* New York: Dekker.
Perlman, R. L., Telser, A. & Dorfman, A. (1964). The biosynthesis of chondroitin sulphate by a cell-free preparation. *J. biol. Chem.*, **239**, 3623–9.
Perricone, E., Palmoski, M. & Brandt, K. (1977). Failure of proteoglycans to form aggregates in morphologically normal aged human hip cartilage. *Arthritis Rheum.*, **20**, 1372–80.
Person, P. & Philpott, D. E. (1969). The nature and significance of invertebrate cartilage. *Biol. Rev.*, **44**, 1–16.
Peterkofsky, B. & Udenfriend, S. (1965). Enzymatic hydroxylation of proline in microsomal polypeptides leading to formation of collagen. *Proc. natn. Acad. Sci. USA*, **53**, 335–42.
Peters, T. J. & Smillie, I. S. (1971). Studies on chemical composition of menisci from the human knee joint. *Proc. R. Soc. Med.*, **64**, 261–2.
Phelps, C. F. (1973). The biosynthesis of glycosaminoglycans. *Biochem. Soc. Trans.*, **1**, 814–19.
Phelps, C. F. & Stevens, R. (1975). Nucleotide sugar metabolism in glycosaminoglycan biosynthesis. *Ann. rheum. Dis.*, **34**, Suppl. 2, 48–51.
Philpott, D. E. & Person, P. (1970). The biology of cartilage. II. Invertebrate cartilages: squid head cartilage. *J. Morph.*, **131**, 417–30.
Piez, K. A. (1967). Soluble collagen and the components resulting from its denaturation. In *Treatise on collagen*, **1**, ed. G. N. Ramachandran, pp. 207–52. London & New York: Academic Press.
Piez, K. A. & Gross, J. (1960). The amino acid composition of some fish collagens: the relation between composition and structure. *J. biol. Chem.*, **235**, 995–8.
Piez, K. A. & Trus, B. L. (1977). Microfibrillar structure and packing of collagen: hydrophobic interactions. *J. molec. Biol.* **110**, 701–4.
Pita, J. C., Muller, F. & Howell, D. S. (1975). Disaggregation of proteoglycan aggregate during endochondral ossification: physiological role of lysozyme. In

Dynamics of connective tissue macromolecules, ed. P. M. C. Burleigh & A. R. Poole, pp. 247–58. Amsterdam: North-Holland.

Podrazky, V., Steven, F. S., Jackson, D. S., Weiss, J. B. & Leibovich, S. J. (1971). Interaction of tropocollagen with proteinpolysaccharide complexes. *Biochim. biophys. Acta*, **229**, 690–7.

Poole, A. R. (1975). Immunocytochemical studies of the secretion of a proteolytic enzyme, cathepsin D, in relation to cartilage breakdown. In *Dynamics of connective tissue macromolecules*, ed. P. M. C. Burleigh & A. R. Poole, pp. 357–79. Amsterdam: North-Holland.

Poole, A. R., Barratt, M. E. J. & Fell, H. B. (1973). The role of soft connective tissue in the breakdown of pig articular cartilage cultivated in the presence of complement-sufficient antiserum to pig erythrocytes. II. Distribution of immunoglobulin G (IgG). *Int. Archs Allergy appl. Immun.*, **44**, 469–88.

Poole, A. R., Hembry, R. M. & Dingle, J. T. (1974). Cathepsin D in cartilage: the immuno-histochemical localisation of extracellular enzyme in normal and pathological conditions. *J. Cell Sci.*, **14**, 139–61.

Prader, A. (1947*a*). Die frühembryonal Entwicklung der menschlichen Zwischenwirbelscheibe. *Acta anat.*, **3**, 68–83.

Prader, A. (1947*b*). Die Entwicklung der Zwischenwirbelscheibe beim menschlichen Keimling. *Acta anat.*, **3**, 115–52.

Pratt, C. W. M. (1957). Observations on osteogenesis in the femur of the foetal rat. *J. Anat.*, **91**, 533–44.

Preston, B. N., Davies, M. & Ogston, A. G. (1965). The composition and physico-chemical properties of hyaluronic acids prepared from ox synovial fluid and from a case of mesothelioma. *Biochem. J.*, **96**, 449–74.

Priest, R. E. & Koplitz, R. M. (1962). Inhibition of synthesis of sulfated mucopolysaccharides by estradiol. *J. exp. Med.*, **116**, 565–74.

Pritchard, J. J. (1952). A cytochemical and histochemical study of bone and cartilage in the rat. *J. Anat.*, **86**, 259–77.

Prives, M. G., Funshein, L. V., Shcherban, E. I. & Shishova, V. G. (1959). Significance of a method of labeled compounds for the investigation of the arterial system of the bone in experiments *in vivo*. *Arkh. Anat. Gistol. Embriol.*, **37**, 56–64.

Prockop, D. J., Dehm, P., Olsen, B. R., Berg, R. A., Grant, M. E., Vitto, J. & Kivirikko, K. I. (1973). Recent studies on the biosynthesis of collagen. In *Biology of the fibroblast*, ed. E. Kulonen & J. Pikkarainen, pp. 311–20. London & New York: Academic Press.

Prockop, D. J., Kaplan, A. & Udenfriend, S. (1962). Oxygen-18 studies on the conversion of proline to hydroxyproline. *Biochem. biophys. Res. Commun.*, **9**, 162–6.

Pugh, D. & Walker, P. G. (1961). Localisation of *N*-acetyl-β-glucosaminidase in tissues. *J. Histochem. Cytochem.*, **9**, 242–50.

Putschar, W. (1931). Über Fett im Knorpel unter normalen und pathologischen Verhaltnissen. *Beitr. path. Anat.*, **87**, 526–39.

Quintarelli, G. & Dellovo, M. C. (1966). Age changes in the localization and distribution of glycosaminoglycans in human hyaline cartilage. *Histochemie*, **7**, 141–67.

Rabe, F. (1910). Experimentalle Untersuchungen über den Gehalt des Knorpels an Fett und Glykogen. *Beitr. path. Anat.*, **48**, 554–75.

Rabinovitch, A. L. (1974). Autoradiographic studies of lipid synthesis and transport in growth plate cartilage of mice. *Fedn Proc. Fedn Am. Socs. exp. Biol.* **33**, 617.

Radin, E. L., Paul, I. L. & Rose, R. M. (1972). Role of mechanical factors in pathogenesis of primary osteoarthritis. *Lancet*, **1**, 519–22.

Ramachandran, G. N. (1967). Structure of collagen at the molecular level. In *Treatise on collagen*, **1**, ed. G. N. Ramachandran, pp. 103–83. London & New York: Academic Press.

Ramachandran, G. N., Bansal, M. & Bhatnagar, R. S. (1973). A hypothesis on the role of hydroxyproline in stabilizing collagen structure. *Biochim. biophys. Acta*, **322**, 166–71.

Ramachandran, G. N. & Kartha, G. (1955). Structure of collagen. *Nature, Lond.*, **176**, 593–5.

Ranvier, L. (1875). *Traité technique d'histologie*. Paris: Savy.

Rauterberg, J., Fietzek, P., Rexrodt, F., Becker, U., Stark, M. & Kuhn, K. (1972). The amino acid sequence of the carboxyterminal non-helical cross-link region of the α_1 chain of calf skin collagen. *FEBS Lett.*, **21**, 75–9.

Redfern, P. (1851). On the healing of wounds in articular cartilage. Reprinted in 1969 in *Clin. Orthop.*, **64**, 4–6.

Redler, I. (1975). Scanning electron microscopy of human osteoarthrotic cartilage. *Ann rheum. Dis.*, **34**, Suppl. 2, 23–5.

Redler, I., Mow, van C., Zimny, M. L. & Mansell, J. (1975). The ultrastructure and biomechanical significance of the tidemark of articular cartilage. *Clin. Orthop.*, **112**, 357–62.

Reith, E. J. (1968). Collagen formation in developing molar teeth of rats. *J. Ultrastruct. Res.*, **21**, 383–414.

Repo, R. U. & Mitchell, N. (1971). Collagen synthesis in mature articular cartilage of the rabbit. *J. Bone Jt Surg.*, **53*B***, 541–8.

Revel, J. P. & Hay, E. D. (1963). An autoradiographic and electron microscopic study of collagen synthesis in differentiating cartilage. *Z. Zellforsch. mikrosk. Anat.*, **61**, 110–44.

Reynolds, J. J., Murphy, G., Sellers, A. & Cartwright, E. (1977). A new factor that may control collagen resorption. *Lancet*, **2**, 333–5.

Rhoads, R. E. & Udenfriend, S. (1968). Decarboxylation of α-ketoglutarate coupled to collagen proline hydroxylase. *Proc. natn. Acad. Sci. USA*, **60**, 1473–8.

Rhoads, R. E., Udenfriend, S. & Bornstein, P. (1971). *In vitro* enzymatic hydroxylation of prolyl residues in the α1-CB2 fragment of rat collagen. *J. biol. Chem.*, **246**, 4138–42.

Rich, A. & Crick, F. H. C. (1961). The molecular structure of collagen. *J. molec. Biol.*, **3**, 483–506.

Rigal, W. M. (1962). The use of tritiated thymidine in studies of chondrogenesis. In *Radioisotopes and bone*, ed. F. C. McLean, P. Lacroix & A. M. Budy, pp. 197–225. Oxford: Blackwell.

Robert, L. & Poullain, N. (1963). Etudes sur la structure de l'élastine et le mode d'action de l'élastase. *Bull. Soc. Chim. biol.*, **45**, 1317–26.

Robins, S. P. & Bailey, A. J. (1972). Age-related changes in collagen: the identification of reducible lysine–carbohydrate condensation products. *Biochem. biophys. Res. Commun.*, **48**, 76–84.

Robins, S. P., Shimokomaki, M. & Bailey, A. J. (1973). The chemistry of the collagen cross-links. Age-related changes in the reducible components of intact bovine collagen fibres. *Biochem. J.*, **131**, 771–80.

Robinson, H. C., Brett, M. J., Tralaggan, P. J., Lowther, D. A. & Okayama, M. (1975). The effect of D-xylose, β-D-xyloside and β-D-galactosides on chondroitin sulphate biosynthesis in embryonic chick cartilage. *Biochem. J.*, **148**, 25–34.

Robinson, H. C. & Dorfman, A. (1969). The sulfation of chondroitin sulfate in embryonic chick cartilage epiphyses. *J. biol. Chem.*, **244**, 348–52.

Robinson, H. C., Telser, A. & Dorfman, A. (1966). Studies on biosynthesis of the linkage region of chondroitin sulfate-protein complex. *Proc. natn. Acad. Sci. USA*, **56**, 1859–66.

Robinson, R. A. & Cameron, D. A. (1956). Electron microscopy of cartilage and bone matrix at the distal epiphyseal line of the femur in the newborn infant. *J. biophys. biochem. Cytol.*, **2** (Suppl.), 253–60.

Robison, R. (1923). The possible significance of hexosephosphoric esters in ossification. *Biochem. J.*, **17**, 286–93.

Rodan, G. A., Mensi, T. & Harvey, A. (1975). A quantitative method for the application of compressive forces to bone in tissue culture. *Calcif. Tissue Res.*, **18**, 125–31.

Roden, L. (1968). The protein–carbohydrate linkages of acid mucopolysaccharides. In *The chemical physiology of mucopolysaccharides*, ed. G. Quintarelli, pp. 17–29. Boston: Little Brown.

Roden, L. & Schwartz, R. B. (1973). The biosynthesis of chondroitin sulphate. *Biochem. Soc. Trans.*, **1**, 227–30.

Rokosova, B. & Bentley, J. P. (1973). The uptake of (^{14}C) glucose into rabbit ear cartilage proteoglycans isolated by differential extraction and by collagenase digestion. *Biochim. biophys. Acta*, **297**, 473–85.

Romer, A. S. (1942). Cartilage an embryonic adaptation. *Am. Nat.*, **76**, 394–404.

Ropes, M. W. & Bauer, W. (1953). *Synovial fluid changes in joint disease*. Cambridge, Mass.: Harvard University Press.

Rosenberg, L., Hellman, W. & Kleinschmidt, A. K. (1970). Macromolecular models of protein-polysaccharides from bovine nasal cartilage based on electron microscopic studies. *J. biol. Chem.*, **245**, 4123–30.

Rosenberg, L., Hellmann, W. & Kleinschmidt, A. K. (1975). Electron microscopic studies of proteoglycan aggregates from bovine articular cartilage. *J. biol. Chem.*, **250**, 1877–83.

Rosenberg, L., Johnson, B. & Schubert, M. (1965). Protein polysaccharides from human articular and costal cartilage. *J. clin. Invest.*, **44**, 1647–56.

Rosenberg, L. C., Pal, S. & Beale, R. J. (1973). Proteoglycans from bovine proximal humeral articular cartilage. *J. biol. Chem.*, **248**, 3681–90.

Rosenberg, M. J. & Caplan, A. I. (1974). Nicotinamide adenine dinucleotide levels in cells of developing chick limbs: possible control of muscle and cartilage development. *Devl Biol.*, **38**, 157–64.

Rosenberg, M. J. & Caplan, A. I. (1975). Nicotinamide adenine dinucleotide levels in chick limb mesodermal cells *in vitro*: effects of 3-acetylpyridine and nicotinamide. *J. Embryol. exp. Morph.*, **33**, 947–56.

Rosenbloom, J. & Prockop, D. J. (1971). Incorporation of *cis*-hydroxyproline into protocollagen and collagen. *J. biol. Chem.*, **246**, 1549–55.

Rosenthal, O., Bowie, M. A. & Wagoner, G. (1941). Studies on the metabolism of articular cartilage. I. Respiration and glycolysis in relation to its age. *J. cell. comp. Physiol.*, **17**, 221–33.

Rosenthal, O., Bowie, M. A. & Wagoner, G. (1942*a*). The nature of the dehydrogenatic ability of bovine articular cartilage. *J. cell. comp. Physiol.*, **19**, 15–28.

Rosenthal, O., Bowie, M. A. & Wagoner, G. (1942*b*). The dehydrogenatic ability of bovine articular cartilage in relation to its age. *J. cell. comp. Physiol.*, **19**, 333–40.

Ross, R. & Benditt, E. P. (1965). Wound healing and collagen formation. V. Quantitative electron microscope autoradiographic observations of proline H^3 utilization by fibroblasts. *J. Cell Biol.*, **27**, 83–106.

Ross, R. & Bornstein, P. (1969). The elastic fibre. I. The separation and partial characterization of its macromolecular components. *J. Cell Biol.*, **40**, 366–81.

Roth, T. F. & Porter, K. R. (1964). Yolk protein uptake in the oocyte of the mosquito *Aedes aegypti*. *J. Cell Biol.*, **20**, 313–32.

Rothwell, A. G. & Bentley, G. (1973). Chondrocyte multiplication in osteoarthritic articular cartilage. *J. Bone Jt Surg.*, ***55B***, 588–94.

Rouget, C. (1859). Des substances amyloides; de leur rôle dans la constitution des tissus des animaux. *J. Physiol. Paris*, **2**, 308–25.

Roughley, P. J. & Mason, R. M. (1976). The electrophoretic heterogeneity of bovine nasal cartilage proteoglycans. *Biochem. J.*, **157**, 357–67.

Roy, S. & Meachim, G. (1968). Chondrocyte ultrastructure in adult human articular cartilage. *Ann. rheum. Dis.*, **27**, 544–58.

Ruggeri, A. (1972). Ultrastructural, histochemical and autoradiographic studies on the developing chick notochord. *Z. Anat. EntwGesch.*, **138**, 20–33.

Ruggeri, A., Dell'Orbo, C. & Quacci, D. (1975). Electron microscopic visualization of proteoglycans with alcian blue. *Histochem. J.*, **7**, 187–97.

Ruggeri, A., Dell'Orbo, C. & Quacci, D. (1977). Electron microscopic visualization of proteoglycans with ruthenium red. *Histochem. J.*, **9**, 249–52.

Saaty, T. L. & Alexander, J. M. (1975). Optimization and the geometry of numbers: packing and covering. *Soc. ind. app. Maths Rev.*, **17**, 475–519.

Sacerdotti, C. (1900). Ueber das Knorpelfett. *Virchows Arch. path. Anat. Physiol.*, **159**, 152–72.

Sajdera, S. W. (1974). Discussion in *Normal and osteoarthrotic articular cartilage*, ed. S. Y. Ali, M. W. Elves & D. H. Leaback, p. 63. London: Institute of Orthopaedics.

Sajdera, S. W., Franklin, S. & Fortuna, R. (1976). Matrix vesicles of bovine fetal cartilage: metabolic potential and solubilization with detergents. *Fedn Proc. Fedn Am. Socs. exp. Biol.*, **35**, 154–5.

Sajdera, S. W. & Hascall, V. C. (1969). Proteinpolysaccharide complex from bovine nasal cartilage. A comparison of low and high shear extraction procedures. *J. biol. Chem.*, **244**, 77–87.

Salmon, W. D. (1972). Investigation with a partially purified preparation of serum sulfation factor: lack of specificity for cartilage sulphation. In *Growth and growth hormone*, ed. A. Pecile & E. E. Müller, pp. 180–91. Amsterdam: Excerpta Medica.

Salmon, W. D., Bower, P. H. & Thompson, E. Y. (1963). Effect of protein anabolic steroids on sulfate incorporation by cartilage of male rats. *J. Lab. clin. Med.*, **61**, 120–8.

Salmon, W. D. & Daughaday, W. H. (1957). A hormonally controlled serum factor which stimulates sulfate incorporation by cartilage *in vitro*. *J. Lab. clin. Med.*, **49**, 825–6.

Salpeter, M. M. (1968). H^3-proline incorporation into cartilage: electron microscope autoradiographic observations. *J. Morph.*, **124**, 387–421.

Salter, R. B. & Field, P. (1960). The effects of continuous compression on living articular cartilage: an experimental investigation. *J. Bone Jt Surg.*, **42*A***, 31–49.

Saltzman, H. A., Sicker, H. O. & Green, J. (1963). Hexosamine and hydroxyproline content in human bronchial cartilage from aged and diseased lungs. *J. Lab. clin. Med.*, **62**, 78–83.

Sames, K. (1975). Histochemical studies on the distribution of acidic glycosaminoglycans in human rib cartilage during the ageing process. *Mech. Ageing Dev.*, **4**, 431–48.

Saunders, J. W. (1948). The proximo-distal sequence of origin of the parts of the chick wing and the role of the ectoderm. *J. exp. Zool.*, **108**, 363–403.

Saunders, J. W. (1966). Death in embryonic systems. *Science*, **154**, 604–12.

Saunders, J. W. (1972). Developmental control of three-dimensional polarity in the avian limb. *Ann. NY Acad. Sci.*, **193**, 29–42.

Saunders, J. W. & Gasseling, M. T. (1968). Ectodermal–mesenchymal interactions in the origin of limb symmetry. In *Epithelial–mesenchymal interactions*, ed. R. Fleischmajer & R. E. Billingham, pp. 78–97. Baltimore: Williams & Wilkins.

Schaffer, J. (1930). Die Stutzgewebe. In *Handbuch der mikroskopischen Anatomie des Menschen*, **2**, 2nd edn, ed. W. von Mollendorff, pp. 1–390. Berlin: Springer.

Schallock, G. (1942). Untersuchungen zur Pathogenese von Aufbrauchveranderungen an den knorpeligen Anteilen des Kniegelenkes. *Veröff. Konstit. u. Wehrpath.*, **49**, 1–68.

Schatton, J. & Schubert, M. (1954). Isolation of a mucoprotein from cartilage. *J. biol. Chem.*, **211**, 565–73.

Schenk, R., Merz., W. A., Muhlbauer, R., Russell, R. G. G. & Fleisch, H. (1973). Effect of ethane-1-hydroxy-1,1-diphosphonate (EHDP) and dichloromethylene diphosphonate (Cl_2MDP) on the calcification and resorption of cartilage and bone in the tibial epiphysis and metaphysis of rats. *Calcif. Tissue Res.*, **11**, 196–214.

Schenk, R. K., Spiro, D. & Wiener, J. (1967). Cartilage resorption in the tibial epiphyseal plate of growing rats. *J. Cell Biol.*, **34**, 275–91.

Schenk, R. K., Wiener, J. & Spiro, D. (1968). Fine structural aspects of vascular invasion of the tibial epiphyseal plate of growing rats. *Acta anat.*, **69**, 1–17.

Scherft, J. P. & Daems, W. (1967). Single cilia in chondrocytes. *J. Ultrastruct. Res.*, **19**, 546–55.

Schiller, S. (1966). Connective and supporting tissues: mucopolysaccharides of connective tissues. *A. Rev. Physiol.*, **28**, 137–58.

Schiller, S. & Dorfman, A. (1963). The distribution of acid mucopolysaccharides in skin of diabetic rats. *Biochim. biophys. Acta*, **78**, 371–73.

Schmiedeberg, O. (1891). Ueber die Chemische Zusammensetzung des Knorpels. *Arch. exp. Path. Pharmak.*, **28**, 355–404.

Schneider, E. L., Chaillet, J. R. & Tice, R. R. (1976). *In vivo* labelling of mammalian chromosomes. *Expl Cell Res.*, **100**, 396–9.

Schofield, B. H., Williams, B. R. & Doty, S. B. (1975). Alcian blue staining of cartilage for electron microscopy. Application of the critical electrolyte concentration principle. *Histochem. J.*, **7**, 139–49.

Schumacher, H. R. (1976). Ultrastructural findings in chondrocalcinosis and pseudo-gout. *Arthritis Rheum.*, **19** (Suppl.), 413–25.

Scott, J. E. (1962). Separation and measurement of the products of reaction of phenyl isothiocyanate with galactosamine and glucosamine. *Biochem. J.*, **82**, 43*P*.

Scott, J. E. (1968). Ion binding in solutions containing acid mucopolysaccharides. In *The chemical physiology of mucopolysaccharides*, ed. G. Quintarelli, pp. 171–84. Boston: Little Brown.

Scott, J. E. (1973). Distribution of acid mucopolysaccharides in cartilage. In *Current developments in rheumatology*, ed. O. Lövtren, H. Boström, B. Olhagen & R. Sannerstedt, pp. 10–29. Mölndal: Lindgren.

Scott, J. E. (1974). The histochemistry of cartilage proteoglycans in light and electron microscopes. In *Normal and osteoarthrotic articular cartilage*, ed. S. Y. Ali, M. W. Elves & D. H. Leaback, pp. 19–32. London: Institute of Orthopaedics.

Scott, J. E. (1975). Physiological function and chemical composition of pericellular proteoglycan (an evolutionary view). *Phil. Trans. R. Soc.*, *B*, **271**, 235–42.

Scott, J. E., Conochie, L., Faulk, W. P. & Bailey, A. J. (1975). Passive agglutination method using collagen-coated tanned sheep erythrocytes to demonstrate collagen-glycosaminoglycan interaction. *Ann. rheum. Dis.*, **34**, Suppl. 2, 38–9.

Scott, J. E. & Dorling, J. (1965). Differential staining of acid glycosaminoglycans (mucopolysaccharides) by alcian blue in salt solutions. *Histochemie*, **5**, 221–33.

Scott, J. E., Dorling, J. & Stockwell, R. A. (1968). Reversal of protein blocking of basophilia in salt solutions: implications in the localization of polyanions using alcian blue. *J. Histochem. Cytochem.*, **16**, 383–6.

Scott, J. E. & Stockwell, R. A. (1967). On the use and abuse of the critical electrolyte concentration approach to the localization of tissue polyanions. *J. Histochem. Cytochem.*, **15**, 111–13.

Scott, J. E., Tigwell, M. J. & Sajdera, S. W. (1972). Loss of basophilic (sulphated) material from sections of cartilage treated with periodate. *Histochem. J.*, **4**, 155–67.
Scott, J. H. (1953). The cartilage of the nasal septum. *Br. dent. J.*, **95**, 37–43.
Searls, J. C. (1976). A radioautographic study of chondrocyte proliferation in nasal septal cartilage of the prenatal rat. *Cleft Palate J.*, **3**, 330–41.
Searls, R. L. (1965*a*). An autoradiographic study of the uptake of S^{35}-sulfate during the differentiation of limb bud cartilage. *Devl Biol.*, **11**, 155–68.
Searls, R. L. (1965*b*). Isolation of mucopolysaccharide from the precartilaginous embryonic limb bud. *Proc. Soc. exp. Biol. Med.*, **118**, 1172–6.
Searls, R. L. (1967). The role of cell migration in the development of the embryonic chick limb bud. *J. exp. Zool.*, **166**, 39–50.
Searls, R. L. (1973). Chondrogenesis. In *Developmental regulation: aspects of cell differentiation*, ed. S. J. Coward, pp. 219–51. New York & London: Academic Press.
Searls, R. L., Hilfer, S. R. & Mirow, S. M. (1972). An ultrastructural study of early chondrogenesis in the chick wing bud. *Devl Biol.*, **28**, 123–37.
Searls, R. L. & Janners, M. Y. (1969). The stabilization of cartilage properties in the cartilage-forming mesenchyme of the embryonic chick limb. *J. exp. Zool.*, **170**, 365–76.
Seedhom, B. B., Takeda, T., Wright, V. & Tsubuku, M. (1978). Mechanical stress and patello-femoral osteoarthrosis. *Ann. rheum. Dis.*, **37**, 486.
Seno, N. & Anno, K. (1961). The presence of hyaluronic acid in whale cartilage. *Biochim. biophys. Acta*, **49**, 407–8.
Seno, N., Meyer, K., Anderson, B. & Hoffman, P. (1965). Variations in keratosulphates. *J. biol. Chem.*, **240**, 1005–10.
Sensenig, E. C. (1949). Early development of the human vertebral column. *Contr. Embryol.*, **33**, 23–41.
Serafini-Fracassini, A. & Smith, J. W. (1966). Observations on the morphology of the proteinpolysaccharide complex of bovine nasal cartilage and its relationship to collagen. *Proc. R. Soc., B*, **165**, 440–9.
Serafini-Fracassini, A. & Smith, J. W. (1974). *The structure and biochemistry of cartilage*. Edinburgh & London: Churchill Livingstone.
Sewell, A. C. & Pennock, C. A. (1976). The chemistry of human neonatal femoral epiphysial cartilage. *Clinica chim. Acta*, **68**, 123–6.
Shands, A. R. (1931). The regeneration of hyaline cartilage in joints. *Archs Surg.*, **22**, 137–78.
Shapiro, F., Holtrop, M. E. & Glimcher, M. J. (1977). Organization and cellular biology of the perichondrial ossification groove of Ranvier. A morphological study in rabbits. *J. Bone Jt Surg.*, **59*A***, 703–23.
Shaw, N. E. & Lacey, E. (1973). The influence of corticosteroids on normal and papain-treated articular cartilage in the rabbit. *J. Bone Jt Surg.*, **55*B***, 197–205.
Shaw, N. E. & Martin, B. F. (1962). Histological and histochemical studies on mammalian knee joint tissues. *J. Anat.*, **96**, 359–73.
Sheehan, J. F. (1948). A cytological study of the cartilage of developing long bones of the rat with special reference to the Golgi apparatus, mitochondria, neutral red bodies and lipid inclusions. *J. Morph.* **82**, 151–99.
Sheehan, J. K., Atkins, E. D. T. & Nieduszynski, I. A. (1975). X-ray diffraction studies on the connective tissue polysaccharides. *J. molec. Biol.*, **91**, 153–63.
Sheldon, H. & Robinson, R. A. (1958). Studies on cartilage: I. Electron microscopy observations on normal rabbit ear cartilage. *J. biophys. biochem. Cytol*, **4**, 401–6.
Sheldon, H. & Robinson, R. A. (1960). Studies on cartilage: II. Electron microscope observations on rabbit ear cartilage following administration of papain. *J. biophys. biochem. Cytol.*, **8**, 151–63.

Shetlar, M. R. & Masters, Y. F. (1955). Effect of age on the polysaccharide composition of cartilage. *Proc. Soc. exp. Biol. Med.*, **90**, 31–3.

Shimomura, Y., Wezeman, F. H. & Ray, R. D. (1973). The growth cartilage plate of the rat and rib: cellular proliferation. *Clin. Orthop.*, **90**, 246–54.

Shipp, D. W. & Bowness, J. M. (1975). Insoluble non-collagenous cartilage glycoproteins with aggregating sub-units. *Biochim. biophys. Acta*, **379**, 282–94.

Shulman, H. J. & Meyer, K. (1968). Cellular differentiation and the aging process in cartilaginous tissues. *J. exp. Med.*, **128**, 1353–62.

Shulman, H. J. & Opler, A. (1974). The stimulatory effect of calcium on the synthesis of cartilage proteoglycan. *Biochem. biophys. Res. Commun.*, **59**, 914–19.

Siegel, R. C. & Martin, G. R. (1970). Collagen cross-linking. Enzymatic synthesis of lysine-derived aldehydes and the production of cross-linked components. *J. biol. Chem.*, **245**, 1653–8.

Siegling, J. A. (1941). Growth of the epiphyses. *J. Bone Jt Surg.*, **3**, 23–36.

Silberberg, M. & Silberberg, R. (1961). Ageing changes in cartilage and bone. In *Structural aspects of ageing*, ed. G H. Bourne, pp. 85–108. London: Pitman Medical.

Silberberg, M. & Silberberg, R. (1971). Steroid hormones and bone. In *The biochemistry and physiology of bone*, **3**, 2nd edn, ed. G. H. Bourne, pp. 401–84. New York & London: Academic Press.

Silberberg, M., Silberberg, R. & Hasler, M. (1964). Ultrastructure of articular cartilage in mice treated with somatotrophin. *J. Bone Jt Surg.*, **46*A***, 766–80.

Silberberg, M., Silberberg, R. & Hasler, M. (1966). Fine structure of articular cartilage in mice receiving cortisone acetate. *Archs Path.*, **82**, 569–82.

Silberberg, R. (1968). Ultrastructure of articular cartilage in health and disease. *Clin. Orthop.*, **57**, 233–57.

Silberberg, R. & Hasler, M. (1971). Submicroscopic effects of hormones on articular cartilage of adult mice. *Archs Path.*, **91**, 241–55.

Silberberg, R. & Hasler, M. (1972). Effect of testosterone proprionate on the ultrastructure of articular cartilage in mice. *Growth*, **36**, 17–33.

Silberberg, R., Hasler, M. & Silberberg, M. (1966). Response of articular cartilage of dwarf mice to insulin: electron microscopic studies. *Anat. Rec.*, **155**, 577–90.

Silbert, J. E. (1964). Incorporation of ^{14}C and ^{3}H from nucleotide sugars into a polysaccharide in the presence of a cell-free preparation from cartilage. *J. biol. Chem.*, **239**, 1310–15.

Silbert, J. E. (1973). Catabolism of mucopolysaccharides and protein-polysaccharides. In *Molecular pathology of connective tissues*, ed. R. Perez-Tamayo & M. Rojkind, pp. 355–73. New York: Dekker.

Silbert, J. E. & Deluca, S. (1969). Biosynthesis of chondroitin sulphate. III. Formation of a sulfated glycosaminoglycan with a microsomal preparation from chick embryo cartilage. *J. biol. Chem.*, **244**, 876–81.

Silpananta, P., Dunstone, J. R. & Ogston, A. G. (1968). Fractionation of a hyaluronic acid preparation in a density gradient. *Biochem. J.*, **109**, 43–9.

Silver, I. A. (1975). Measurement of pH and ionic composition of pericellular sites. *Phil. Trans. R. Soc.*, *B*, **271**, 261–72.

Simmons, D. J. (1964). Circadian mitotic rhythm in epiphysial cartilage. *Nature, Lond.*, **202**, 906–7.

Simmons, D. J. & Kunin, A. S. (1967). Autoradiographic and biochemical investigations of the effect of cortisone on the bones of the rat. *Clin. Orthop.*, **55**, 201–15.

Simmons, D. P. & Chrisman, O. D. (1965). Salicylate inhibition of cartilage degeneration. *Arthritis Rheum.*, **8**, 960–9.

Simon, W. H. (1970). Scale effects in animal joints. I. Articular cartilage thickness and compressive stress. *Arthritis Rheum.*, **13**, 244–56.

Simon, W. H. (1971). Scale effects in animal joints. II. Thickness and elasticity in the deformability of articular cartilage. *Arthritis Rheum.*, **14**, 493–502.
Simon, W. H., Friedenberg, S. & Richardson, S. (1973). Joint congruence. *J. Bone Jt Surg.*, **55*A***, 1614–20.
Simunek, Z. & Muir, H. (1972). Changes in the protein-polysaccharides of pig articular cartilage during prenatal life, development and old age. *Biochem. J.*, **126**, 515–23.
Sissons, H. A. (1953). Experimental determination of rate of longitudinal bone growth. *J. Anat.*, **87**, 228–36.
Sissons, H. A. (1956). Experimental study of the effect of local irradiation on bone growth. In *Progress in radiobology*, Proc. IVth Int. Conf. Radiobiology, ed. J. S. Mitchell, B. E. Homes & C. L. Smith, pp. 436–48. Edinburgh & London: Oliver & Boyd.
Sissons, H. A. (1971). The growth of bone. In *The biochemistry and physiology of bone*, **3**, 2nd edn, ed. G. H. Bourne, pp. 145–80. New York & London: Academic Press.
Skoog, T. & Johansson, S. H. (1976). The formation of articular cartilage from free perichondrial grafts. *Plast. reconstr. Surg.*, **57**, 1–6.
Slavkin, H. C., Croissant, R. D., Bringas, P., Matosian, P., Wilson, P., Mino, W. & Guenther, H. (1976). Matrix vesicle heterogeneity: possible morphogenetic functions for matrix vesicles. *Fedn Proc. Fedn Am. Socs exp. Biol.*, **35**, 127–34.
Sledge, C. B. (1973). Growth hormone and articular cartilage. *Fedn Proc. Fedn Am. Socs exp. Biol.* **32**, 1503–5.
Smillie, I. S. (1943). Observations on the regeneration of the semilunar cartilages in man. *Br. J. Surg.*, **31**, 398–401.
Smith, A. U. (1965). Survival of frozen chondrocytes isolated from cartilage of adult mammals. *Nature, Lond.*, **205**, 782–4.
Smith, B. D., Byers, P. H. & Martin, G. R. (1972*a*). Production of procollagen by human fibroblasts in culture. *Proc. natn. Acad. Sci. USA*, **69**, 3260–2.
Smith, D. W., Brown, D. M. & Carnes, W. H. (1972*b*). Preparation and properties of salt-soluble elastin. *J. biol. Chem.*, **247**, 2427–32.
Smith, J. W. (1968). Molecular pattern in native collagen. *Nature, Lond.*, **219**, 157–8.
Smith, J. W. (1970). The disposition of protein polysaccharide in the epiphysial plate cartilage of the young rabbit. *J. Cell Sci.*, **6**, 843–64.
Smith, J. W. & Frame, J. (1969). Observations on the collagen and proteinpolysaccharide complex of rabbit corneal stroma. *J. Cell Sci.*, **4**, 421–36.
Smith, J. W., Peters, T. J. & Serafini-Fracassini, A. (1967). Observations on the distribution of the proteinpolysaccharide complex and collagen in bovine articular cartilage. *J. Cell Sci.*, **2**, 129–36.
Smith, J. W. & Walmsley, R. (1951). Experimental incision of the intervertebral disc. *J. Bone Jt Surg.*, **33*B***, 612–25.
Smith, T. W. D., Duckworth, T., Bergenholtz, A. & Lemperg, R. K. (1975). Role of growth hormone in glycosaminoglycan synthesis by articular cartilage. *Nature, Lond.*, **253**, 269–71.
Sokoloff, L. (1973). A note on the histology of cement lines. In *Perspectives in biomedical engineering*, ed. R. M. Kenedi, pp. 135–8. London: Macmillan.
Sokoloff, L. (1976). Articular chondrocytes in culture. *Arthritis Rheum.*, **19** (Suppl.), 426–9.
Solursh, M., Vaerewyck, S. A. & Reiter, R. S. (1974). Depression by hyaluronic acid of glycosaminoglycan synthesis by cultured chick embryo chondrocytes. *Devl Biol.*, **41**, 233–44.
Sood, S. C. (1971). A study of the effects of experimental immobilization on rabbit articular cartilage. *J. Anat.*, **108**, 497–507.

Sorgente, N., Hascall, V. C. & Kuettner, K. E. (1972). Extractability of lysozyme from bovine nasal cartilage. *Biochim. biophys. Acta*, **284**, 441–50.

Souter, W. A. & Taylor, T. K. F. (1970). Sulphated acid mucopolysaccharide metabolism in the rabbit intervertebral disc. *J. Bone Jt Surg.*, ***52B***, 371–84.

Speakman, P. T. (1971). Proposed mechanism for the biological assembly of collagen triple helix. *Nature, Lond.*, **229**, 241–3.

Spiro, R. G. (1970). Glycoproteins. *A. Rev. Biochem.*, **39**, 599–638.

Spiro, M. J. & Spiro, R. G. (1971). Studies on the biosynthesis of the hydroxylysine-linked disaccharide units of basement membranes and collagens. *J. biol. Chem.*, **246**, 4910–18.

Sprinz, R. (1961). Further observations on the effect of surgery on the meniscus of the mandibular joint in rabbits. *Archs oral Biol.*, **5**, 195–201.

Sprinz, R. & Stockwell, R. A. (1976). Changes in articular cartilage following intra-articular injection of tritiated glyceryl trioleate. *J. Anat.*, **122**, 91–112.

Sprinz, R. & Stockwell, R. A. (1977). Effect of experimental lipoarthrosis on the articular fibro-cartilage of the rabbit mandibular condyle. *Archs oral biol.*, **22**, 313–16.

Srivastava, V. M., Malemud, C. J. & Sokoloff, L. (1974). Chondroid expression by rabbit articular cells in spinner culture following monolayer culture. *Conn. Tissue Res.*, **2**, 127–36.

Stanescu, V., Maroteaux, P. & Sobczak, E. (1977). Proteoglycan populations of baboon (*Papio papio*) articular cartilage. *Biochem. J.*, **163**, 103–9.

Stedman, E. & Stedman, E. (1943). Probable function of histone as a regulator of mitosis. *Nature, Lond.*, **152**, 556–7.

Stein, G. S., Spelsberg, T. C. & Kleinsmith, L. J. (1974). Non-histone chromosomal proteins and gene regulation. *Science*, **183**, 817–24.

Stellwagen, R. H. & Tomkins, G. M. (1971). Preferential inhibition by 5-bromodeoxyuridine of the synthesis of tyrosine amino-transferase in hepatoma cell cultures. *J. molec. Biol.*, **56**, 167–82.

Stidworthy, G., Masters, Y. F. & Shetlar, M. R. (1958). The effect of ageing on mucopolysaccharide composition of human costal cartilage as measured by hexosamine and uronic acid content. *J. Geront.*, **13**, 10–13.

Stockwell, R. A. (1965). Lipid in the matrix of ageing articular cartilage. *Nature, Lond.*, **207**, 427–8.

Stockwell, R. A. (1966). The ageing of cartilage. PhD thesis, University of London.

Stockwell, R. A. (1967*a*). The cell density of human articular and costal cartilage. *J. Anat.*, **101**, 753–63.

Stockwell, R. A. (1967*b*). The lipid and glycogen content of rabbit articular and non-articular hyaline cartilage. *J. Anat.*, **102**, 87–94.

Stockwell, R. A. (1967*c*). Lipid content of human costal and articular cartilage. *Ann. rheum. Dis.*, **26**, 481–6.

Stockwell, R. A. (1970). Changes in the acid glycosaminoglycan (mucopolysaccharide) content of the matrix of ageing human articular cartilage. *Ann. rheum. Dis.*, **29**, 509–15.

Stockwell, R. A. (1971*a*). The ultrastructure of cartilage canals and the surrounding cartilage in the sheep fetus. *J. Anat.*, **109**, 397–410.

Stockwell, R. A. (1971*b*). The interrelationship of cell density and cartilage thickness in mammalian articular cartilage. *J. Anat.*, **109**, 411–21.

Stockwell, R. A. (1971*c*). Cell density, cell size and cartilage thickness in adult mammalian articular cartilage. *J. Anat.*, **108**, 584.

Stockwell, R. A. (1974). Fine structure and macromolecular organization of connective tissue. *Trans. ophthal. Soc. UK*, **94**, 648–60.

Stockwell, R. A. (1975*a*). Stain distribution with alcian blue in articular cartilage. *Ann. rheum. Dis.*, **34**, Suppl. 2, 17–18.
Stockwell, R. A. (1975*b*). Structural and histochemical aspects of the pericellular environment in cartilage. *Phil. Trans. R. Soc.*, *B*, **271**, 243–5.
Stockwell, R. A. (1978*a*). Matrix histochemistry of the deep zone of human articular cartilage. *Ann. R. Coll. Surg.*, **60**, 136.
Stockwell, R. A. (1978*b*). Chondrocytes. *J. clin. Path.*, **31**, Suppl. 12, 7–13.
Stockwell, R. A. & Barnett, C. H. (1964). Changes in permeability of articular cartilage with age. *Nature, Lond.*, **201**, 835–6.
Stockwell, R. A. & Meachim, G. (1973). The chondrocytes. In *Adult articular cartilage*, ed. M. A. R. Freeman, pp. 51–99. London: Pitman Medical.
Stockwell, R. A. & Scott, J. E. (1965). Observations on the acid glycosaminoglycan (mucopolysaccharide) content of ageing cartilage. *Ann. rheum. Dis.*, **24**, 341–50.
Stockwell, R. A. & Scott, J. E. (1967). Distribution of acid glycosaminoglycans in human articular cartilage. *Nature, Lond.*, **215**, 1376–8.
Stoolmiller, A. C. & Dorfman, A. (1969). The metabolism of glycosaminoglycans. In *Comprehensive biochemistry*, **17**, ed. M. Florkin & E. H. Stotz, pp. 241–75. Amsterdam: Elsevier.
Strangeways, T. S. P. (1920). Observations on the nutrition of articular cartilage. *Br. med. J.*, **1**, 661–3.
Strawich, E. & Nimni, M. E. (1971). Properties of a collagen molecule containing three identical components extracted from bovine articular cartilage. *Biochemistry*, **10**, 3905–11.
Streeter, G. L. (1949). Developmental horizons in human embryos (4th issue). A review of the histiogenesis of cartilage and bone. *Contr. Embryol.*, **220**, 149–67.
Strickland, A. L. & Sprinz, H. (1973). Studies of the influence of estradiol and growth hormone on the hypophysectomised immature rat epiphysial cartilage growth plate. *Am. J. Obstet. Gynaec.*, **115**, 451–7.
Strom, C. M. & Dorfman, A. (1976). Distribution of 5-bromodeoxyuridine and thymidine in the DNA of developing chick cartilage. *Proc. natn. Acad. Sci. USA*, **73**, 1019–23.
Strominger, J. L., Maxwell, E. S., Axelrod, J. & Kalckar, H. M. (1957). Enzymatic formation of uridine diphosphoglucuronic acid. *J. biol. Chem.*, **224**, 79–90.
Strudel, G. (1973). Relationship between the chick periaxial metachromatic extracellular material and vertebral chondrogenesis. In *Biology of the fibroblast*, ed. E. Kulonen & J. Pikkarainen, pp. 93–101. London & New York: Academic Press.
Stump, C. W. (1925). The histiogenesis of bone. *J. Anat.*, **59**, 136–54.
Summerbell, D. (1974). Interaction between the proximo-distal and antero-posterior coordinates of positional value during the specification of positional information in the early development of the chick limb-bud. *J. Embryol. exp. Morph.* **32**, 227–37.
Summerbell, D. & Wolpert, L. (1972). Cell density and cell division in the early morphogenesis of the chick wing. *Nature New Biol.*, **239**, 24–6.
Suzuki, S. & Strominger, J. L. (1960). Enzymatic sulfation of mucopolysaccharides in hen oviduct. I. Transfer of sulfate from 3′-phospho-adenosine-5′-phosphosulphate to mucopolysaccharide. *J. biol. Chem.*, **235**, 257–66.
Swann, D. A. (1968). Studies on hyaluronic acid. II. The protein component of rooster comb hyaluronic acid. *Biochim. biophys. Acta*, **160**, 96–105.
Swann, D. A. (1975). Purification and properties of articular lubricant. *Ann. rheum. Dis.*, **34**, Suppl. 2, 91–3.
Swanson, S. V. (1973). Lubrication. In *Adult articular cartilage*, ed. M. A. R. Freeman, pp. 247–76. London: Pitman Medical.
Szirmai, J. A. (1963). Quantitative approaches in the histochemistry of mucopolysaccharides. *J. Histochem. Cytochem.*, **11**, 24–34.

Szirmai, J. A. (1969). Structure of cartilage. In *Ageing of connective and skeletal tissue*, ed. A. Engel & T. Larsson, pp. 163–84. Stockholm: Nordiska Bokhandelns.

Szirmai, J. A. & Doyle, J. (1961). Quantitative histochemical and chemical studies on the composition of cartilage. *J. Histochem. Cytochem.*, **9**, 611.

Szirmai, J. A., Van Boven-de Tyssonsk, E. & Gardell, S. (1957). Microchemical analysis of glycosaminoglycans, collagen, total protein and water in histological layers of nasal septum cartilage. *Biochim. biophys. Acta*, **136**, 331–50.

Tanner, J. M. (1962). *Growth at adolescence*, 2nd edn. Oxford: Blackwell.

Tanzer, M. L., Church, R. L., Yaeger, J. A., Wampler, E. & Park, E.-D. (1974). Procollagen: intermediate forms containing several types of peptide chains and non-collagen peptide extensions at NH_2 and COOH ends. *Proc. natn. Acad. Sci. USA*, **71**, 3009–13.

Tanzer, M. L., Fairweather, R. & Gallop, P. M. (1972). Collagen cross-links: isolation of reduced N^{ε}-hexosylhydroxylysine from borohydride-reduced calf skin insoluble collagen. *Archs Biochem. Biophys.*, **151**, 137–41.

Taylor, J. R. (1973). Growth and development of the human intervertebral disc. PhD thesis, University of Edinburgh.

Teitz, C. C. & Chrisman, O. D. (1975). The effect of salicylate and chloroquine on prostaglandin-induced articular damage in the rabbit knee. *Clin. Orthop.*, **108**, 264–74.

Tell, G. P. E., Cuatrecasas, P., Van Wyk, J. J. & Hintz, R. L. (1973). Somatomedin: inhibition of adenylate cyclase activity in subcellular membranes of various tissues. *Science*, **180**, 312–15.

Telser, A., Robinson, H. C. & Dorfman, A. (1965). The biosynthesis of chondroitin sulfate protein complex. *Proc. natn. Acad. Sci. USA*, **54**, 912–19.

Telser, A., Robinson, H. C. & Dorfman, A. (1966). The biosynthesis of chondroitin sulfate. *Archs Biochem. Biophys.*, **116**, 458–65.

Termine, J. D. & Conn, K. M. (1976). Inhibition of apatite formation by phosphorylated metabolites and macromolecules. *Calcif. Tissue Res.*, **22**, 149–57.

Thanassi, N. M. & Newcombe, D. S. (1974). Cyclic AMP and thyroid hormone: inhibition of epiphyseal cartilage cyclic-3′,5′-nucleotide phosphodiesterase activity by L-triiodothyronine. *Proc. Soc. exp. Biol. Med.*, **147**, 710–14.

Thomas, J., Elsden, D. F. & Partridge, S. M. (1963). Partial structure of two major degradation products from the cross-linkages in elastin. *Nature, Lond.*, **200**, 651–2.

Thompson, R. C. & Bassett, C. A. L. (1970). Histological observations on experimentally induced degeneration of articular cartilage. *J. Bone Jt Surg.*, **52*A***, 435–43.

Thorngren, K.-G. & Hansson, L. I. (1973). Cell kinetics and morphology of the growth plate in the normal and hypophysectomised rat. *Calcif. Tissue Res.*, **13**, 113–29.

Thorogood, P. V. & Hinchliffe, J. R. (1975). An analysis of the condensation process during chondrogenesis in the embryonic chick hind limb. *J. Embryol. exp. Morph.*, **33**, 581–606.

Thyberg, J. & Friberg, U. (1970). Ultrastructure and acid phosphatase activity of matrix vesicles and cytoplasmic dense bodies in the epiphysial plate. *J. Ultrastruct. Res.*, **33**, 554–73.

Thyberg, J., Lohmander, S. & Friberg, H. (1973). Electron microscopic demonstration of proteoglycans in guinea pig epiphyseal cartilage. *J. Ultrastruct. Res.*, **45**, 407–27.

Thyberg, J., Lohmander, S. & Heinegard, D. (1975). Proteoglycans of hyaline cartilage. Electron microscopic studies on isolated molecules. *Biochem. J.*, **151**, 157–66.

Tickle, C., Summerbell, D. & Wolpert, L. (1975). Positional signalling and specification of digits in chick limb morphogenesis. *Nature, Lond.*, **254**, 199–202.

Todd, R. B. & Bowman, W. (1845). *The physiological anatomy and physiology of man.* London: Parker.
Tonna, E. A. (1961). The cellular complement of the skeletal system studied autoradiographically with tritiated thymidine (H^3TDR) during growth and ageing. *J. biophys. biochem. Cytol.*, **9**, 813–24.
Tonna, E. A. & Cronkite, E. P. (1964). A study of the persistence of the H^3-thymidine label in the femora of rats. *Lab. Invest.*, **13**, 161–71.
Toole, B. P. (1969). Solubility of collagen fibrils formed *in vitro* in the presence of sulphated acid mucopolysaccharide-protein. *Nature, Lond.*, **222**, 872.
Toole, B. P. (1972). Hyaluronate turnover during chondrogenesis in the developing chick limb and axial skeleton. *Devl Biol.*, **29**, 321–9.
Toole, B. P. (1973). Hyaluronate inhibition of chondrogenesis: antagonism of thyroxine, growth hormone and calcitonin. *Science*, **180**, 302–3.
Toole, B. P., Jackson, G. & Gross, J. (1972). Hyaluronate in morphogenesis: inhibition of chondrogenesis *in vitro*. *Proc. natn. Acad. Sci. USA*, **69**, 1384–6.
Toole, B. P. & Lowther, D. A. (1968). The effect of chondroitin sulphate-protein on the formation of collagen fibrils *in vitro*. *Biochem. J.*, **109**, 857–66.
Towe, K. M. (1970). Oxygen-collagen priority and the early Metazoan fossil record. *Proc. natn. Acad. Sci. USA*, **65**, 781–8.
Toynbee, J. (1841). Researches tending to prove the non-vascularity and the peculiar uniform mode of organization and nutrition of certain animal tissues. *Phil. Trans. R. Soc., B*, **131**, 159–92.
Traub, W. & Piez, K. A. (1971). The chemistry and structure of collagen. *Adv. Protein Chem.*, **25**, 243–352.
Traub, W., Yonath, A. & Segal, D. M. (1969). On the molecular structure of collagen. *Nature, Lond.*, **221**, 914–17.
Trelstad, R. L. (1975). Collagen fibrillogenesis *in vitro* and *in vivo*: the existence of unique aggregates and the special state of the fibril end. In *Extracellular matrix influences on gene expression*, ed. H. Slavkin & R. Greulich, pp. 331–9. New York: Academic Press.
Trelstad, R. L., Kang, A. H., Cohen, A. M. & Hay, E. D. (1973). Collagen synthesis *in vitro* by embryonic spinal cord epithelium. *Science*, **179**, 295–6.
Treuhaft, P. S. & McCarty, D. J. (1971). Synovial fluid pH, lactate, oxygen and carbon dioxide partial pressure in various joint diseases. *Arthritis Rheum.*, **14**, 475–84.
Trias, A. (1961). Effect of persistent pressure on the articular cartilage. *J. Bone Jt Surg.*, **43*B***, 376–86.
Trinick, E. (1974). Wound repair in the sternal cartilage of the domestic fowl. *J. Anat.*, **118**, 367.
Trueta, J. & Amato, V. P. (1960). The vascular contribution to osteogenesis. III. Changes in the growth cartilage caused by experimentally induced ischaemia. *J. Bone Jt Surg.*, **42*B***, 571–87.
Trueta, J. & Harrison, M. H. M. (1953). The normal vascular anatomy of the femoral head in adult man. *J. Bone Jt Surg.*, **35*B***, 442–61.
Trueta, J. & Little, K. (1960). The vascular contribution to osteogenesis. II. Studies with the electron microscope. *J. Bone Jt Surg.*, **42*B***, 367–76.
Trueta, J. & Morgan, J. D. (1960). The vascular contribution to osteogenesis. I. Studies by the injection method. *J. Bone Jt Surg.*, **42*B***, 97–109.
Tsiganos, C. P. & Muir, H. (1969). Studies on protein-polysaccharides from pig laryngeal cartilage: extraction and purification. *Biochem. J.*, **113**, 879–84.
Tsiganos, C. P. Hardingham, T. E. & Muir, H. (1972). Aggregation of cartilage proteoglycans. *Biochem. J.*, **128**, 121*P*.
Tushan, F. S., Rodnan, G. P., Altman, M. & Robin, E. D. (1969). Anaerobic

glycolysis and lactate dehydrogenase LDH isoenzymes in articular cartilage. *J. Lab. clin. Med.*, **73**, 649–56.

Ubermuth, H. (1929). Die Bedeutung der Altersveränderungen der menschliche Bandscheiben für die Pathologie der Wirbelsaule. *Arch. klin. Chir.*, **156**, 567–77.

Urban, J., Holm, S., Maroudas, A. & Nachemson, A. (1978). Nutrition of the intervertebral disk. An *in vivo* study of solute transport. *Clin. Orthop.*, **129**, 101–14.

Vane, J. R. (1971). Inhibition of prostaglandin synthesis as a mechanism of action for aspirin-like drugs. *Nature New Biol.*, **231**, 232–5.

Vaubel, E. (1933). Form and function of synovial cells in tissue culture. *J. exp. Med.*, **58**, 63–95.

Veis, A., Anesey, J. & Mussell, S. (1967). A limiting microfibril model for the three-dimensional arrangement within collagen fibres. *Nature, Lond.*, **215**, 931–4.

Venn, M. F. (1978). Variation of chemical composition with age in human femoral head cartilage. *Ann. rheum. Dis.*, **37**, 168–74.

Venn, M. & Maroudas, A. (1977). Chemical composition and swelling of normal and osteoarthrotic femoral head cartilage. I. Chemical composition. *Ann. rheum. Dis.*, **36**, 121–9.

Vignon, E., Arlot, M., Patricot, L. M. & Vignon, G. (1976). The cell density of human femoral head cartilage. *Clin. Orthop.*, **121**, 303–8.

Virchow, R. (1863). *Die krankhaften Geschwulste*, **1**. Berlin: Hirschwald.

Vuust, J. & Piez, K. A. (1972). A kinetic study of collagen biosynthesis. *J. biol. Chem.*, **247**, 856–62.

Walker, K. V. R. & Kember, N. F. (1972*a*). Cell kinetics of growth cartilage in the rat tibia. I. Measurements in young male rats. *Cell Tissue Kinet.*, **5**, 401–8.

Walker, K. V. R. & Kember, N. F. (1972*b*). Cell kinetics of growth cartilage in the rat tibia. II. Measurements during ageing. *Cell Tissue Kinet.*, **5**, 409–19.

Walker, P. S., Dowson, D., Longfield, M. D. & Wright, V. (1968). Boosted lubrication in synovial joints by fluid entrapment and enrichment. *Ann. rheum. Dis.*, **27**, 512–20.

Walker, P. S., Sikorski, J., Dowson, D., Longfield, M. D., Wright, V. & Buckley, T. (1969). Behaviour of synovial fluid on surfaces of articular cartilage. A scanning electron microscope study. *Ann. rheum. Dis.*, **28**, 1–14.

Walmsley, R. (1953). The development and growth of the intervertebral disc. *Edinb. med. J.*, **60**, 341–64.

Walmsley, R. & Bruce, J. (1938). The early stages of replacement of the semilunar cartilages of the knee joint in rabbits after operative excision. *J. Anat.*, **72**, 260–3.

Wasteson, A. & Lindahl, U. (1971). The distribution of sulphate residues in the chondroitin sulphate chain. *Biochem. J.*, **125**, 903–8.

Watermann, R. (1961). *Die Gefässkanäle der Kniegelenksnahen Wachstumszonen.* Köln: Universitätsverlag.

Watterson, R. L., Fowler, I. & Fowler, B. J. (1954). The role of the neural tube and notochord in development of the axial skeleton of the chick. *Am. J. Anat.*, **95**, 337–99.

Waugh, W. (1958). The ossification and vascularisation of the tarsal navicular and their relation to Köhlers disease. *J. Bone Jt Surg.*, **40*B***, 765–77.

Weber, M. M. & Kaplan, N. O. (1957). Flavoprotein-catalysed pyridine nucleotide transfer reactions. *J. biol. Chem.*, **225**, 909–20.

Weidenreich, F. (1930). Das Knochengewebe. In *Handbuch der mikroskopischen Anatomie des Menschen*, 2nd edn, ed. W. von Mollendorf, pp. 391–520. Berlin: Springer-Verlag.

Weightman, B., Chappell, D. J. & Jenkins, E. A. (1978). A second study of tensile fatigue properties of human articular cartilage. *Ann. rheum. Dis.*, **37**, 58–63.

Weightman, B. O., Freeman, M. A. R. & Swanson, S. A. V. (1973). Fatigue of articular cartilage. *Nature, Lond.*, **244**, 303–4.

Weinstock, M. & Leblond, C. P. (1974). Synthesis, migration and release of precursor collagen by odontoblasts as visualized by radioautography after (^{3}H)-proline administration. *J. Cell Biol.*, **60**, 92–127.

Weis-Fogh, T. & Anderson, S. O. (1970). New molecular model for the long-range elasticity of elastin. *Nature, Lond.*, **227**, 718–21.

Weiss, C., Rosenberg, L. & Helfet, A. J. (1968). An ultrastructural study of normal young adult human articular cartilage. *J. Bone Jt Surg.*, **50*A***, 663–74.

Weiss, L. & Dingle, J. T. (1964). Lysosomal activation in relation to connective tissue disease. *Ann. rheum. Dis.*, **23**, 57–63.

Weiss, P. A. (1933). Functional adaptation and the role of ground substances in development. *Am. Nat.*, **67**, 322–40.

Wells, P. J. & Serafini-Fracassini, A. (1973). Molecular oraganization of cartilage proteoglycan. *Nature New Biol.*, **243**, 266–8.

Werb, Z. (1975). The role of endocytosis in controlling the secretion of non-lysosomal proteinases from connective tissue cells. In *Dynamics of connective tissue macromolecules*, ed. P. M. C. Burleigh & A. R. Poole, pp. 159–66. Amsterdam: North-Holland.

Whitehead, R. G. & Weidman, S. M. (1957). Fractionation of phosphorus compounds in ossifying cartilage. *Nature, Lond.*, **180**, 1196–7.

Whitehead, R. G. & Weidman, S. M. (1959). Oxidative enzyme systems in ossifying cartilage. *Biochem. J.*, **72**, 667–72.

Wiebkin, O. W., Hardingham, T. E. & Muir, H. (1975). The interactions of proteoglycans and hyaluronic acid and the effect of hyaluronic acid on proteoglycan synthesis by chondrocytes of old cartilage. In *Dynamics of connective tissue macromolecules*, ed. P. M. C. Burleigh & A. R. Poole, pp. 81–98. Amsterdam: North-Holland.

Wiebkin, O. W. & Muir, H. (1973). The inhibition of sulphate incorporation in isolated adult chondrocytes by hyaluronic acid. *FEBS Lett.*, **37**, 42–6.

Wiebkin, O. W. & Muir, H. (1975). The effect of hyaluronic acid on proteoglycan synthesis and secretion by chondrocytes of adult cartilage. *Phil. Trans. R. Soc.*, *B*, **271**, 283–91.

Wild, A. E. (1975). Role of the cell surface in selection during transport of proteins from mother to foetus and newly born. *Phil. Trans. R. Soc.*, *B*, **271**, 395–410.

Wilkins, J. F. (1968). Proteolytic destruction of synovial boundary lubrication. *Nature, Lond.*, **219**, 1050–1.

Wilsman, N. J. & Van Sickle, D. C. (1970). The relationship of cartilage canals to the initial osteogenesis of secondary centers of ossification. *Anat. Rec.*, **168**, 381–92.

Wilsman, N. J. & Van Sickle, D. C. (1971). Histochemical evidence of a functional heterogeneity in neonatal canine epiphyseal chondrocytes. *Histochem. J.*, **3**, 311–18.

Wilsman, N. J. & Van Sickle, D. C. (1972). Cartilage canals, their morphology and distribution. *Anat. Rec.*, **173**, 79–94.

Winterburn, P. J. & Phelps, C. F. (1971). Studies on the control of hexosamine biosynthesis by gluosamine synthetase. *Biochem. J.*, **121**, 711–20.

Winterburn, P. J. & Phelps, C. F. (1972). The significance of glycosylated proteins. *Nature, Lond.*, **236**, 147–51.

Wlassics, T. (1930). Über den Zusammenhang der Anderung des Fettgehaltes der Zellen des hyalinen Knorpels mit dem Ernahrungsgrad des individuums. *Z. mikrosk.-anat. Forsch.*, **22**, 220–6.

Wolpert, L. (1976). Mechanism of limb development and malformation. *Br. med. Bull.*, **32**, 65–70.

Wolpert, L., Lewis, J. & Summerbell, D. (1975). Morphogenesis in the vertebrate limb. In *Cell patterning*, ed. R. Porter & J. Rivers, pp. 95–130. CIBA Foundation Symposium 29 (N.S.). Amsterdam: Elsevier.

Wood, G. C. (1960). The formation of fibrils from collagen solutions. 2. A mechanism of collagen fibril formation. *Biochem. J.*, **75**, 598–605.

Woods, C. G., Greenwald, A. S. & Haynes, D. W. (1970). Subchondral vascularity in the human femoral head. *Ann. rheum. Dis.*, **29**, 138–42.

Woodward, C. & Davidson, E. A. (1968). Structure–function relationships of protein polysaccharide complexes: specific ion-binding properties. *Proc. natn. Acad. Sci. USA*, **60**, 201–5.

Woolley, D. E., Glanville, R. W., Lindberg, K. A., Bailey, A. J. & Evanson, J. M. (1973). Action of human skin collagenase on cartilage collagen. *FEBS Lett.*, **34**, 267–9.

Wuthier, R. E. (1975). Lipid composition of isolated epiphysial cartilage cells, membranes and matrix vesicles. *Biochim. biophys. Acta*, **409**, 128–43.

Wuthier, R. E. (1976). Lipids of matrix vesicles. *Fedn Proc. Fedn Am. Socs exp. Biol.*, **35**, 117–21.

Wuthier, R. E. (1977). Electrolytes of isolated epiphyseal chondrocytes, matrix vesicles and extra-cellular fluid. *Calcif. Tissue Res.*, **23**, 125–33.

Wuthier, R. E. & Eanes, E. D. (1975). Effect of phospholipids on the transformation of amorphous calcium phosphate to hydroxyapatite *in vitro*. *Calcif. Tissue Res.*, **19**, 197–210.

Wuthier, R. E., Majeska, R. J. & Collins, G. M. (1977). Biosynthesis of matrix vesicles in epiphyseal cartilage. I. *In vivo* incorporation of ^{32}P-orthophosphate into phospholipids of chondrocyte membrane and matrix vesicle fractions. *Calcif. Tissue Res.*, **23**, 135–9.

Wyburn, G. M. (1944). Observations on the development of the human vertebral column. *J. Anat.*, **78**, 94–102.

Yee, R. Y., Englander, S. W. & von Hippel, P. H. (1974). Native collagen has a two-bonded structure. *J. molec. Biol.*, **83**, 1–16.

Young, M. H. & Crane, W. A. J. (1964). Effect of hydrocortisone on the utilization of tritiated thymidine for skeletal growth in the rat. *Ann. rheum. Dis.*, **23**, 163–8.

Zarek, J. M. & Edwards, J. (1963). The stress–structure relationship in articular cartilage. *Med. Electron Biol. Engng*, **1**, 497–507.

Zbinden, G. (1953). Über Feinstruktur und Altersveranderungen des hyalinen Knorpels im elektron-mikroskopischen Schnittpraparat. *Schweiz Z. allg. Path. Bakt.*, **16**, 165–89.

Zwilling, E. (1961). Limb morphogenesis. *Adv. Morphogen.*, **1**, 301–30.

INDEX